REUSE-BASED METHODOLOGIES AND TOOLS IN THE DESIGN OF ANALOG AND MIXED-SIGNAL INTEGRATED CIRCUITS

Reuse-Based Methodologies and Tools in the Design of Analog and Mixed-Signal Integrated Circuits

by

R. CASTRO-LÓPEZ
IMSE-CNM-CSIC,
Spain

F.V. FERNÁNDEZ
University of Seville, IMSE-CNM-CSIC,
Spain

O. GUERRA-VINUESA
University of Seville,
Spain

and

Á. RODRÍGUEZ-VÁZQUEZ
University of Seville,
Spain

A C.I.P. Catalogue record for this book is available from the Library of Congress.

ISBN-13 978-90-481-7289-4
ISBN-10 1-4020-5139-5 (e-book)
ISBN-13 978-1-4020-5139-5 (e-book)

Published by Springer,
P.O. Box 17, 3300 AA Dordrecht, The Netherlands.

www.springer.com

Printed on acid-free paper

Rafael Castro López

To my family and friends

Francisco V. Fernández

To Eli, Judit, and Nuria

Óscar Guerra Vinuesa

To my wife and daughters

Ángel Rodríguez Vázquez

To my former doctoral students and current friends

Contents

Preface

Whether the widely cited Moore's Law –forecasting that the number of transistors that can be fit into a chip roughly doubles every two years– has actually represented a roadmap the semiconductor industry has struggled to comply with or a long-term prediction proven true, the fact is that this industry has accomplished spectacular breakthroughs in past decades, pervasively impacting most aspects of everyday life.

Despite these breakthroughs, the spiraling cost of integrated circuit (IC) design is slowly but surely wrapping a noose around the neck of the semiconductor industry. The economics of building today's even-more-complex ICs under even-more-stringent time-to-market requirements (perhaps the most impelling forces in modern semiconductor industry) are already so daunting that the 2003 ITRS report singled out the cost of chip design as "the greatest threat to the continuation of the semiconductor roadmap". The resulting design productivity gap –the gulf between what is possible to manufacture and what is possible to design– will certainly widen, slowing down this industry's phenomenal growth.

In the past, the industry has extracted itself from design cost traps by finding a way to automate portions of the IC design process, allowing designers to become more productive and driving costs back down. Today, the problem cannot be tackled by still relying on 20-year-old design automation technology or by simply hiring more qualified engineers. The design community believes that powerful computer-aided design (CAD) tools and capable CAD-based methodologies do not suffice in order to successfully and utterly bridge the design gap, but that some kind of design paradigm shift must be urgently put on stage.

In this sense, reuse-based design practices are regarded as a promising solution, and concepts such as IP Block, Virtual Component, and Design

Reuse have become commonplace thanks to the significant advances in the digital arena. Although far from being completely settled, an important market has flourished around digital reuse that furnishes design companies with solutions to noticeably improve their productivity rate.

When it comes to analog and mixed-signal (AMS) design, the scenario is, unfortunately, not that optimistic. The current level of AMS CAD, lagging several generations behind digital design automation partly because of the very nature of AMS design –more subtle, hierarchically loose, and handicraft-demanding–, partly because of the comparatively smaller amount of R&D dedicated to AMS CAD, and the huge heterogeneity of AMS circuits, has so far hindered a similar level of consensus and development on AMS reuse-based design, frequently influencing the idea that inheriting digital reuse concepts is impractical or simply unrealizable. It is necessary to remark, however, the importance of improving AMS design productivity: despite the relatively smaller silicon area dedicated to AMS circuitry, the time needed to design this circuitry dominates, in most cases, the total design time. Therefore, any research ultimately targeted at the improvement of the design productivity of ICs should consider AMS design productivity as a goal priority as well. Otherwise, design productivity will eventually get stuck on the AMS design bottleneck.

In this scenario, the research reported in this book tries to demonstrate not only that reuse-based design in the AMS arena is possible, but also that by following such a design paradigm and making use of appropriate CAD tools, techniques, and methods, it is possible to break through the bottlenecks of AMS design and enhance the design productivity. The concept of reuse here cannot be simply based on plug-in pre-designed, fixed circuit blocks out of a design repository, but rather on recycling these blocks; that is, adopting a flexible methodology by which a circuit can be easily and seamlessly adapted to different design specifications, different environments, and different technology nodes and foundries, thereby completing a AMS design project in time.

This book presents a framework for the reuse-based design of AMS integrated circuits. This framework is founded on three key elements:

- first, a CAD-supported hierarchical design flow that facilitates the incorporation of AMS reusable blocks. Thanks to this design reuse flow, overall design time can be reduced and increasing AMS design complexity can be efficiently managed;

- second, a complete and clear definition of the AMS reusable block. Such definition is structured into three separate facets or views: the behavioral, structural, and layout facets. Throughout block reuse, design information flows from one facet to another, progressively adapting it to the targeted performance and technology. Each facet is devised to suit a stage of the design reuse flow, at its corresponding hierarchical level. In this way, the behavioral and structural facets are used for top-down electrical synthesis and bottom-up verification, and the layout facet is used for bottom-up physical synthesis;

- third, the set of methods, tools, and guidelines composing the design for reusability methodology, which allows producing fully reusable AMS blocks. This methodology relies on intensive facet parameterization as well as on the capture and encapsulation of design knowledge within each facet.

Although the book undertakes the problem from a general perspective, covering all different stages of the design flow, it makes special emphasis on AMS physical design reuse, as this is one of the most (if not the most) crucial, knowledge-intensive stages of the AMS design flow, thus posing a greater challenge to reuse-based design.

The framework is completed with a synthesis technique that aims at speeding up the design process of AMS ICs by reducing the time-consuming, error-prone iterations between electrical and physical synthesis, traditionally considered as non-miscible design stages. In this so-called layout-aware electrical synthesis, a simulation-based optimization algorithm explores the design space while specific and detailed information of the circuit layout –its geometric features and its layout-induced degradation on the circuit's performance– is used to improve the synthesized solution, yielding a correct-by-construction physical implementation of the circuit during the first pass.

The framework has been put into practice and assessed on a well-known, commercial design environment (*Design Framework II* from Cadence®). Furthermore, the framework has been validated through an industrial-scale, functional silicon prototype, consisting in an universal IQ transmit interface for wireless communications.

The contents of this book are organized in seven chapters as follows.

Chapter 1 introduces the problem rationale by examining the evolution of the semiconductor industry, analyzing the current challenges, and delving into the causes of the design productivity gap. To set the background of the research, the chapter then proceeds to clearly define the problem by resorting to several key concepts such as hierarchy, abstraction level, and circuit view, and answering the question of why traditional AMS design methodologies cannot solve it.

Chapter 2 reviews the current state of AMS design automation technology and, in the light of this revision, presents the reuse-based design paradigm. The digital reuse scenario is then examined in order to give insight into the differing requirements of AMS reuse. Afterwards, the chapter surveys the state-of-the-art of AMS reuse-based design. Last, the reuse-based design framework proposed in this book is described.

Chapters 3, 4, and 5 respectively describe the behavioral, structural, and layout facets of the AMS reusable block. The description of each facet follows a three-part structure: what is and what is the facet used for, what requirements does reuse-based design impose on the facet, and how reusability can be built on the facet. Accompanying the descriptions, each chapter contains detailed illustrative examples.

Chapter 6 reports the experimental demonstration of the validity of the reuse-based design framework. This chapter comprises several design experiments, as well as the description and experimental verification results of the silicon prototype mentioned above, whose analog section has been designed under the proposed framework.

Finally, Chapter 7 presents and demonstrates the layout-aware synthesis technique.

The considerations presented in Chapters 4, 5, and 7 are complemented in Appendix A.

The work presented in this book has been partially supported by the TEC2004-01752 Project (funded by the Spanish Ministry of Education and Science with support from the European Regional Development Fund) and the MEDEA+ 2A101-SPIRIT Project.

Chapter 1

Introduction

1 PROBLEM OVERVIEW: THE DESIGN GAP

Nowadays, the semiconductor industry and the design community are facing some very exciting and difficult challenges. For this industry to continue with its phenomenal historical growth and the well-known Moore's law, advances in all fronts are necessary. Although the integration of more and more functionalities onto a single chip is being proven as an effective strategy in terms of fabrication costs, the design effort has been continuously increasing. Both tightening time-to-market pressures and increasing design complexity are widening the gap between the available number of transistors and the ability to design them. This is even more pronounced in the area of analog and mixed-signal design, since design automation is still very far from its digital counterpart. Furthermore, analog and mixed-signal design methodologies are unable to cope with time-to-market and design complexity, the two fundamental forces driving the semiconductor industry.

In this chapter, the problem is investigated and properly defined and for that purpose this section provides the main motivations of the research reported in the book. First, an overview of the evolution of the semiconductor industry is given to set the background. Then, the design gap problem is discussed and, finally, the impact on analog and mixed-signal design is analyzed.

1.1 Evolution of the semiconductor industry

In the nineteenth century, there were more technology achievements than in the nine centuries preceding it. Then, in the first twenty years of the twentieth century, we saw more advancement than in all of the nineteenth century. Now, paradigm shifts occur in only a few years' time. In the twenty-first century, it is expected that there will be almost 1000 times greater technological changes than in its predecessor. This fact also holds true for the industry of microelectronics, which, in the past 40 years, has experimented an incredible and rapid improvement in its products. Multiple evidences of this development are all around us. Semiconductor devices are becoming smaller, almost disappearing into the background. Computational power derived is being applied to many areas of human experience: communications, data storage, medicine, genomics, and so on. The electronic industry is now one of the largest industries in terms of output as well as employment in many nations. The importance of electronics in the economic, social, and even political development throughout the world will no doubt continue to increase.

Semiconductor devices have long been used in electronics. By 1947, the physics of semiconductors was sufficiently understood to allow Brattain and Bardeen to create an amplifying circuit utilizing a point-contact "transfer resistance" device that later became known as a transistor. In 1958, Kilby created the first integrated circuit (IC), ushering in the era of modern semiconductor industry.

The sustained growth of electronics has resulted principally from the industry's ability to decrease exponentially the minimum feature size it uses to fabricate integrated circuits, commonly referred as Moore's Law[1]. Gordon Moore made his famous observation in 1965, just six years after the first planar integrated circuit was completed. The press called his analysis the "Moore's Law", and the name has stuck. In his original paper [Moore65], Moore observed an exponential growth in the number of transistors per integrated circuit and predicted that this trend would continue. In his own words:

> *The complexity for minimum component costs has increased at a rate of roughly a factor of two per year [...] Certainly over the short term this rate can be expected to continue, if not to increase. Over the longer term, the rate of increase is a bit more uncertain, although there is no reason to believe it will not remain nearly constant for at least 10 years. That means by 1975, the number of components per integrated circuit for minimum cost will be 65,000.*

That is, Gordon Moore predicted that the number of transistors that can be fit into a chip would roughly double every year. Later, in 1975, he updated this figure, so the prediction was that the number of transistors would double every two years [Moore75]. The plot in Fig. 1 illustrates this progress. It

---

1. This Law actually refers to digital circuits implementing dynamic memories (DRAM), whose topological regularity allows a higher integration capacity, thus giving an idea of the maximum number of transistors that can be integrated in a given fabrication technology.

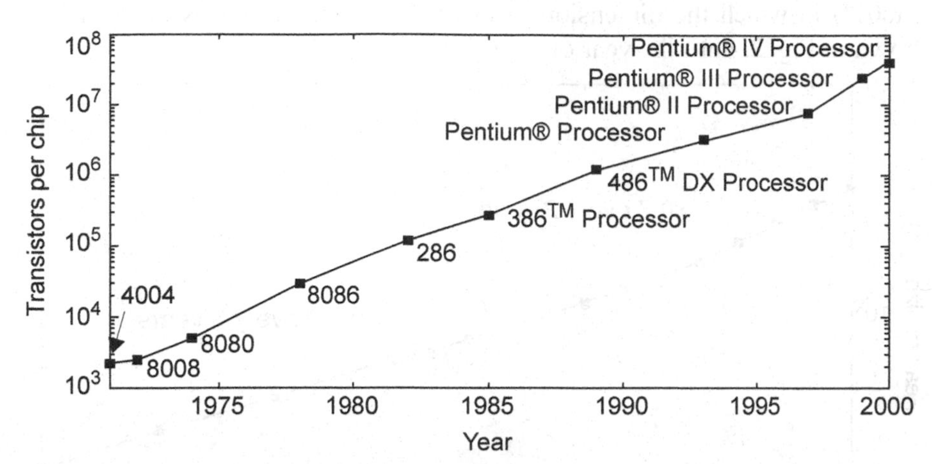

Figure 1. The Moore's Law with Intel® processors.

shows how the number of transistors of some of the Intel® microprocessors has evolved through time.

As said above, this huge progress in the chip density has been driven by the downsizing of the semiconductor components, such as the MOSFET[2] (also referred as just MOST). By such a downsizing, the number of transistors in the chip is enabled to increase and, therefore, the functionality of the product can be improved. This shrinking of the minimum feature size basically pursues two different goals [Hu93]. First, it is aimed at increasing the density, i.e., the maximum number of devices that can be integrated within a single chip. This requires both shorter transistor channel length and smaller channel width. Second, it is devised to increase the current density in order to enhance the speed of switching transistors, since the larger the transistor current, the smaller the time required to charge or discharge the parasitic capacitors associated to the circuit nodes. This implies short channels and high gate oxide field for higher inversion layer charge densities.

As integrated circuit technology has advanced, semiconductor industry has moved from single devices to small scale integration (SSI), with 5 to 20 logic blocks per chip, to medium scale integration (MSI), with 20 to 200 logic blocks per chip, to large scale integration (LSI), when thousands of transistors were put into a single integrated circuit, and, finally, to very large scale integration (VLSI), making integration of hundreds of thousands of transistors onto a single chip possible. In 2001, the Semiconductor Industry Association (SIA) published a technology roadmap of semiconductors (the ITRS report

---

2. The MOSFET, acronym for metal oxide semiconductor field effect transistor, is by far the most common field effect transistor in both digital and analog circuits.

[Itrs01]) in which the dimension of every future generation is predicted, as shown in Fig. 2, until the year of 2016.

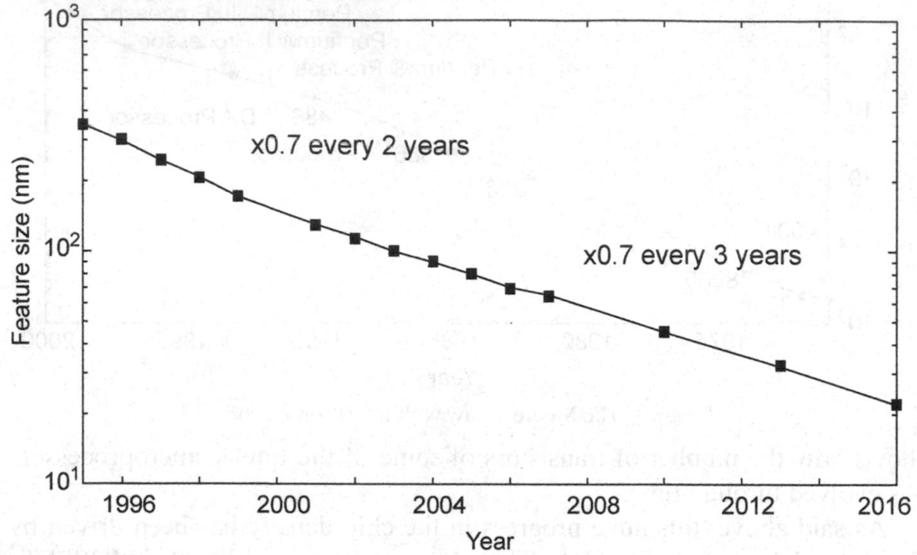

Figure 2. Minimum feature size evolution.

Note that the feature sizes used now and predicted for future chip fabrication are incredibly small: the human hair is 100,000-nm wide, a red blood cell is 'just' 7000-nm wide, and the human immunodeficiency virus (HIV) is 100-nm wide. From the plot in Fig. 2, it can be observed that the minimum feature size has shrunk at a rate of x0.7 every 2 years, until 2001. Since then, the forecast is that such feature will be downsized by a factor of x0.7 every 3 years. This represents the creation of a new technology node (i.e., a significant technology progress governed by the smallest feature printed) every 3 years.

Thanks to this capacity to integrate more functionality onto less silicon area, designers are putting entire systems on a single chip. Thus, chips that are self-contained systems, including processing, memory, and input-output functions, known as systems-on-a-chip (SoCs) [Abid99], are providing us with an overwhelming capability to create new products that will impact the way we work, live, learn, and play.

Basically, two types of SoC design exist: one coming from the application-specific integrated circuit, or ASIC[3], and the other from the custom integrated circuit world. The former type of design is mostly digital, comprising programmable elements, such as control processors and digital signal

---

[3] An ASIC is a circuit designed to suit a customer's particular requirement, as opposed to general-purpose designs, such as DRAMs or microprocessors.

processors or DSPs, memory sub-systems, complex bus architectures, an input/output interface to the external system, software elements, complex bus architectures, clock and power distributions, test structures, etc. The processor can be anything from an 8-bit 8051 to a 64-bit RISC. The memory sub-system could include SRAM and/or DRAM. Analog and mixed-signal blocks may also be present, but it is likewise possible that some of these AMS blocks, such as radio frequency components, remain as a separate chip for this type of design. The other type of SoC design is known as analog and mixed-signal SoC (AMS-SoC). This kind of SoCs inherits the custom analog and mixed-signal design style. These chips are high-performance circuits and typically have complex signal paths through both analog and digital components. Examples of these designs include xDSL front-ends, disk drive controllers, and RF front-ends. Unlike the case of ASIC-SoCs, the analog and mixed-signal part is no longer an "option": it is the critical and probably the differentiating part of these integrated circuits with the digital part optional as to whether or not it is integrated [Kund00].

There are many reasons for using SoC concepts. Apart from the enhanced portability due to smaller system size, the system performance is improved, and the system power dissipation can be reduced as well. Despite these benefits, however, designers have to deal with a challenging issue: the increasing cost of designing SoCs. The following section explains the causes behind this increase and its impact on the design community, and, especially, on the analog design arena.

1.2 The design gap

Although the wafer fabrication costs grow significantly between technology nodes, the fabrication cost per function has been traditionally dropping exponentially, as it is illustrated in Fig. 3 [Itrs01], mainly because more functions can be integrated into the same wafer. The historical trend has been reducing the cost per function ratio by 25 to 30 percent per year.

Unfortunately, the cost of a chip in terms of design procedures, design tools, and human resources, does not remain constant. Actually, while designer productivity has slightly improved with time, there are factors causing this productivity to be insufficient to cope with advances in chip density. Consequently, the cost of the chip design is growing with the complexity of the circuit: although the integration of a system onto a single chip may have attractive manufacturing costs, it also has "frightening" design costs.

As a consequence, there exists a widening design productivity gap, in which the number of available transistors grows faster than the ability to design them meaningfully, as illustrated in Fig. 4 [Itrs99]. In this figure, it is visible that the separation between the chip density and designer's productivity

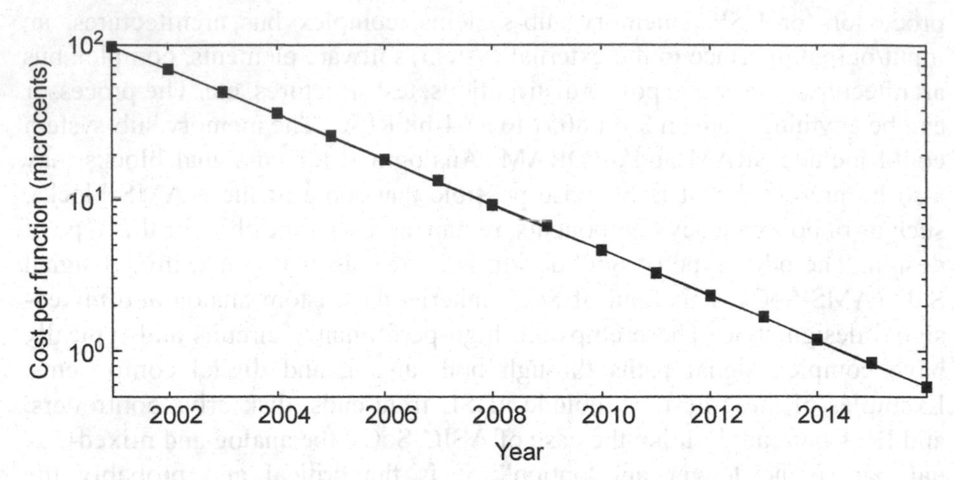

Figure 3. Exponential drop of the cost per function.

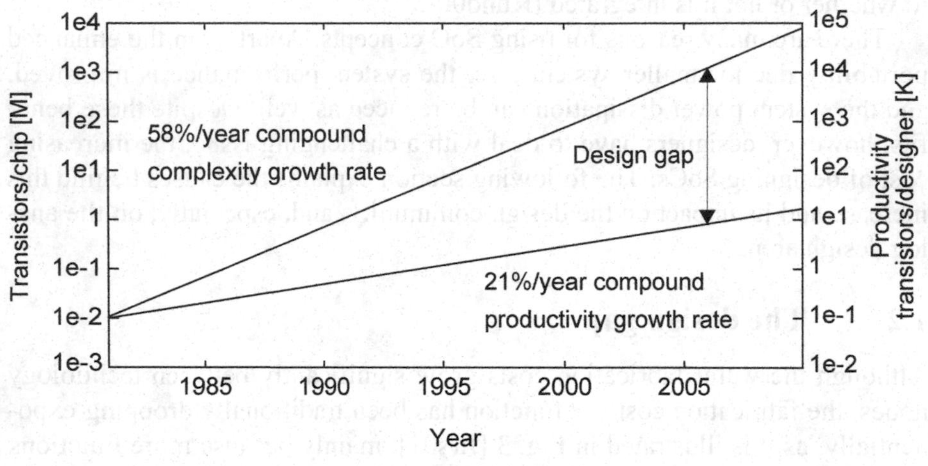

Figure 4. Design Gap.

is growing as fabrication technology evolves. Actually, the overriding message in the 2001 ITRS report (still present in the 2003 report [Itrs03]) is that design cost is the greatest threat to continuation of the semiconductor industry's phenomenal growth [Itrs01].

In order to understand the causes of this design gap, that is, why the design community is unable to cope with the continuously increasing chip density, and, therefore, discern what possible solutions can be provided, it is first necessary to analyze the challenges and driving forces guiding the evolution of the design community. The first driving force is the availability of shrinking process technologies, introducing new problems and effects that need to be accounted for during the design process, being its more direct consequence the rise of design complexity. The other fundamental driving force is the stress on reducing the time-to-market figure of a new chip product. Let us now examine in detail each one of these driving forces.

1.2.1 Time-to-market

The time-to-market (TTM) can be defined as the time it takes to get a product from the concept to the marketplace. The TTM is a very critical factor for ASICs and SoCs that eventually end up in any market place: if the vendor misses the initial market window, prices, and, therefore, profit can be seriously eroded. The first vendor to market with a new product makes a profit, and obtains immediate feedback to improve the next product generation. The second vendor to market will only break even, but the third vendor will surely lose money. The importance of dealing with TTM constraints is illustrated in Fig. 5 [Levi92].

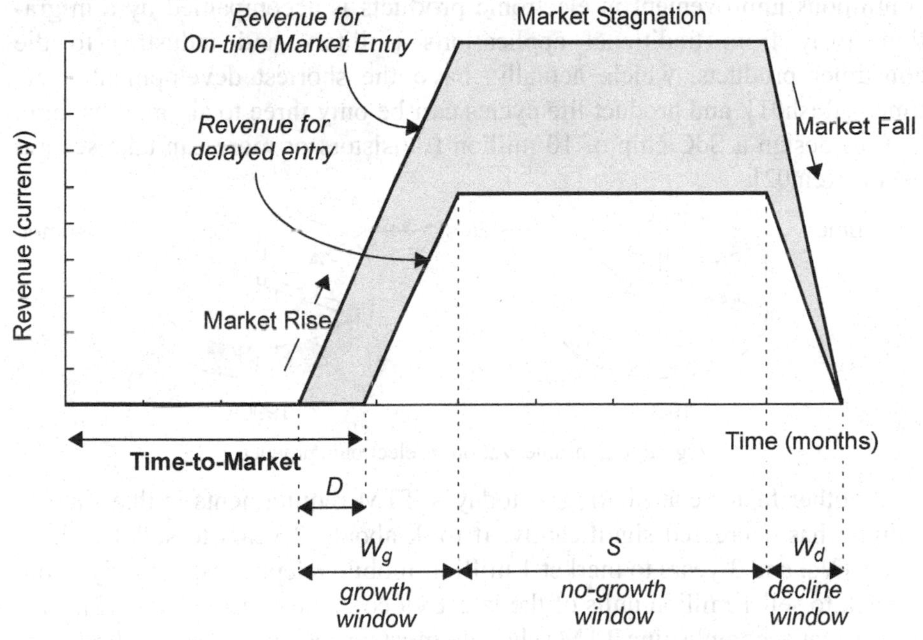

Figure 5. Importance of delayed market entry.

Every product has a market growth window, W_g, a period of no-growth or stagnation, S, and a market decline window, W_d, thus being the lifetime of the product equal to $W_g + S + W_d$. The product begins to gain market acceptance and sales grow rapidly as the product reaches the market. Then, as the product matures, the competence grows as other vendors enter the stage and the product becomes thereby depreciated, all resulting in a market stagnation. Finally, as technological process advances and superior products are launched, product sales start declining. For an on-time market entry (that is, zero time delay D), there is a total expected revenue, R_o. For a delayed product, however, there is a total revenue loss, R_L, which can be easily calculated by computing the area of the shaded region in Fig.5. The relative revenue loss, $r_L = R_L / R_o$, is then expressed by the following equation:

$$r_L = \left[\frac{D(2W_g - D + 2S + W_d)}{W_g(W_g + W_d + 2S)}\right] \tag{1}$$

For instance, according to Eq. (1), a typical four-month delay over a growth window of 1 year, a stagnation period of 1 year, and a market decline window of half a year, corresponds to a relative loss of 40 percent of the total revenue corresponding to a product with zero time delay.

A major reason for hardening TTM pressures is the continued "consumerization" of the electronic marketplace [Kund00]. As shown in Fig.6, the continuous improvement in electronic products is accompanied by a migration away from traditional applications (military and industry) to the consumer products, which, actually, have the shortest development cycle times [Rash01], and product life cycles can be only three to six months long. Yet, to design a SoC chip of 10 million transistors or more can take several years [Rein02].

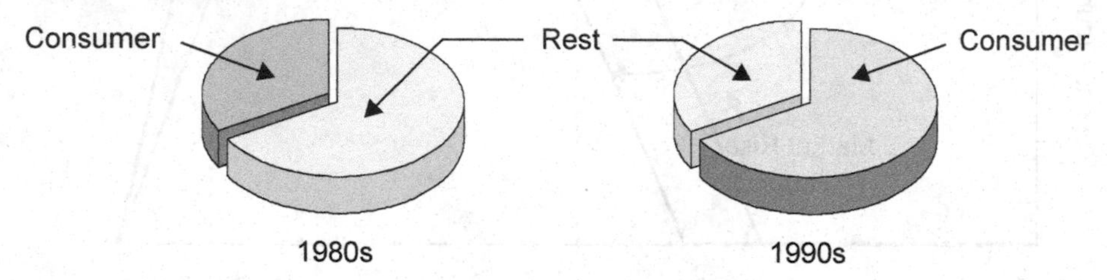

Figure 6. Consumerization of electronic products.

Another factor contributing to today's TTM requirements is that time to volume has decreased significantly. It took about 12 years to sell 1 million color TVs, and 3 years to market 1 million mobile telephones, but only about 1 week to sell 1 million units of the latest video game console [Rein02]. This means that accomplishing TTM schedule must be accompanied by a first-time correct product. That is, circuit designers have to rely on methods and tools that guarantee a rapid transition from idea to market, and avoid, as much as possible, the presence of mistakes or bugs in the final product.

1.2.2 Design complexity

Design complexity means different things to different people. The first question to answer is what metric must be used to measure design complexity: the number of transistors or components, the area size, the size of the design team, or even the number of several disciplines and types of expertise involved in a project. If we take the transistor count to measure design complexity, there is a significant difference upon the type of integrated circuit considered. For example, the difference in design cost or effort between a 256-kBytes and a 256-MBytes memory chip is not all that great, since there are many repetitive elements on a memory chip [Rein02]. However, ASIC-SoC and, especially AMS-SoC designs, may have many different components, digital, analog, and mixed-signal, the latter two requiring major design efforts. Therefore, design complexity for SoCs cannot be compared just in terms of the number of transistors. Actually, the main factors prompting an increase of the design complexity are the following [Giel00] [Itrs01]:

- Increasing number of transistors per chip: there are both analog and digital components that need to be co-designed, with numerous design issues (such as increasing leakage and crosstalk) involved.

- New signal processing algorithms and system architectures: emerging applications require new to-be-developed functionalities.

- Larger design teams: designing a SoC require larger design teams, simply because of the different nature the SoC components, which often require many kinds of expert knowledge.

- Shrinking processes: due to the fast evolution of the fabrication technology and provided that a circuit is planned to be implemented in upcoming fabrication technologies, the expectation for a change in the process parameters should be taken into account during its design phase.

As it can be easily noted, design time will surely increase when design complexity grows. This is in clear contrast with the need to market the product as quick as possible. Therefore, designers have to face the design challenge of managing the TTM-design complexity trade-off. The key to deal with this trade-off, i.e., correctly managing design complexity while meeting TTM goals, lies in adopting appropriate and well-structured design methodologies which must be supported by efficient computer-aided design (CAD) tools.

1.3 Analog design automation

According to the SIA definition, a CAD[4] resource is a sophisticated, computerized workstation and software used to design integrated circuit chips [Glos03], which encompasses not only separate and specific-purpose tools, but also global design methodologies with which human designers can build functionality while satisfying intended performance specifications [Chang97]. The goal of CAD is to reduce the manual design time required for circuit design. In this book, CAD tools and CAD methodologies will be referred as the two fundamental concepts of CAD. The CAD methodology must be consistent enough to properly guide the CAD tools towards the complete satisfaction of all chip performance requirements [Chang97].

In the digital circuit domain, where information is represented as numbers with discrete (non-continuous) values, usually expressed as a sequence of binary digits (ones and zeros), CAD tools are fairly well developed, widely used, and commercially available today. There are several reasons explaining this. The most important one comes as a consequence of the very nature of digital circuits. These circuits work by "flipping" transistors on and off to the 1 and 0 states. Thanks to this Boolean representation of the digital systems, their functionality can be easily translated to algorithmic constructs using programming languages. This eases the automation of many of the design tasks involved, certainly for the lowest levels of the design. In this way, digital systems can be described through high-level languages, and digital CAD tools perform the task of synthesizing this description into a structural representation. This representation can be then transformed into the final physical representation of the digital circuit, for the selected technological process. In addition, digital circuits have a great immunity to noise, because the signal values are associated to just discrete 1's and 0's. This noise margin further facilitates automation of every design step. Therefore, the design effort can be focused on creating more and better electronics, directing it towards trade-offs between power consumption, speed, and area. Lastly, it is necessary to note that the digital integrated circuit market is quite large, with numerous established vendors investing on CAD development, resulting in a steady steam of efficient digital CAD tools. Although the level of computer digital design automation is far from the push-button stage as design complexity is rapidly increasing, the development of digital CAD tools and methodologies are moderately keeping pace with chip density. Nevertheless, a limit is being reached in many areas (e.g., verification) and, consequently, the digital design gap is slowly but steadily widening.

4. CAD for microelectronics is also known with the name of electronic design automation (EDA).

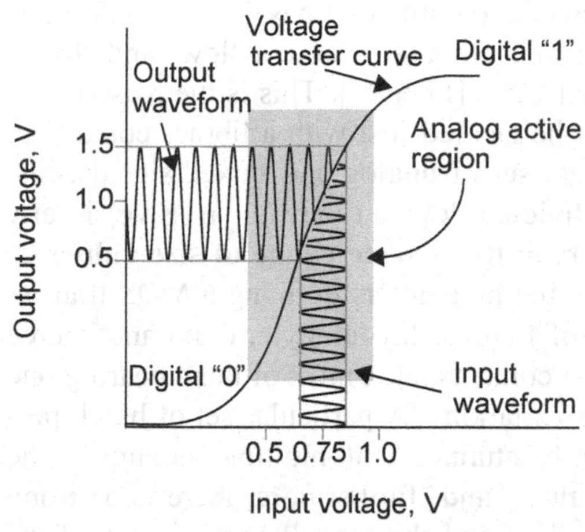

Figure 7. Analog vs. digital behavior.

Unfortunately, for analog and mixed-signal designs, the picture, so far, is even worst. As illustrated in Fig. 7, an analog system carries signals in the form of, for instance, voltages, currents, and charges, which are continuous functions of the continuous time variable [Greg86]. Thus, in contrast to digital circuits, analog systems work by biasing the transistor somewhere between what digital designers would call fully on ("1"), and fully off ("0"). Analog and mixed-signal design entails specialized knowledge and almost handicraft design skills, usually acquired after many years of experience. Some of the characteristics of analog and mixed-signal design making the automation of common analog design tasks so difficult are the following:

1. **Heterogeneity**. Analog and mixed-signal functionalities can be implemented through a **wide range of circuit blocks**, and the same function can be accomplished with several different circuit topologies, each one suited to specific applications. There are hundreds of circuit topologies for analog and mixed-signal circuits and systems (containing 10-5000 devices) in use today. Besides, the **spectrum of performance specifications** for any analog and mixed-signal circuit is much **larger** than for digital circuits. For instance, a simple operational amplifier (opamp) has, to say the least, 10 performance requirements which are typically continuous (such as small signal DC gain, bandwidth, phase margin, slew rate, settling time, output voltage swing, input offset, noise, area, power consumption, and so forth [Koh90]). On the other hand, digital standard[5] cells usually have 2 performance parameters, power and timing. A digital standard cell library probably contains around 1000 cells, arranged as the rough product of 10 logic functions (NAND, NOR, flip-flop, etc.), 10 different input/output alternatives (NAND2, NAND4, and so on), and 10 different timing and power alternatives per cell, while even a small

5. A standard cell is a predefined circuit element that may be used to create an integrated circuit more easily than through design.

analog cell with 10 performance parameters, has $2^{10} = 1024$ variants, even limiting the value of each parameter to a "low" and "high" value (which is not the usual case) [Leen01]. This is the reason why this analog diversity cannot be tackled just with a library containing all alternatives of a complete set of analog and mixed-signal cells. Last but not least, each individual device (active or passive) in any analog and mixed-signal circuit has a **wide range of size values** as well. Consider, for instance, the parameters defining a MOS transistor: length, width, number of fingers, layout style, drain and source area capacitances, number of contacts, etc., most of them bearing one or two orders of magnitude variations. A particular set of block performance specifications can be attained with multiple variants of the whole set of device size values, and, furthermore, there is a strong interaction between device sizes and the overall performance of the analog and mixed-signal circuit. Therefore, fine tuning of this set to obtain highly optimized circuits (e.g., in terms of area or power consumption) is a crucial task of analog and mixed-signal circuits.

2. **Sensitivity**. Analog and mixed-signal circuits are extremely sensitive to second and higher-order non-ideal effects. Some of these effects are caused by the imperfect nature of the fabrication process and materials involved: device mismatch, parasitic capacitances, and temperature or biasing variations, among others, may critically jeopardized the foreseen performance of the integrated circuit [Chang97]. Another source of high-order effects is the sensitivity of analog and mixed-signal to system-level interactions. Due to capacitive and inductive crosstalk and substrate and power-supply noise coupling, and the peaking nature of digital circuits, the desired performance of the analog and mixed-signal part of the chip can be easily worsened. A direct consequence is that layout precautions should be taken in order to isolate the analog and mixed-signal parts from the digital part. Another consequence is that the design of the digital and the analog and mixed-signal parts should not be separated. Besides, since design decisions must, to a large extent, change as the technological process is changed, direct process-to-process reuse of an integrated circuit is not feasible.

3. **Hierarchy**. As opposed to digital circuits, hierarchy is not so strictly and clearly defined for analog and mixed-signal designs. As it is well known, different hierarchical levels are described using different levels of abstraction in any digital application, whereas, in analog circuits, voltages, currents, and impedances must all be considered at all hierarchical levels. It is thus not possible to establish higher levels of

abstraction that shields all the device-level and process-level details when designing at higher levels of the hierarchy. This close interaction between hierarchical levels forces the designer to take into account how a single change in one level affects the higher or lower levels of the hierarchy. Consequently, no straight automated means to switching from one level to another exist.

4. **High-performance**. As increasingly demanding electronic applications call for more aggressive digital signal processing, the performance of the analog and mixed-signal components (progressively pushed to interface the inherently analog world with the digital signal processing cores) is consequently required to be higher. In most cases, such higher performance is achieved by taking into account the correlation between electrical parameters and their variations due to statistical fluctuations of the chip fabrication process [Chang97]. This requires extremely careful design, which cannot be easily translated into computation algorithms, and, therefore, easily automated.

Due to all these features, analog and mixed-signal design is still regarded as less systematic, far more knowledge-intensive, and heuristic[6] than digital design. Consequently, a large amount of analog and mixed-signal circuits are still designed without any really robust and commercial CAD tools but a SPICE-like simulation shell and an interactive and scarcely-automated layout environment [Giel00]. Therefore, analog design turns out to be a time-consuming, prone-to-errors effort. As fabrication technology progresses and new technology nodes appear (and feature size shrinks as well), analog and mixed-signal designers are unable to keep the pace and deal with increasing TTM pressures and design complexity.

In AMS-SoCs, the design of the analog and mixed-signal components is a bottleneck of the overall design process, as shown in Fig. 8(a). Current estimates indicate that 60% of today's ICs include some analog or mixed-signal content, as illustrated in Fig. 8(b). In addition, the time required to design this analog or mixed-signal content usually dominates the total design time. For the semiconductor industry to maintain its design productivity and keep providing high-quality yet cheap electronic products, design time and cost of the analog circuitry, from the description of its specifications to first-pass silicon, has to be reduced. To this end, it is clear that an efficient CAD methodology and adequate CAD tools to design both analog and mixed-signal circuits are urgently required.

---

6. The word *heuristic* is derived from the Greek word *heuriskein* (ευρισκειν), which means *to discover*. It refers to a common-sense set of rules intended to increase the probability of solving some problem.

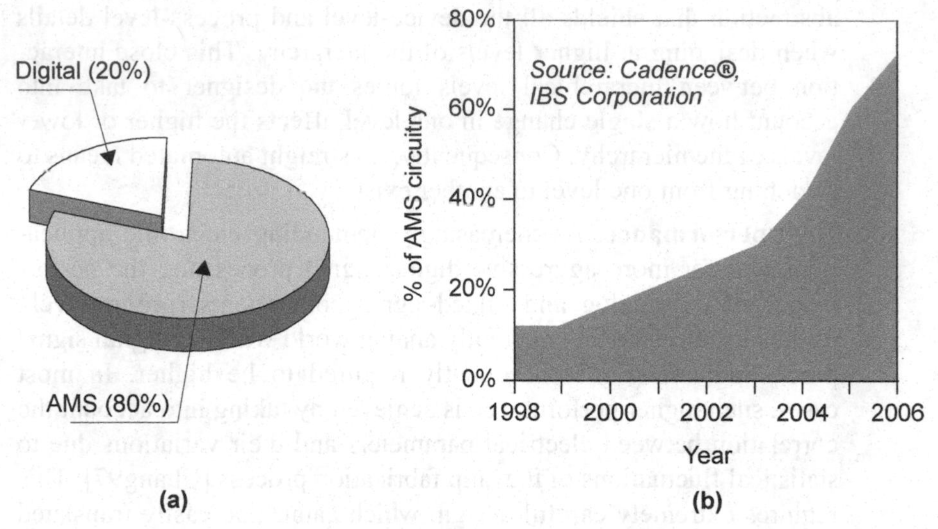

Figure 8. (a) Relative design time for the analog and digital parts of an AMS-SoC. (b) Percentage of chips with some AMS content.

2 PROBLEM DEFINITION

As explained in the previous section, there exists an urgent need for CAD tools and methodologies specifically tailored for analog and mixed-signal integrated circuits that render a design process significantly enhanced in these two aspects:

- **Rapidness**, by reducing the overall design time, thus improving design productivity and meeting TTM requirements.
- **Correctness**, by increasing the likelihood of first-pass right silicon with an adequate management of the increasing complexity of analog and mixed-signal designs.

In order to correctly define the problem, it is necessary to examine the currently available methodologies and tools in detail, and then analyze where do they fail to cope with the goals of rapidness and correctness. But prior to such analysis it is useful to revisit some important concepts.

2.1 Hierarchy, abstraction, and views

To master the ever increasing design complexity of both the digital and analog parts of integrated circuits, there are two concepts, **hierarchy** and

abstraction, that are commonly used throughout the entire design process [Donn94a]. Hierarchical design is commonly used in programming as a procedure in which the whole list of primitive statements is managed through proper calls to simpler procedures, possibly involving many of the primitive statements. Each procedure breaks down the task into smaller operations until each step is refined into a procedure simple enough to be written directly. This *divide-and-conquer* technique reduces the complexity of the whole process by recursively breaking it into manageable pieces. Chip designers also use this approach: it is then easier to understand a 1,000,000-transistors hierarchical SoC design than the same design expressed directly as a million transistors wired together. In this way, the required specifications for the complete design behavior can be passed down the hierarchy, reducing the overall complexity of the chip design process, as it is illustrated in Fig. 9.

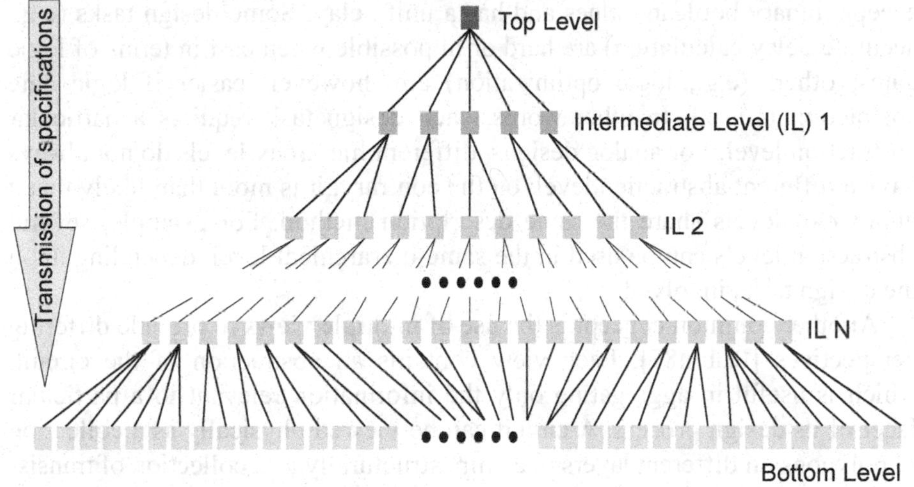

Figure 9. Design hierarchy.

A crucial benefit from hierarchical design is that, as it will be explained in following chapters, hierarchy permits reusability of design knowledge [Toum95]. For instance, the knowledge and expertise used to design a voltage comparator (a building block in the hierarchy of, for instance, an analog-to-digital converter) can be used later when designing any other circuit using that comparator. An important related idea is that a modification that takes place at a certain level of the hierarchy is effectively propagated to all the hierarchical levels above the modified one.

But, as it was said in the previous section, there is a relevant difference of digital and analog hierarchies. Whereas digital hierarchy is well defined (with RTL, structural, and gate levels), the analog hierarchy has no strictly defined levels (and, certainly, not generally accepted). The main reason for this lack of any rigid, formal hierarchy is that the different levels are not based on different levels of signal abstraction, but rather on different and not always standardized structural decompositions over different levels.

Therefore, design abstraction turns out to be a critical concept to integrated circuit design, too. The level of abstraction measures the magnitude of the link between the model and each one of the physical devices contained in it. As physical effects become more important and increase design complexity, the total range of abstraction levels encountered in a single design flow is continuously growing and pulling abstraction and detail in opposite directions [Kund00]. Like programmers, chip designers use multiple levels of design abstraction to meet performance goals for very large designs. For digital designs, the simplest example of abstraction is the logic gate, which is just a simplification of the nonlinear circuits used to build the gate: the logic gates accepts binary boolean values and has a unit delay. Some design tasks (e.g., accurate delay calculation) are hard or impossible when cast in terms of logic gates; others (e.g., logic optimization) are, however, easier if logic gate abstraction is used. In other words, each design task requires a particular abstraction level. For analog designs, different hierarchy levels do not always have a different abstraction level; on the contrary, it is more than likely that a number of levels share the same description method. For example, several abstraction levels can be used in the same hierarchical level, depending upon the design tasks involved.

Another common concept is the use of multiple **views** to provide differing perspectives [Rubin87]. Each view contains an abstraction of the circuit, which is useful in aggregating only the information relevant to a particular facet of its design process. A circuit can be viewed physically as a collection of polygons on different layers of a chip, structurally as a collection of transistors, resistors, capacitors, etc., or behaviorally as a set of operational restrictions in a behavioral-description language. It is useful to be able to flip among these views when building a circuit because each has its own merit in aiding design.

2.2 The AMS design flow

Figure 10 shows a general design flow illustrating the main stages of the creation of integrated circuits, both digital and analog [Donn94a] [Chang97].

The five stages of the flow are described below:

1. **Electrical synthesis**: the design process starts with the specification of the circuit's performance that has to be realized. A typical example is to specify the maximum power consumption or minimize the area occupation, minimizing thus the overall production cost. The output

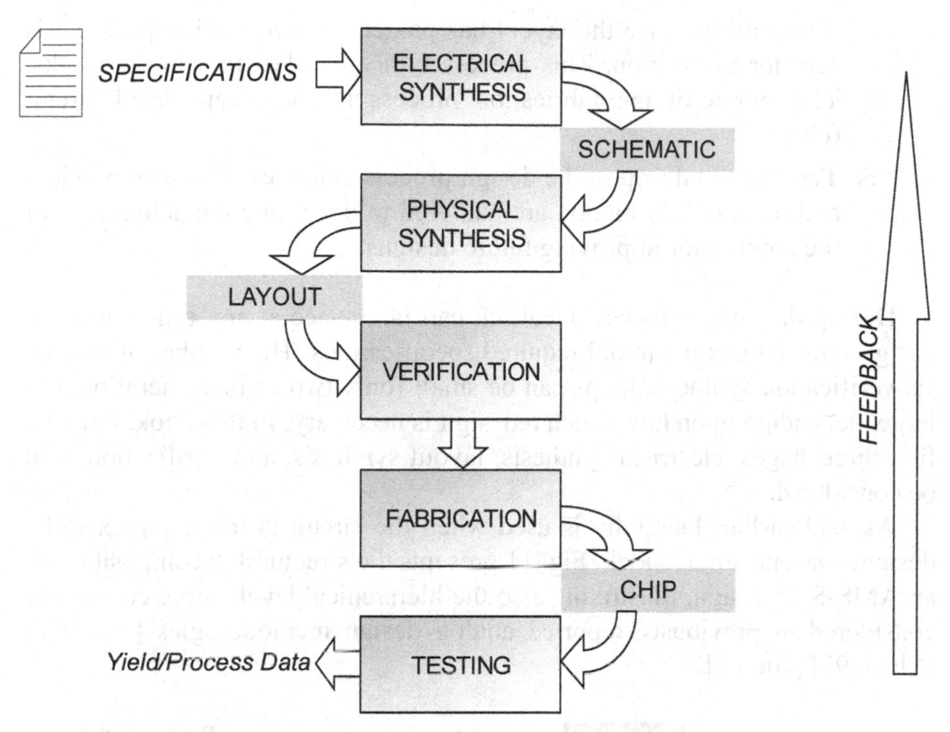

Figure 10. Typical design flow of an integrated circuit.

to this process is a *schematic*, a list of all the devices (transistors, resistors, capacitors, etc.) composing the circuit which specifies not also how they are connected (a list of nets), but also the characteristics of each single device (e.g., transistor width and length).

2. **Physical synthesis**: the goal of this phase is to obtain the physical representation of the circuit schematic, known as layout, a collection of geometric shapes and layers, later used in the fabrication process. Layouts can be generated either manually or automatically. The layout must be checked for errors with respect to the collection of design rules dictated by the technological process, and for its complete correspondence with the intended schematic.

3. **Verification**: after completion of the layout, it must be verified that the initial specifications have been met. This verification is usually carried out through simulation of the whole circuit, which may include the unavoidable layout-induced degradation[7].

[7] Strictly speaking, verification is also required at many other places of the design flow, for instance, after completing the electrical synthesis of every circuit's building block. The aim of Fig. 10, however, is just to highlight the various tasks of the IC design flow, which, to keep things simple enough, have been clearly separated. The role of verification will be specified in the next pages.

4. **Fabrication**: once the layout has passed the verification phase, it is sent for fabrication. This process typically takes two to six weeks. The output of the fabrication process is the encapsulated circuit (chip).
5. **Test**: the final step of the design process is the test of the chip. Yield and process information are gathered to determine the actual cost of the product for improving future designs.

During the entire process, feedback can take place at any time when the design fails to meet the initial required specifications. The number of synthesis-verification-synthesis loops can be small (only two or three iterations) or large, depending upon how much redesign is necessary. In this book, only the first three stages, electrical synthesis, layout synthesis, and verification, will be considered.

As said earlier, hierarchy is used when the circuit is too complex to be designed as one single block. Fig. 11 presents the structural decomposition of an AMS-SoC design, illustrating also the hierarchical levels more commonly considered in previously reported analog design methodologies [Dess01a] [Chang97] [Giel00].

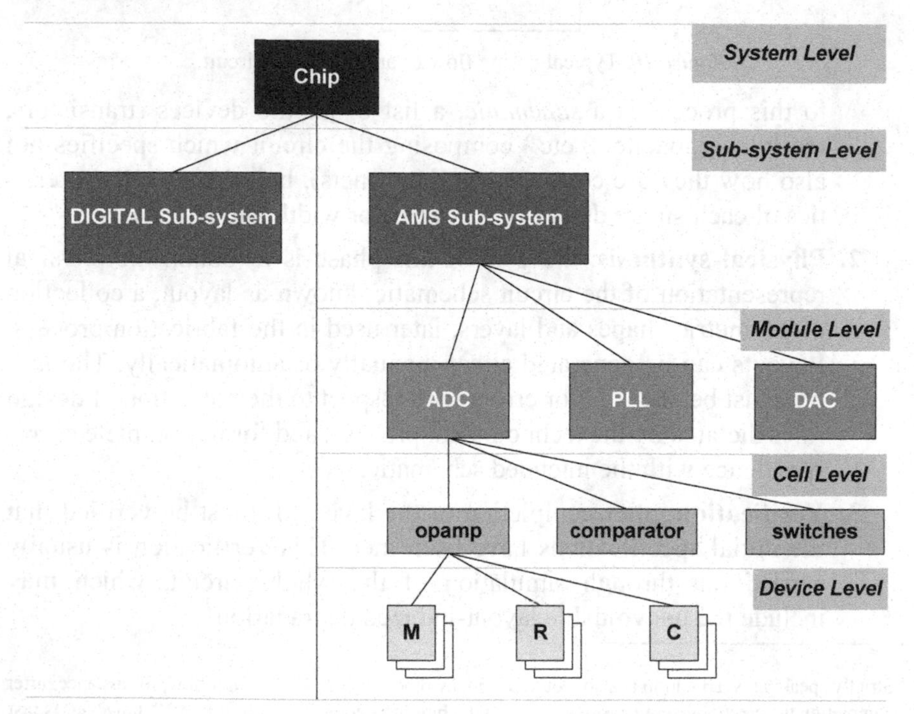

Figure 11. Hierarchical structure of an AMS-SoC design.

These levels are:

- Digital and mixed-signal **sub-systems**[8], placed at the top level of the hierarchy.
- **Module** level, defined as stand-alone functions with robust interface that can be clearly distinguished from its environment [Dess01a].
- **Cell** level, where blocks are basic functions (e.g., opamp, comparator, or switches) used to build the module-level blocks and are solely composed of device-level components.
- **Device** level, composed of active and passive devices, described by means of specific technological models.

Although not explicitly indicated in Fig. 11, module-level blocks may also comprise device-level components that are not part of any cell-level block (e.g., the RC components of a biquad filter –the module-level block– whose active components are opamps –the cell-level blocks–). What differentiates cell from module-level blocks is that all components of a cell-level block are devices (e.g., transistors, resistors, capacitors, and so on) whereas a module-level block can be composed of cell-level blocks as well as of individual devices. In the same way, a system or sub-system level may comprise instances of each of the lower hierarchical levels (i.e., module-level blocks and separate cell and device-level components). For the sake of generality, we will additionally suppose that the hierarchy between the system (or sub-system) and the cell levels can be further divided into one or more module levels. That is, whereas there is only one system (or sub-system) level –to define the highest hierarchical level and whose performance requirements we will consider as externally set–, only one cell level, and only one device level –from which the physical implementation can be realized–, each module may include other modules as well as other cell and device building blocks.

Many analog design methodologies exist, but despite different on the surface, most current design paradigms for analog circuits are similar. The first issue to consider is that of **top-down** versus **bottom-up** design. In the former, a circuit is designed with successively more detail; in the latter, a circuit is viewed with successively less detail. Fig. 12 illustrates the top-down and bottom-up concepts. Top-down design starts with the decomposition of the system specifications into a sub-set of specifications for each system's building block, which is known as *translation* or *mapping* of specifications. In doing so, the interactions between the blocks are approximated to allow a certain

[8] Hereinafter, the word *sub-system* is to be replaced by *system* provided that the analog or mixed-signal circuit is the top-most hierarchical level (i.e., there is no digital section).

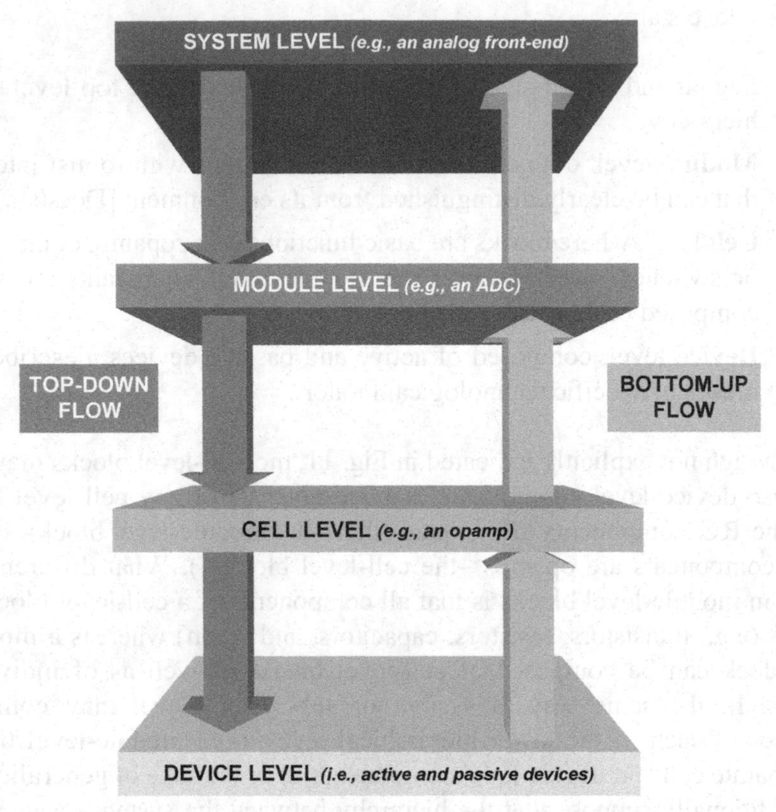

Figure 12. Top-down vs. bottom-up design processes.

independence of the design [Chang97]. The top-down process ends at the bottom level, where active and passive devices (transistors, resistors, and capacitors) compose the lower level of abstraction of the blocks. The reverse style, the bottom-up design process, begins by designing the individual cell-level blocks and ends with the assembly of the system-level circuit.

The main advantage of the top-down design flow described above is that system performance is verified early in the process, attaining thus a higher chance of first-time success, while obtaining a better overall system design. On the other hand, if some circuit specifications cannot be met at any hierarchical level, the designer has to go back and change the system architecture, which can be a costly operation at that point. Clear disadvantages of bottom-up approaches are that the architecture cannot be optimized for best system performance and that system performance cannot be verified until all the blocks have been designed, which can lead to major design changes late in the design process.

Thereby, a hierarchical design flow combining the top-down and bottom-up paths, together with redesign or backtracking iteration loops is preferred [Chang97] [Giel00]. As illustrated in Fig. 13, in between any two hierarchical levels i and i + 1 , this design methodology consists of a sequence of steps.

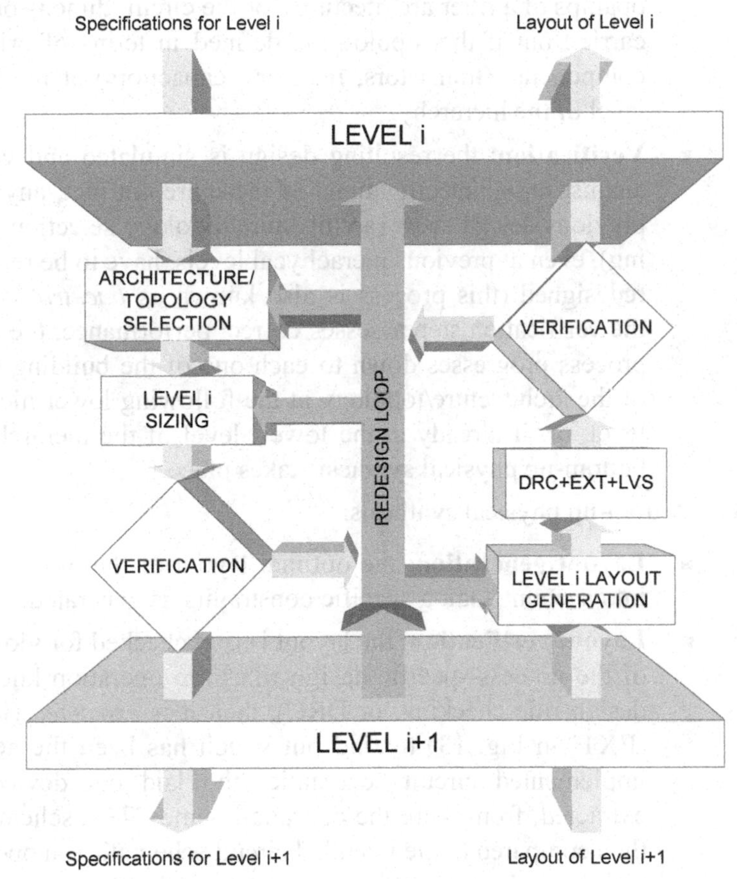

Figure 13. Hierarchical top-down bottom-up design flow.

These steps are:

a. Top-down electrical synthesis:

- **Architecture/topology selection**[9]: starting with the performance specifications from the immediately higher level, the user selects the most promising architecture/topology that can address these specifications.

[9]. *Architecture* is hereinafter referred as the internal structure and connectivity of module, sub-system, and system-level circuits; the term *topology* is related to the cell-level circuits.

- **Sizing**: the architecture/topology is optimally sized such that the circuit meets the required specifications. This means that either the specifications from level i – 1 are mapped into specifications for the building blocks of the architecture (e.g., opamps of a filter architecture), or the circuit dimensioning is carried out if the topology is defined in terms of primitive components (transistors, resistors, capacitors) at the lowest level of the hierarchy.
- **Verification**: the resulting design is simulated and verified against initial specifications. If these are not met, any of the previous design steps (architecture/topology selection or sizing), even at previous hierachycal levels, have to be repeated/ redesigned (this process is also known as *backtracking*). If the verification step assesses correct performance, the design process progresses down to each one of the building blocks of the architecture/topology at the following lower hierarchy level, or, if already at the lowest level of the hierarchy, the bottom-up physical synthesis takes place.

b. Bottom-up physical synthesis:

- **Layout generation**: the optimal layout of the block, taking into account analog specific constraints, is generated.
- **Layout verification**: the layout is first checked for violations of the process-specific design rules (an operation known as design rule checking, or DRC), then it is *extracted* (labeled 'EXT' in Fig. 13) to find out which has been the actually implemented circuit schematic (the laid out devices are extracted, from there the operation name). This schematic is then compared to the intended circuit schematic (an operation known as layout versus schematic, or LVS). Extraction is also used to obtain the layout-induced effects in terms of added devices –typically capacitances and resistances–, known as layout *parasitics*[10]. The whole process (i.e., DRC+EXT+LVS) is known as *formal verification* in contrast to the subsequent performance verification.
- **Performance verification**: the extracted layout (i.e., the circuit schematic plus the layout parasitics) is then simulated and, as in the top-down verification phase, if unacceptable

10. In general, a layout *parasitic* can be defined as every cause of performance degradation which is not intended by the circuit designer and whose value is determined by the circuit layout [Lamp99].

deviations from the required performance are uncovered, the design process is repeated.

The illustration of the hierarchical design methodology for analog and mixed-signal circuits, with the levels of abstraction explained above is given in Fig. 14. It is worth noting that every step of the hierarchical top-down bottom-up synthesis methodology (architecture/topology selection, sizing or specification translation, layout generation, simulation, and verification) can be, in principle, automated with CAD approaches. Such an automation can range from merely relieving the designer from long, repetitive tasks to performing the complete design flow with minimal designer intervention.

Although hierarchical design represents a substantial improvement of the time-to-market figure and the ability to handle complexity when compared with plain bottom-up or top-down approaches, there still are some points where this methodology needs improvement in order for it to cope entirely with those two design challenges. In this sense, a number of goals have been defined in the 2001 ITRS report [Itrs01]. Among them, the work presented in this book aims at the following:

1. **Avoid iterations**. Iterations between layout generation and sizing –a direct consequence of the complete separation of the bottom-up layout generation and the top-down electrical synthesis–, incur repeating the related synthesis processes and other interfacing costs, which altogether may eventually lead into product-to-market failure. Therefore, these spins should be reduced as much as possible.
2. **Improve synthesis**. AMS circuit design typically requires very specific expert knowledge that should be incorporated into the various automated synthesis processes in order to enhance their efficiency.
3. **Reuse**. To extend the notion of library-based approaches, and thereby reduce overall design time, reuse of captured designers' expertise should be also a major concern. Therefore, methods for reusing stored expertise must be defined and developed that improve the hierarchical methodology for AMS circuits.

The problem, and, hence, the goal of the research here reported, consists in the improvement of the analog and mixed-signal design process by means of the definition, development, and demonstration of appropriate CAD methodologies and tools, which, attending to the issues mentioned above, help to reduce the design cost and, hereby, to bridge the design gap described in Section 1.2.

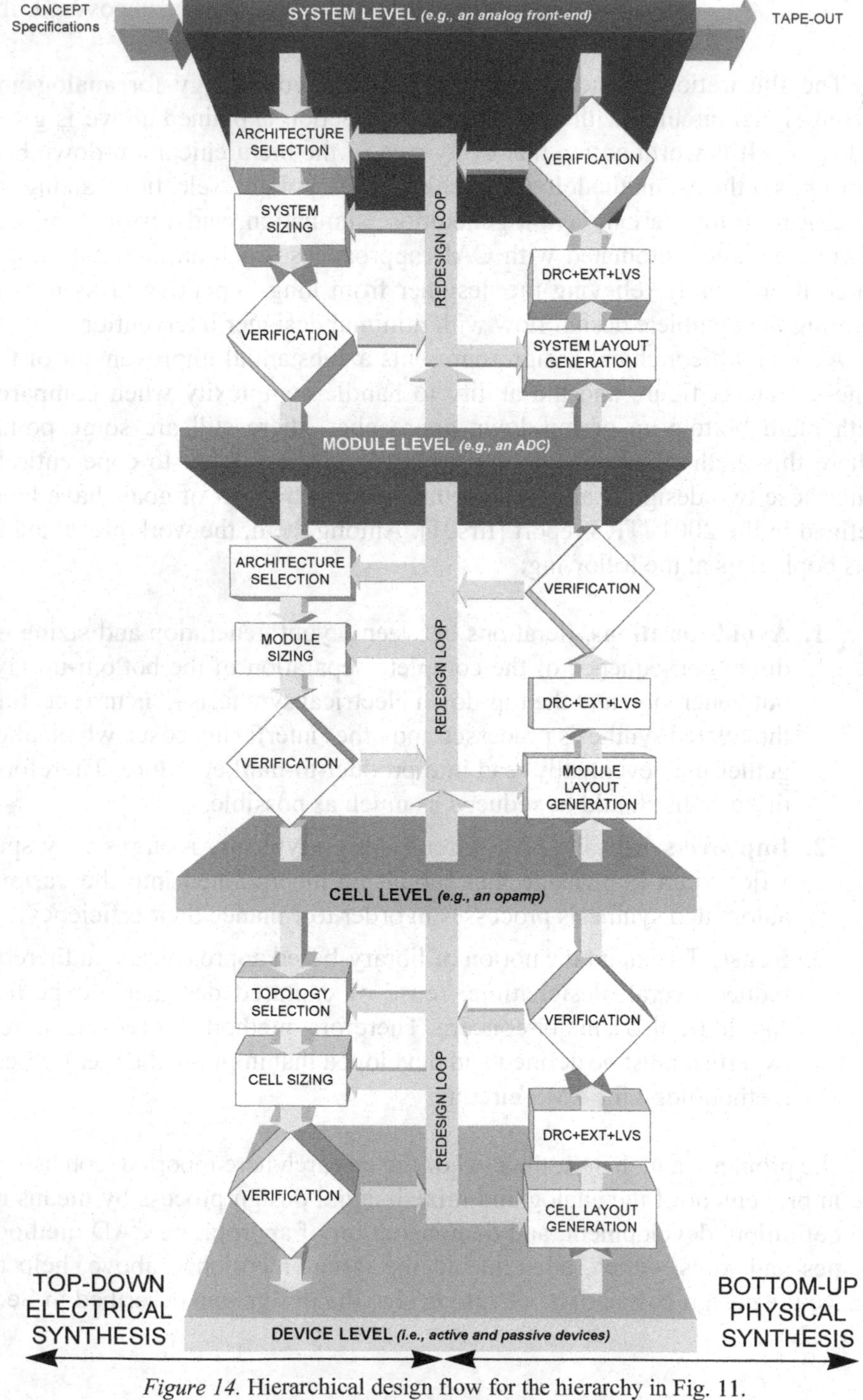

Figure 14. Hierarchical design flow for the hierarchy in Fig. 11.

3 SUMMARY

In the first part of this chapter, the motivations driving the research reported in this book have been introduced. The ever-widening gap between the silicon transistor capacity and the available design rate is posing a major, serious challenge in the semiconductor industry. Although the design capacity has been moderately improving thanks to design methodologies and computer-aided design tools so far, the situation is becoming critical as time-to-market pressures tighten and design complexity increases. If this scenario is worrying enough in the digital arena, where design methodologies and tools are relatively well-advanced, it turns alarming with regard to analog or mixed-signal circuits. It is concluded that further development of the design methodologies and design tools, and even the working out of new design paradigms in the area of analog and mixed-signal integrated circuits are, altogether, urgently required.

In the second part of the chapter, the problem is carefully analyzed. Concepts involved such as abstraction, hierarchy, and views have been revisited in the context of the analog and mixed-signal design flow. This flow has also been examined in detail in order to find out where it fails to cope with present design requirements.

In the following chapter, several approaches attempting to bridge the design gap are revised.

Chapter 2

A Reuse-based Design Framework for Analog ICs

The preceding chapter introduced the motivation for further improvement in modern analog and AMS circuit design methodologies. A first important effort towards such enhancement is automating the several steps involved in the AMS synthesis problem, especially the sizing and layout generation design steps. Commonly known as design automation, it can be carried out by following either a knowledge-based approach or an algorithmic, optimization-based approach. Both have their advantages and limitations. Increasing the efficiency of synthesis, however, can be achieved if a way to capture and at a later time use the reliable design knowledge that analog and mixed-signal designers insist on is devised. Circuit reuse, a new paradigm, is seen as a promising solution that, combined with the appropriate synthesis tools, can bridge the increasing design gap between what we can fabricate and what we can design and that calls for advances in AMS design methodologies.

This chapter presents a novel reuse-based design framework for AMS ICs. The cornerstone of such framework is the analog and mixed-signal reusable block and a clear, structured definition of this type of blocks is thus provided here. Furthermore, the development of these special blocks is supported with a systematic methodology for reusability. Last but not least, this reuse-based framework comprises a design reuse flow, by which reusable blocks are incorporated into the hierarchical top-down bottom-up flow introduced in the previous chapter.

1 DESIGN AUTOMATION

An important conclusion of the previous chapter was that there are still some aspects where the hierarchical methodology can be improved in order to bridge the existing design gap and, in doing so, efficiently cope with more and more stringent time-to-market pressures and increased analog design complexity.

The work presented in this book, aiming at three of these aspects (i.e., avoid iterations, improve synthesis, and exploit reuse) starts off by analyzing the existing synthesis approaches for AMS designs. A thorough understanding of their advantages and limitations is essential to find out why synthesis alone does not suffice to bridge the design gap, and how new design paradigms can help closing the gap.

1.1 Preliminary definitions

As explained in Chapter 1, a typical design flow for AMS circuits is composed of a number of design steps repeated top-down, from the system hierarchical level to the device level, and bottom-up for composition and verification. The steps between any two of these hierarchical levels are:

- architecture/topology selection,
- sizing,
- layout generation, and
- verification.

A **computer-aided design methodology** for AMS circuits aims at automating the different steps of the design methodology. Such a CAD methodology entails the integration of one or more CAD tools into a flow providing a specific level of automation. Here, automation refers to the ratio of the time it takes to manually complete the entire design for the first time, to the time it takes with the CAD methodology [Ocho98]. The CAD tools that can be used in any CAD methodology are explained below[1] (see Fig. 1):

- **Point** tools, to help the designer in a single particular task, relieving him from tedious, error-prone jobs. Examples are simulators (e.g., HSPICE® [Hspi04]), layout tools (e.g., Cadence® Virtuoso® [Virt00a]), and verification tools (e.g., Cadence® Diva® [Diva05]).
- **Design flow** (DF) tools, to manage the design database and the transformations the database undergoes from concept to *tapeout*. The scope is not restricted to a single design step, or between two point tools, but it may rather spread over the whole design flow as well.
- **Design automation** (DA) tools, supporting the automation of a particular design step, such as a decision-making algorithm for circuit sizing. These tools are also known as synthesis tools, and can be used for (system, module, or cell) sizing and layout generation.

---

1. The definitions given here have been adapted from the ones found in [Daems02].

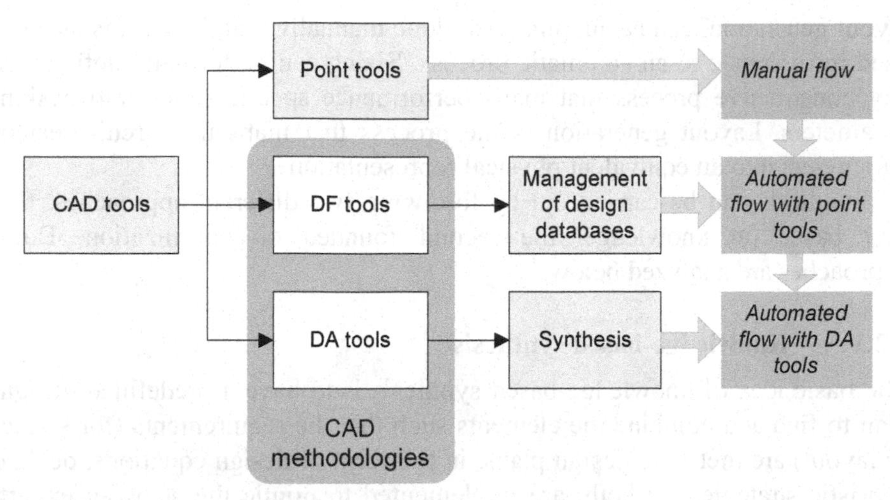

Figure 1. Design automation concepts.

A CAD methodology aims at automating either the design flow or a design step by means of DF tools and DA tools, respectively. A flow solely supported by point tools is called **manual design flow**, as the only existing degree of automation is that of the separate point tools used. When DF tools manage the input and output data of the point tools, the flow is known as an **automated design flow with point tools**. Examples of these flows are electronic database management systems such as Cadence® *Design Framework II* (www.cadence.com) and the *Falcon Framework*® from Mentor Graphics® (www.mentor.com). Last, when DA tools are incorporated, the resulting flow, denoted here as **automated design flow with DA tools**, achieves a higher degree of automation, enabling the designer to focus on other key tasks, such as the working out of new, better AMS architectures and topologies.

For the remainder of this book, the term *automation* is used to mean design automation, i.e., automation achieved by means of synthesis tools.

1.2 The two sides of automation

Synthesis is the process of creating a low-level representation (low-level abstraction) from a higher-level (more abstract) representation [Vsia01]. The synthesized representation should have the same function as the higher-level representation. Synthesis literally means combining constituent entities to form a whole unit. In circuit design, synthesis refers to the process of finding a set of elements and a way of combining them, such that when so combined to form a circuit, the circuit meets its requirements. Synthesis, both sizing and layout generation, can be automated or done manually, but the term is usually used in reference to an automatic process. Sizing can be formally defined as any constructive process that maps performance specifications into design parameters. Layout generation is the process that maps the circuit design parameters into an equivalent physical representation.

Synthesis can be carried out by following two different approaches, the first based on knowledge, the second founded on optimization. Both approaches are analyzed below.

1.2.1 Knowledge-based synthesis

The basic idea of knowledge-based synthesis is to have a predefined design plan to find and combine the elements such that the requirements (for sizing or layout) are met. The design plans, in the form of design equations, design heuristic strategies, or both, are implemented to mimic the steps an expert designer could take to reach an optimum solution of the design problem. Then, the underlying principle is to capture the expertise (knowledge) of a designer, so the synthesis method can reach an optimum solution using such captured reasoning.

Used for sizing, the design equations are formulated in such a way that given the performance characteristics, the components characteristics can be calculated. This means that for sizing at the system level, given the performance characteristics of the system-level circuit, the performance characteristics of the building modules can be calculated. Equivalently for module sizing, design plans allow the calculation of the required performance of the building cell or module-level blocks from the performance of the module-level circuit. Last, design plans are used to obtain the device sizes from the cell-level circuit performances. When some design parameters (i.e., module or cell performance characteristics, or device sizes) cannot be easily obtained by using equations, a specific design strategy is used. Such a strategy usually comprises information on how these design parameters are to be heuristically chosen in order to achieve correct performance.

Examples of knowledge-based sizing tools are IDAC [Degr87], OASYS [Harj89], BLADES [ElTu89], ISAID [Makr95], CADICS [Jusuf90], MIDAS [Been93], and COMDIAC [Porte97].

Used for layout synthesis, the intended captured knowledge refers to the procedures that expert layout designers use to improve the quality of the layout, spanning a wide variety of techniques, from specific placement[2]

[2] Roughly, the phases of layout generation are placement, where all circuit components are distributed over the layout plane according to certain quality rules (such as symmetry and orientation), and routing, where all components are interconnected. Chapter 5 delves deeper into these two phases.

strategies to improve device matching and minimize the layout area, to routing techniques to minimize the loading effects. Knowledge-driven approaches are developed to generate the layout of fixed-architecture/topology circuits, for which placement and routing are specified in advance. There are two types of knowledge-driven approaches, namely rule-based and template-based approaches. Rule-based approaches store the layout knowledge in a customizable rule set (defining a "good" analog layout) which are to be obeyed by whichever the layout placement and routing algorithms are used during circuit layout generation. Template-based tools are also developed to best use layout designers' expertise. The underlying idea is to capture this expertise in a pattern or template that specifies all necessary component-to-component and component-to-wiring spatial relationships. Besides, it must capture analog specific constraints like symmetry, device matching, and parasitic minimization. To generate a circuit layout from this template, the value of a set of electrical and geometrical parameters (e.g., the transistor width and length, or the maximum current density allowed to flow on certain layer type) need to be provided. Capture of layout knowledge through templates can be done by using procedural generators, actually the most mature layout technique used for analog circuits [Kuhn87] [Lamp99]. The mechanisms to describe these procedural generators can be specific layout languages or common spreadsheet interfaces, but both approaches are intended to code the analog-specific layout knowledge into the software itself.

Examples of layout generation tools based on knowledge are ALSYN [Bext93], the design-by-example approach presented in [Conw92], the procedural layout generator in [Kuhn87], BALLISTIC [Owen95], ICEWATER [Harv92], and [Hend93], as well as special-purpose layout generators for digital-to-analog data converters (DAC) [Neff95], switched-capacitor analog-to-digital data converters (ADC) [Jusuf90], and filters [Yagh88].

1.2.2 Optimization-based synthesis

In these tools, the synthesis problem is translated into a function minimization problem that can be solved through numerical methods [Temes67]. Optimization techniques were used for analog design in the 1950s [Aaron56], and they were rediscovered when faster computers and improved simulators made them more practical for industrial design and research [Nye88].

For sizing, the basics of optimization-based synthesis are illustrated in Fig. 2. The optimization process is an iterative procedure, design variables being updated at each iteration, until an equilibrium point is reached. The degree of compliance of the design performance with the optimization goals at each iteration is quantified through a cost function. In general, any implementation of this technique relies on two separate modules, design information flowing between them. These modules are the performance evaluation tool and the optimization tool itself. Whereas the latter deals with the cost function and explores the available design space to minimize such function, the former provides a way to evaluate the circuit optimality with regard to the intended requirements. Depending upon the type of performance evaluation as well as the optimization technique used, different approaches can be distinguished.

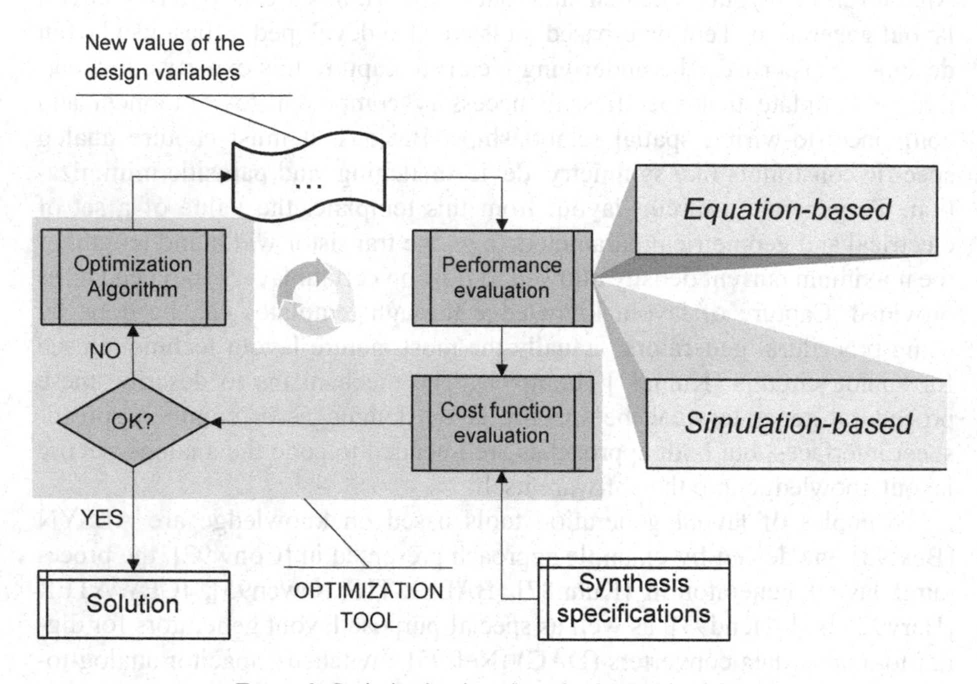

Figure 2. Optimization-based synthesis used in sizing.

With respect to performance evaluation, a possibility is to use equations, which can be derived either manually or by using symbolic analyzers [Fern97] [Giel90]. In this case, the optimization is known as equation-based. There also exists the possibility of evaluating the performance with the help of a simulation tool. The optimization is then known as simulation-based.

With respect to the optimization techniques, there exist two approximations: deterministic and statistical. Both techniques are described below:

- **Deterministic methods**, where updating requires information on the cost function and on their derivatives. An important disadvantage is that only changes of design parameters that make the value of the cost function decrease are permitted –the optimization process is quickly trapped in a local minimum of the cost function, so the usefulness of these techniques concentrates on the fine-tuning of sub-optimal sizings. This dependency on the starting point could be palliated if several sizing processes are started on several different points of the design space, thus increasing the likelihood of arriving at the global minimum. However, the sizing time increases as well.

- **Statistical methods**, where design parameters are varied randomly and hence information on the derivatives of the cost function is not required. The main advantage of the statistical techniques over the deterministic ones is the capability to escape from local minima, thanks to a non-zero probability of accepting movements that may increase the cost function. An example is the simulated annealing technique [Laar87]. This powerful optimization algorithm permits the global minimization of a multi-variable function at the price of a lower convergence speed. Therefore, it is suitable for the search in large design spaces. The price to pay is, however, a larger computational cost. Genetic algorithms[3] [Holl92] are also able to jump out of local minima, albeit an intensive computation capability is also required.

Examples of optimization-based sizing tools using simulation are DELIGHT.SPICE [Nye88], AZTECA [Horta91], CATALYST [Vital93], MAELSTROM [Kras99], ANACONDA [Phel00a], and FRIDGE [Mede99]. Examples of tools using equations are OPASYN [Koh90], OPTIMAN [Giel90], STAIC [Harv92], the constrained optimization technique developed by Maulik et al. [Maul93], ASTRX/OBLX [Ocho96], GPCAD [Hers98], and SD-OPT [Mede99].

When optimization-based synthesis is used for layout synthesis, placement and routing stages of the layout generation are determined by an optimization tool according to a certain cost function. This cost function typically considers minimization of some design aspects such as area and net length, while penalizing violation of some analog design constraints, such as device mismatch, loading capacitances, and crosstalk. The quality of these optimization-driven tools is mainly determined by the set-up of this cost function.

Depending upon the way of deriving the cost function and dealing with constraints on analog performance, two categories are usually considered [Lamp99]. The first group is composed of heuristic-based approaches, where layout-induced degradation on the circuit performance is taken into account

3. These algorithms are part of evolutionary computing techniques [Bäck93] which, simply said, try to solve optimization problems by an evolutionary process starting from a population of trial solutions and resulting in a best (fittest) solution (survivor).

by classifying nets according to their sensitivity and circuit function. A notable improvement is accomplished by the other group of approaches, whose operation is based on performance-driven optimization of the circuit layout (also known as constraint-driven optimization). Unlike heuristic approaches, where no quantification of the performance degradation is done during the layout synthesis, performance-driven tools try to measure the layout-induced degradation and keep it below desired margins. In this way, the impact of each layout parasitic is weighed out according to its effect on the circuit performance [Mala96]. Early reported contributions modeled the effect of layout parasitics by firstly using sensitivities and then mapping the performance constraints for the circuit to a set of constraints on the layout parasitics [Chou90a] [Chou90b] [Chou90c]. Later approaches have shown that this intermediate mapping (i.e., resorting to constraints on parasitics) can be skipped altogether [Lamp99].

Examples of heuristic-based layout generators tools are ILAC [Rijm89], ANAGRAM [Garr88], KOAN/ANAGRAMII [Cohn91], and LADIES [Moga89]. First contributions of performance-driven tools were to perform channel routing [Chou90a] [Chou90b] [Chou90c]. Other examples of performance-driven layout generators are ROAD [Mala90], [Mala96], ANAGRAM III [Basa93], PUPPY [Char92] [Char94], SPARCS-A [Felt93], LAYLA [Lamp99], and GELSA [Prie01].

1.2.3 Quality metrics for analog synthesis

There are several metrics for measuring the quality of the above-described synthesis approaches [Neff95] [Ocho98] [Hjal03]. Three categories are here distinguished:

- **Quality of the solution**. Possibly the most importance metric, it must be particularized for sizing and layout synthesis.
  - For sizing, quality refers to **accuracy** and **robustness** of the solution. Accuracy reflects the discrepancy between the synthesis tool's internal performance prediction mechanisms and the actual performance of the obtained solution, which may include the layout-induced degradation. Robustness refers to the ability of the sizing tool to create circuits that are tolerant of manufacturing line and operating point variations.
  - Regarding layout synthesis, quality, somewhat more difficult to measure, is typically considered as a comparison between the layout generated by the tool and a custom layout generated by an analog layout expert. Quality metrics such as layout area, matching,

symmetries, and minimization of parasitics, are typically compared.

- **Degree of automation**. As said above, it refers to the ratio of the time it takes to manually design a circuit for the first time to the time it takes to design the same circuit with the synthesis tool. Considering this, the following aspects are of concern:
  - **Running time**. This metric refers to the time elapsed between the start and the end of a synthesis run. Running time is continuously being affected by the increase in computer capability: today's developers of analog sizing and layout synthesis tools have an advantage over their predecessors, considering methods that, only 15 years ago were intractable and too costly 5 years ago. Ever increasing circuit complexity, however, counteracts this advantage so running time remains a meaningful metric figure.
  - **Design entry time/effort**. It refers to the preparatory time and effort required to render the problem in a form suitable for input to the synthesis tool. This time is usually longer than the synthesis running time itself. Usually, the time/effort required is considerably longer than the time/effort required to do the full custom design. Note that the time/effort also depends on the level of knowledge that the user has on the synthesis tool requirements.
- **Breadth**. It refers to the set of analog design problems the tool can deal with. This is particularly important for analog design, where there are many different types of architectures, each one posing different synthesis problems. An analog synthesis tool that allows ease of incorporation of new design problems (usually known as an open tool) may succeed in the long run, whereas tools devised for one problem but which are difficult to apply to similar problems (i.e., a closed tool) quickly become obsolete.

1.3 Knowledge versus optimization-based synthesis

Based on the set of quality metrics described above, it is possible to make a comparison between synthesis based on knowledge and synthesis based on optimization. Sizing and layout generation are considered apart.

Although knowledge-based sizing does reach solutions quickly, provided the design plan has been already derived, this approach suffers from several drawbacks. First, the quality of the solutions in terms of both accuracy and robustness is not acceptable. The very concept of knowledge-based sizing, i.e., to explicitly solve the design equations and provide strategies and additional rules of thumb, and the need for easy reformulation of the design objectives, forces the design equations to be simple. Therefore, modern accurate device models cannot be used, as large deviations of the real performance from the predicted one may result. In a worst-case scenario the error is several hundred percent. A second disadvantage is that they are, in general, closed tools, that is, they are limited to a reduced number of architectures/topologies and design objectives. Faced with changes in the circuit architecture/topology or even in the specifications, the design plans should be remade, which is very expensive. Major drawbacks are, however, the large preparatory time/effort required to develop a design plan (taking four times longer –if not much more– than the time/effort required to manually design a circuit [Been93]) and the likely difficulty in using them in a different technology.

Optimization-based sizing circumvents the need for a detailed design plan. The incentive of using equations lies in the computation speed, though the results are always approximate, even more when the equations are obtained manually, so accuracy may be compromised. In this case, and just like knowledge-based sizing, the tool is closed because new equations have to be obtained for each new architecture or topology. Furthermore, design entry time/effort is large. When equations are derived by a symbolic analyzer, the sizing approach becomes more open, because the equations are generated automatically. The complexity of the circuit (in terms of the number of circuit devices and nodes) is however limited by the possibilities of the analyzer. Besides, large-signal, strongly nonlinear specifications cannot be easily handled [Fern97]. Preparatory time/effort is slightly improved with respect to manual development of equations; however, the designer has to “instruct” the symbolic analyzer to derive appropriate expressions, which can be time-consuming as well. It is also worth noting that symbolic expressions are always attained at the expense of accuracy, since an approximation is required in order to reduce the huge complexity of the analysis problem. With optimization-based sizing using simulation, better accuracy can be attained as long as accurate simulation models are used, but the running times are typically much longer than sizing with equations. Furthermore, the capabilities of the simulator determine whether the tool is closed or open. In principle, preparatory time/effort of optimization-based sizing approaches using simulation is similar to that spent in verifying the design in the handcrafted sizing approach (a simulation testbench, from which the circuit performance is obtained, must be also derived). Definition of the circuit’s explorable design space as well as other possible inputs, whose complexity depends on the sizing tool, requires additional effort/time before the sizing can be carried out.

The quality of the obtained layout when using rule-based techniques largely depends on the set of rules, which must be specifically developed and adjusted for each new circuit. Besides, creating an adequate set of rules can be very time-consuming. Knowledge-based layout synthesis features a high quality of the produced layouts if the template approach is used. As the experience of the layout expert is encapsulated with the template, techniques such as the use of matching-improving structures or symmetric routing to minimize the impact of layout imperfections can be directly stored and reproduced whenever necessary. In addition, the running times (e.g., a few seconds [Cast02b]) are quite acceptable as no time-consuming algorithm needs to be executed. Nevertheless, two major drawbacks are worth noting. First, since the template must be created specifically for each circuit, a large time/effort is typically required. Second, as a direct consequence of the latter drawback, generality of this type of layout synthesis methods is poor. Note that rule-based approaches suffer from these two drawbacks as well.

The main benefit from using optimization-based layout synthesis is their generality: in principle, they can be applied to any analog circuit. The drawbacks are the complexity of the optimization problem, the difficulty of the cost-function set-up (which largely determines the quality of the obtained solution), and the excessively long turnaround times.

The above comparative analysis is summarized in Table 1.

Based on this analysis, it is difficult to finally find out the best approach to boost the hierarchical top-down bottom-up design methodology (for either

Table 1. Comparison between knowledge and optimization-based analog synthesis.

Metric	*Knowledge-based*			*Optimization-based*		
	Sizing	*Layout synthesis*		*Sizing*		*Layout synthesis*
		Rule-based	*Template*	*Equation*	*Simulation*	
Layout quality	N/A	+	+ +	N/A	N/A	+
Accuracy	–	N/A	N/A	–	+	N/A
Robustness	–	N/A	N/A	–	+	N/A
Run time	+	+ +	+ +	+	–	– –
Design entry time/ effort	– –	– –	– –	– – / –[a]	–	–
Breadth	–	–	–	+ / –	+ / –[b]	+ +

a. Manually-derived equations / Using a symbolic analyzer.

b. It depends on the simulator.

sizing or layout generation), since each method has its own specific strengths and limitations. In this respect, the work presented in this book considers a design paradigm shift that takes into account the following two aspects:

- Design knowledge plays a very important role: for instance, better layout quality can be attained if the expertise of an analog layout designer can be captured within the problem definition. What is more, analog design knowledge, which efficiently drives knowledge-based approaches, can also be used to procure smarter and faster optimization-based sizing tools, by guiding what would otherwise be a "blind" exploration of the design space. Whatever the adopted approach, a compromise between running time and quality of the synthesized solution is vital to cope with main driving forces of today's semiconductor industry (i.e., time-to-market pressures and complexity).
- All synthesis approaches present inadequately long times or require a considerable effort for design entry. This drawback can be largely alleviated if the design entry has been previously developed and it is prepared in an easily reusable form.

2 CIRCUIT REUSE

In this section, the concept of circuit reuse, as a new design paradigm, is introduced and insight into the advances in this field is provided, putting special emphasis on the analog and mixed-signal arena.

As stated in Section 2.2 of the previous chapter, the ITRS report envisages a set of precepts for enhancement of the analog design [Itrs01]. Among these, profiting from already designed circuit blocks has been proposed as fundamental to improve design productivity. A framework that eases the incorporation of these previously designed circuit blocks is considered a promising solution to bridge the widening gap between available number of transistors and available capacity to design them (see Section 1.2 on page 5).

2.1 Preliminary definitions

There are several important concepts related to reuse in the field of integrated circuits: **design reuse**, **design for reusability** (also known as design for reuse), and **reusable block**. A clear definition of these concepts is required.

Design reuse can help the designer to easily manage the complexity of a system. For instance, reuse is quite common among board designers, where many different chips are continuously re-utilized to compose many different

Design reuse. In information technology, design reuse is the inclusion of previously designed components (blocks of logic or data) in software and hardware. The purpose of design reuse is to be able to efficiently use a previous successful design experience in another system and/or a different fabrication processes [Dess01a]. The degree of reuse is measured by the amount of information and experience that is transferred from the successful first design to the subsequent ones.

applications. Furthermore, though activities of innovation and steady thinking about ways to improve performance shall not be discouraged, time-to-market pressures require reusing, whenever possible, existing designs. The savings in time come from several sources: the needed cell (or something close) is already at hand; it is often proven and debugged, with test data available. Design reuse, then, can make it faster and cheaper to design and build a new product, since the reused components will not only be already designed but also tested for reliability.

Design reuse is not new. From the earliest days, the first task of an engineer facing a new design problem has been to look over old designs to find a circuit topology or an innovative idea that might be successfully applied to the problem. Such an informal process has been supported, in many companies, with libraries of previously designed blocks, resulting in a rapid design process provided that no modifications of the existing circuit are required [Neff95]. Note that effective reuse, though, takes much more than just gathering pre-designed components in a library.

When a design has been successfully accomplished for the very first time, it should be well documented. Besides, it should be better to maintain the same design choices, trade-offs, and heuristics used if the block is to be reused in a different project, so as not to re-start the whole design process from scratch. However, this is not likely to happen. A systematic way to create reusable designs should be therefore defined.

Design for Reusability. Methodologies, tools, and services to enable the creation of reusable designs [Keat99]. Design for reusability is the cornerstone of a compelling design reuse framework. Reusability is the degree to which a circuit block may be used again in other instances for which it may or may not have been specifically intended [Vsia01].

The design reuse and design for reusability concepts are in a sense interchangeable, since the design reuse concept includes the design for reusability concept, but the opposite is not true (the final chip created with the help of any design reuse method does not have to be reusable itself). In any case, a reuse-based design framework is composed not only of the design for reusability elements, but of the methods, tools, and services necessary to incorporate the reusable blocks into the design flow. Note that for higher efficiency and better acceptance from the design community, the reuse-based design framework should be supported by DA and DF tools.

Returning to the board example, a key advantage is that interfaces are standardized and proper documentation on the chip is usually available. The goal of the design for reusability in the IC world can be somehow compared with that in the board world. If encapsulation of chips as well as board materials and formats are standards, an analogous situation should be pursued in the IC world: a systematic methodology, accompanied with suitable synthesis tools, must be developed, and a unified description of the reusable IC block must be agreed upon.

Reusable block. Circuit blocks produced through design for reusability. These blocks are also known as *Virtual Components* (VC) or *Intellectual Property* (IP) blocks [Keat99] [Seep01] [Vsia01]. *IP repository* is the central library of IP blocks and the tools, infrastructure, and services to support the library.

In reusing a circuit block there are two possible scenarios:

1. **Design Retargeting**, when any of the performance specifications of the circuit block is required to change to cope with new design objectives.
2. **Design Migration**, when the circuit block is to be implemented in a fabrication process different from that it was firstly designed for. Design migration implies always a **database migration**, i.e., an adaptation of the circuit block associated files (e.g., device models, GDSII[4] layout file, etc.) to the new fabrication process. Design migration may additionally require a design retargeting as changing the process may entail a degradation of the circuit block performance.

4. GDSII (geometrical data stream II) is the standard file format for transferring/archiving two-dimensional graphical design data. It contains a hierarchy of structures, each structure containing elements (boundary/polygon, path/polyline, text, box, structure references, structure array references).

Both scenarios pose many different problems for the design reuse and design for reusability methodologies. The efficiency with which a reusable block deals with the two scenarios determines its reusability.

In this way, a classification of the IP blocks can be made, attending to the degree of reusability and, therefore, to the different design necessity it is devised to fill in. Before focusing on AMS design, it is interesting to have a look at IP classification for digital design.

2.2 Digital design reuse

According to the Virtual Socket Interface Alliance (VSIA)[5] [Vsia01] and the digital-oriented Reuse Methodology Manual [Keat99], IP blocks can be classified into the following categories:

- **Soft IP**. These reusable blocks are delivered in the form of synthesizable HDL (Hardware Description Language). They have the advantage of being more flexible and the disadvantage of not being as predictable in terms of performance and area.
- **Hard IP**. These blocks have been optimized for power, size, or performance, and mapped to a specific technology. Examples include fully placed and routed netlists, optimized for a specific technological library, a custom physical layout (e.g., as a GDSII file), or a combination of both. Hard IPs have the advantage of being much more predictable, but consequently less flexible and portable due to process dependencies. Already optimized, redesign of this type of IP is not feasible. Database migration turns out very difficult and intensive research is being done in this area [Keat99][6].
- **Firm IP**. These reusable blocks have been structurally and topologically optimized for performance and area through floorplanning/ placement using a generic technology library. The level of detail ranges from region placement of RTL sub-blocks, to relatively placed datapaths, to a fully placed netlist. Often a combination of these approaches is used to meet the design goals. Firm IPs offer a compromise

5. VSIA, formed in September 1996, was established to unify the mix and match of IP blocks, by selecting appropriate design representations –formats– for virtual component deliverables and by defining required and recommended design practices for IP blocks, covering logical design, physical design, test, and bus interfaces.

6. A intermediate solution to layout database migration is optical shrink, by which the mask layer size is reduced while the relative device sizes, in the new technology as in the old one, are kept constant. Note also that this solution is only valid when technologies are similar (e.g., between processes from to the same foundry).

between soft and hard, being more flexible, more portable than hard IP blocks, yet have more predictable performance and area than soft IP blocks.

A tabular summary of the salient differences between the three types of IP blocks is provided in Table 2.

Table 2. Digital IP classification.

IP	*Performance*	*Retargeting*	*Database Migration*
Soft	Not predictable	Feasible	Unlimited
Firm	Predictable	Feasible	Through library migration
Hard	Very predictable	Impossible	Very difficult

2.3 Analog design reuse

The notion of analog IP has been the subject of heated debates (e.g., [Ohr02] [Koch03]). Analog IP, like analog design automation, is lagging several generations behind existing digital technology. Digital IP is well supported by synthesis tools, there is an extensive repository of available digital IP blocks, and most well-accepted, commercial design environments are geared up for these digital IPs. Analog IP, on the other hand, still lacks of sufficient availability and dedicated research and, more importantly, of reliable silicon proof, remaining thus as an interesting but still immature alternative. Moreover, and though the interfaces between analog IPs have already been covered by VSIA [Vsia99], there is no consensus on the requirements of analog IP [Vand03].

Therefore, a clear definition of analog IP is the starting point for a productive, capable reuse-based design framework for analog ICs. When compared to digital, it results naive and unrealistic to expect an exact replica of the digital definitions for the analog counterpart. The reasons for such statement are:

- On average, dealing with analog design retargeting is significantly more involved than dealing with digital design retargeting due, largely, to the much more advanced digital CAD tools and better structured methodologies. Consider, for instance, retargeting a 12-bit analog-to-digital converter to 16 bits, which implies a lot of design choices and trade-offs at the module, cell, and device levels. Far from automated, these subtle design decisions are left to the expert analog designer.

- Regarding design migration, compensating for the changes in the process features and device models may require a resizing of the circuit. For example, porting the layout of an opamp from one technology to another may result in an unstable design, caused by changes in the gain and phase margin. However, not only a resize (i.e., retargeting) is to take place, but layout modifications as well (i.e., database migration), which can be an even more laborious, complex task. Besides, optical shrink alone is not appropriate for analog database migration due to the drastic impact that a process change has on the circuit performance [Gilb02] [Dess01a].
- The adoption of someone else's design without really understanding its limitations and the context it was firstly developed for can be hazardous. Without a meticulous assessment of its suitability to the present environment or clear guidelines for the redesign for suitability, the design may give rise to further problems down-stream [Gilb02]. Since analog design is long founded on specialized, subtle heuristics and designer's expertise, reusing the circuit block is usually difficult without a priori knowing those considerations. A clear example is found in analog layout: implementing a particular structure in a particular style (e.g., common-centroid) may result critical to improve device matching; this expert knowledge is essential for future, successful reuse. The traditional analog design flow does not encapsulate the knowledge of the designer (what to tweak, change, or be watchful of in a design), and, as such, it cannot be used later by a less experienced designer.
- Reusing an analog circuit firstly designed to work in a particular physical environment, may entail a change in the achieved performance for a different environment. Digital switching noise, injected through the substrate, is frequently identified as the culprit.
- The effect of layout parasitics on the design performance is a crucial aspect of the analog design; reusing a circuit for a different set of specifications or in a different fabrication process, changes the value of these parasitics, thereby impacting the performance in a different, possibly dangerous way.
- It is much more difficult to approach analog design from a higher level of abstraction, due to the loose form of hierarchy of analog design. Consequently, synthesis is much more difficult when compared to digital. For instance, capturing of the design plan to realize the complete hierarchical design comes at the expense of precious design time of the expert designer [Vand03].

An approach to the analog IP block definition has been described in [Vand03], distinguishing the following two categories:

- **Commodity IP**. These blocks do not fully challenge the capabilities of the technology, and it is suited for full automation from specifications down to layout. Providing that the appropriate synthesis tools become commercially available, commodity IP can be delivered as soft IP. Commodity IP blocks are well suited to synthesis automation, and the number of times they are planned to be reused is high.
- **Star IP**. These are designs on the edge of the technological boundaries; their architecture/topology and achievable performance is very susceptible to technological changes and evolve with technology. It is to be provided as hard IP and its reusability is thus quite limited.

Analog IP can be simple cells such as opamps, OTAs, comparators, etc., or more complex blocks like DACs, ADCs, and PLLs. There is also a sustained debate around which kind of blocks (simpler or more sophisticated) are best suited to promote design reuse among IC designers [Ohr02]. Whereas some authors claim that, due to the many subtle details of analog design, analog reusable blocks should trade properties only in the same process and for the same specific applications, others assure that commodity analog IP trade holds the most promise of succeeding [Ohr03].

Even in the digital arena, quite a few managers are very reluctant to make an investment on effective, systematic reuse. An estimate is that it takes 2-3 times the effort to develop a digital reusable block than it does to design the same block for a single use [Keat99] [Seep99]. This is mainly due to the analysis and considerations made to turn the block reusable, to the extra verification, to the testbench, that increases in complexity to ensure total functionality and validation, and to the documentation. Furthermore, transitioning a whole design team to reuse-based thinking requires an additional investment in retooling and management practices.

The benefits of design reuse are, however, large in the long term. Another estimate is that integrating a reusable block into a design means 10 to 100 times less effort than designing the block from scratch [Keat99]. Note that this benefit is only derived as long as the investment to make the block reusable has been made. Some authors think that there is a compelling argument for design reuse: we simply will not be able to build tomorrow's chips without it. Although the benefit is large but delayed, the needed investment, both in time and resources, is also considerable (this is why books on software reuse claim that reuse is but a management and cultural problem rather than a technical one).

What most authors agree upon is that analog IP is useless unless intents and decisions of the previous designer are captured [Koch03]. Full capture and archiving of the constraints that experienced analog designers insist on, is very important for subsequent reuse [Giel01]. Reuse of this kind of knowledge is thus essential. In fact, another important reason why analog reuse-based strategies are not widely in place today is the extra effort required to package all that knowledge (specifications, models, layout experience, and so on) into a reusable form. Furthermore, design collateral in the form of numerous associated files, such as test setup documents, results, and other documentation not typically managed by existing design environments should be packaged within the analog IP block as well, as it may affect the ability to share and reuse IPs across designs and design teams and between companies.

Last, it is also worth noting that for the reuse-based framework to be widely accepted by the analog design community, it is recommended that both the design reuse flow and the design for reusability methodology be implemented in a widespread-use design environment (e.g., Cadence's® *Design Framework II*, Mentor Graphics's® *Falcon Framework*®, or Synopsys's® Milkyway™). This would indeed minimize the eventual hesitancy towards the analog reuse-based approach.

Boosting existing analog design methodologies to higher levels of efficiency is not possible unless analog synthesis methods and reusable analog blocks are well accepted [Vand03]. For this to happen, intensive research and efforts are required in order to improve current synthesis methods, provide techniques to create reusable blocks, and unify the standards required to enable reuse from multiple sources. Analog design community still awaits silicon proof that automation and reuse do not come at the expense of reduced performance.

2.4 Other approaches to analog reuse

Although analog reuse is still behind its digital counterpart, it is, fortunately, the subject of ongoing activity in industrial and academic research. This section examines reported approaches to analog reuse, some previous, others coincident in time with the development of the approach reported in this book.

VSIA is helping AMS in that it is providing the structure to better define the hand-off interface between IP authors and IP providers; it does not directly help portability and reuse [Vsia99].

Reuse of analog circuits for a change of the fabrication process and for a change in the performance has been investigated in [Fran99b] and [Phel00b]. The two former approaches rely on the existence of a previous working design and on the assumption that there is full access to such design, but no access to the original designer.

In [Fran99b], an example of retargeting for technology porting of analog circuits through qualitative reasoning is presented. The retargeting starts with a guided scaling step to produce a starting point. Afterwards, fine-tune of the circuit to correct for possible non-fulfillment of performance specifications is carried out by using a qualitative reasoning procedure. This is done through a special dependency matrix, which qualitatively describes the dependency of each performance specification on each design parameter. During layout generation, relative positions as well as aspect ratios of the circuit building blocks are kept constant. Assembly of the complete layout is done manually in a bottom-up manner.

In [Phel00b] an equalizer/filter front-end module is retargeted for a different value of only one performance specification, the output noise. An optimization-based approach using simulation is used. Population ideas from ANACONDA [Phel00a] with some of the annealing ideas from MAELSTROM [Kras99] are applied. To cope with analog complexity, the authors propose a hierarchical decomposition in which cell-level macromodels are used to search for an optimal module-level design. To reduce iteration steps, the retargeting is carried out concurrently. That is, a full cell-level sizing evolves for each cell at the same time searching for optimal module-level design is done. This was made possible through a sophisticated workstation-level parallelism using a computer farm of 20 to 30 Sun UltraSparcs, which, however, connot completely handle the exponential complexity of the problem as the number of levels increases.

The original designer is substituted either by another designer who builds the dependency matrix [Fran99b], or by extensive computing optimization [Phel00b]. Therefore, most of the original design intentions and considerations are lost for subsequent reuse. Whereas no performance retargeting is considered in [Fran99b] (i.e., the ported circuit is retargeted to address the same original performance), no technological considerations are taken in [Phel00b]. Furthermore, the above approaches treat only the sizing phase of the design, the layout being considered as a separate phase, handled with dedicated point and synthesis tools. The effect that layout-induced parasitics have on the circuit performance is not explicitly treated during sizing.

More recently, design methodologies based on layout-oriented synthesis approaches that allow capturing design knowledge for eventual reuse with a close interaction between electrical and physical design has been presented in [Dess01a]. The methodology in [Dess01a] uses two knowledge-based synthesis tools. On one side, circuit-sizing procedures are coded in the COMDIAC knowledge-based sizing environment [Porte97]. On the other side, the methodology relies on technology and size-independent layout templates that contain physical layout information related to the circuit, written in the

CAIRO layout language [Dess99], which incorporates procedural device generators with several analog layout constraints and a routine for area optimization. Physical layout constraints including layout parasitics, global aspect ratio, and reliability design rules are taken into account during circuit sizing. A novel technique to avoid sizing-layout iterations is presented that relies in the extraction of layout parasitics during the sizing process, thus allowing exploring the design space in presence of layout parasitics. Despite the incontestable advance that this work brings in towards analog design reuse, several drawbacks limit the application of the methodology.

First, the transmission of high-level specifications down to the cell hierarchical level is still done manually. This lack of automation precludes a clear and standard definition of the analog reusable block. That is, the knowledge required to perform high-level sizing is missing, so it cannot be conveniently reused. Moreover, such lack of automation entails that neither time-to-market pressures nor analog design complexity are better undertaken than with the handcraft-based approach. Second, accuracy of solutions obtained with knowledge-based sizing is inadequate for high-performance analog design (see Section 1.3 of this chapter). Better accuracy comes at the expense of more detailed equations and complex design strategies. Deriving and encapsulating such high-accuracy design plans requires a costly effort. Furthermore, the effort is tailored to each new reusable analog circuit, thereby increasing its development cost and, possibly, reducing its derived benefits. Third, although the design for reusability is illustrated with an example, no general, well-defined requirements on analog reusable blocks are provided. A less important, but still worth noting drawback is that the presented methodology has been implemented with in-house CAD tools (e.g., COMDIAC and CAIRO), thereby minimizing its impact on the analog design community.

In [Plas01], AMGIE, an analog synthesis flow covering from topology selection to layout generation, is presented. Topology selection from a library of predefined circuits is carried out by using a set of three filters applied to a topology database. The sizing system is a compilation of several tools combined into an optimization-based framework using equations. The solution is verified by using simulations. Generation of equations is done by a symbolic analyzer, which automatically derives the small-signal equations required in AC analyses directly from the circuit topology. All other equations (e.g., nonlinear equations) must be entered by the expert designer. The symbolic equations are simplified in these tools in order to reduce the size of the expressions evaluated inside the optimization loop. Automated setup of design equations directly from the circuit topology is only intended to be used by expert users that can create libraries of sizing plans, subsequently used by ordinary users to size a specific topology[7]. The time to create such a sizing plan for an opamp-like circuit is about 8 hours for an expert user. The user can choose among several optimization algorithms such as simulated annealing for global optimization, and local, gradient-based, methods. The layout generator LAYLA [Lamp99] uses a performance-driven problem formulation in order to minimize the performance degradation in the layout step. The simulated annealing optimization algorithm is used to minimize the impact of interconnect parasitics, mismatch, and thermal effects during the floorplaning.

This approach lacks, however, the reusable factor that is deemed essential to boost today's IC design methodologies. Besides, no truly hierarchical design flow is reported. That is, no systematic way for transmitting performance specifications from higher levels of the hierarchy (e.g., system or module levels) down to lower levels (i.e., module or cell levels) is reported.

In [Vand03], an example of cell-level analog soft IP is described. This IP is to be synthesized with AMGIE. No low-level knowledge (e.g., layout details) is, however, provided. The layout is generated partly automatically (with LAYLA), partly manually. The test case presented shows that design automation comes at no expense of reduced performance. However, no clear standardization on analog IP is provided: the work presents but an AMGIE-tailored example of soft IP, without drawing general guidelines for creating reusable analog IP.

A partial solution to design knowledge encapsulation has been presented in [Liu03], where a system to support designer's knowledge as part of the archival circuit representation is described. This system introduces the notion of *active comments*, which are schematic annotations that control simulation files. The active comments are broken into two main forms: measurement comments, which are similar to other integrated systems, and constraint comments, which are like assertions in normal programs. Transforming the constraint from absolute value to relative or circuit-specific value is required to make the comments technology-portable. However, it is not always straightforward and, most of the times, it requires a deep understanding of the underlying circuit.

The work presented in [Hamo03] presents a definition of analog IP, based on the firm IP category inherited from the digital domain. In terms of views, the proposed analog IP is composed of a schematic view, a behavioral view, an analytical view (i.e., computational models for initial configuration selection), a physical view, and a testbench suite (i.e., to generate the performance measures). A *hardening* flow is also roughly described that transforms the

---

7. The sizing plan should not be confused with the design plan used in knowledge-based systems. In this case, the plans consists in design equations for the optimization engine to operate on and additional knowledge added by the designer in order to reduce the size of the problem.

firm analog IP into a GDSII file. Unfortunately, although interesting, the paper lacks of enough detail regarding the description of the analog IP at all proposed views, as well as the hardening flow (for example, neither information on the involved sizing procedures nor relevant instructions on how to make reusable layout views –both for retargeting and migration– are provided).

Above approaches are academic approaches and none of them is commercially available. There are, however, clear signs that the interest in analog design automation is rising[8]. A few companies are presently commercializing several tools for analog synthesis and reuse-based frameworks. For instance, by the time this book is being published, Cadence® is offering design solutions that cover analog cell sizing (NeoCircuit® [Neo04a]) as well as analog layout generation (NeoCell® [Neo04b]).

3 THE REUSE-BASED DESIGN FRAMEWORK

Reuse-based practices, together with adequate synthesis tools, provide one of the most promising solutions to face stringent time-to-market pressures and increasing complexity of AMS design. Reusable analog IP should not be viewed, then, as a technical desire but as a business necessity.

Although important efforts have been and are still done to boost current AMS design methodologies, it is also true that none of the above reviewed approaches has furnished the design community with a systematic approach to analog and mixed-signal reuse. The solutions to date either have been partial, covering only a portion of the design flow, or do not provide a clear and structured definition of what an analog reusable IP is or how it can be attained.

In such scenario, the work developed in this book tries to provide a unified framework for the reuse-based design of AMS circuits. This goal encompasses the following issues:

- A clear description of the analog reusable block.
- A design for reusability methodology by which the analog reusable block can be realized.

---

8. An analyst at Lehman Brothers recently wrote that "With some of today's analog and mixed-signal EDA toolsets currently being based on a very manual methodology, we see automated analog and mixed-signal toolsets representing a huge market opportunity. It is potentially among the hottest market segments in EDA that we have seen".

- A structured design reuse methodology that tries to improve the hierarchical design methodology.

As shown in Fig. 3, both the design reuse and design for reusability components are based on analog synthesis tools and flows. Design for reusability, relying upon CAD methodologies and tools, allows to produce analog reusable blocks; the automated flow with DA tools acting on these reusable blocks is the foundation of the design reuse flow. This implies that our definition of analog reusable block must be compliant with these automation resources. It is also essential that an optimum compromise between optimization-based and knowledge-based automation be reached.

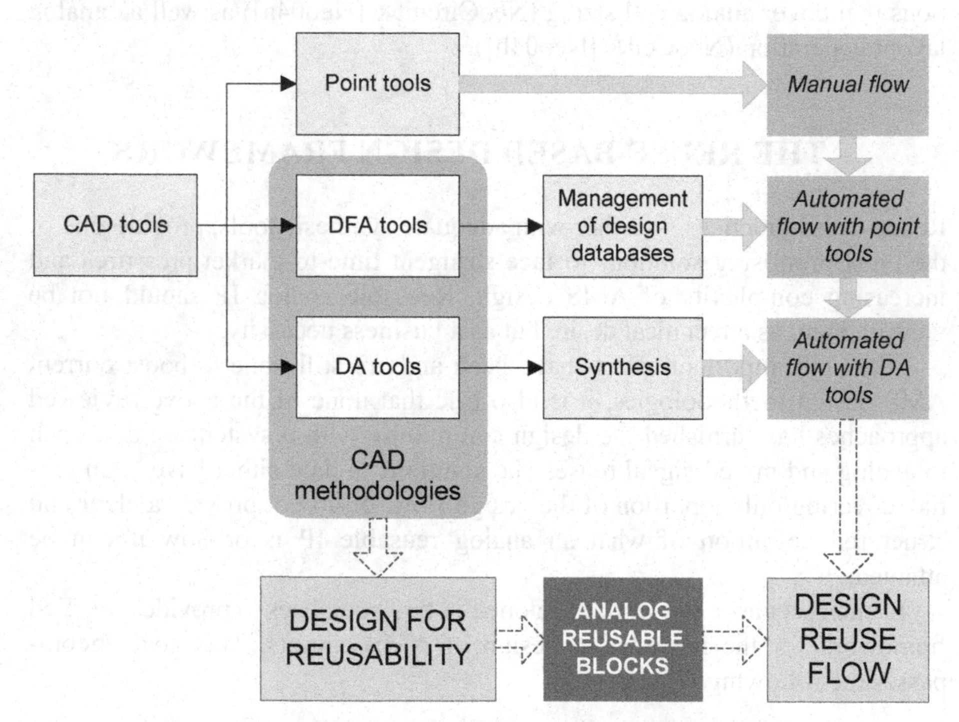

Figure 3. Reuse-based framework for analog and mixed-signal ICs.

With such a reuse-based framework, the main pursued objectives are:

1. Being able to rapidly retarget the analog design to arbitrary new performance goals.
2. Being able to rapidly port the analog design to a new fabrication process.

3. Support full capture, archiving, and reuse of the constraints that practicing analog designers rely on.
4. Support a full front-to-back solution, with automatic circuit and physical synthesis that minimizes redesign (backtracking) iterations, especially in case of critical layout-induced degradation.

To enhance designers' acceptance of the reuse-based framework, the work presented in this book aims at two additional goals:

5. Seamless integration of the reuse-based framework into the custom design flows with which circuit designers have historically been most comfortable.
6. Implementation of the reuse-based framework in a widely used design environment.

Note that by attaining these six objectives it is possible to overcome the drawbacks of the hierarchical design flow described in the preceding chapter.

3.1 The analog reusable block

The first step towards a unified reuse-based analog framework is to clarify the notion of analog reusable block. In this section, a definition and description of the analog reusable block concept is provided.

It is clear that commodity IP provides a higher level of reuse as compared with star IP, since the latter is delivered as a rigid design usually in the form of a GDSII file. The example of commodity IP presented in [Vand03] comprises just a reusable design plan by means of which synthesis can be carried out to obtain the active and passive device sizes. It lacks, however, other aspects such as a reusable layout description making the notion of commodity IP quite limited as far as reusability is concerned.

Nevertheless, it is unrealistic, as said above, to expect that the digital IP classification can be successfully and directly translated into the analog or mixed-signal arenas: neither the digital and analog design problems are the same (see Section 1.3 on page 10), nor the amount and importance of expert, subtle knowledge in analog and digital design reuse is the same. Yet, there is a digital concept that can be used to approach the notion of analog reusable block: this is the firm IP concept.

As illustrated in Table 2, firm IPs provide an intermediate solution between the very flexible, not predictable soft IP and the very predictable but badly flexible hard IP. Analog IP should retain, as it will be explained in Section 3.3, the firm features, namely flexibility and predictability. The purpose is to enable the reuse of the analog block in as many design steps (sizing, layout generation, verification) as possible, thereby profiting from the already embedded design knowledge.

With this idea in mind, the analog reusable block proposed is composed of three separate, yet linked views or facets, each used at a different design step as design information flows from one view to another. Such a reusable structure is shown in Fig. 4.

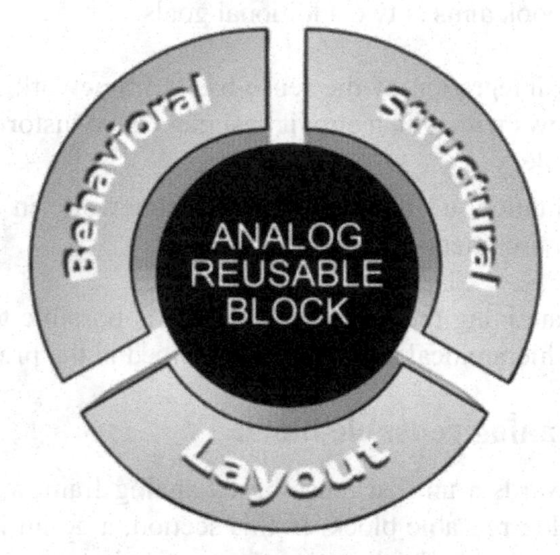

Figure 4. The three facets of the analog reusable block.

The three views of the analog reusable block are:

- **Behavioral** view. This facet comprises a high-level model of the analog block as well as the collateral design knowledge required to perform sizing and verification of the circuit immediately above in the hierarchy.
- **Structural** view. This facet corresponds to a database describing the analog block in terms of its building components as well as the design knowledge required to perform the sizing and verification of the analog block itself.
- **Layout** view. Last but not least, the layout facet contains the valuable expertise for the completion of the analog block's layout.

The facets of the analog reusable block defined above must be worked out in such a way that smoothly performing both design retargeting and design migration is ensured. As it will be explained in Section 3.3, and in full detail demonstrated in Chapters 3 to 5, these reusable views pose several problems that must be solved through a proper design for reusability method. Broadly speaking, whereas the behavioral view is relatively easy to design for reusability, the layout view poses the greatest level of complexity. In a sense, this difference is also present in the digital domain, where hard IPs are very difficult (if not impossible) to reuse for other performance targets or a differing fabrication process. Design for reusability methods presented in the book aim at providing the analog reusable block with the same reusability degree of the three built-in facets. The proposed concept of analog reusable block is much closer to the notion of commodity IP than to star IP as far as a higher reusability degree is concerned. However, the proposed analog reusable block is not limited to low-performance circuit blocks but can be also employed, in contrast to commodity IP, to support the design of state-of-the-art analog circuits.

Note the hierarchical nature of the layout and structural facets of the analog reusable block. While the behavioral facet represents an abstraction of the block performance, the layout facet of a system, sub-system, or module-level reusable block includes the layout facet of their building reusable blocks. For instance, the layout facet of a module-level reusable block comprises the layout facet of any cell-level and module-level reusable blocks below in the hierarchy[9]; a sub-system-level reusable block contains the layout facet of the building module and cell-level reusable blocks. Regarding the structural facet, any (system, sub-system, or module-level) reusable block incorporates the behavioral facet of its building reusable components. For instance, the structural facet of sub-system-level reusable block incorporates the behavioral facet of its module and cell-level building blocks.

As it is explained in the following section, the use of the structural and layout facets of module and system reusable blocks turns out to be essential during bottom-up flow, when module and system layouts are assembled and verified. Nevertheless, analog reusable blocks above the cell hierarchical level should be carefully pondered over. First, because hierarchy should only be used if the decomposed design problem is much easier to synthesize than the overall block as a whole [Vand03]. Second, because these blocks are architecture-fixed; that is, the building components have already been decided upon. Because of the heterogeneity of analog circuits, it is more likely that a particular cell-level reusable block is more often required, for many different applications, than a hierarchically higher, architecture-fixed reusable block is. Therefore, hierarchy should not be considered unless the system or module blocks are to be intensively reused.

---

9. In addition, of course, of the remaining, required elements (such as device-level components, routing wires, shielding guard-rings, etc.) completing the layout of the reusable block.

As discussed in Section 1.2.3 and Section 1.3, design entry time/effort is a negative aspect of both knowledge-based and optimization-based synthesis approaches (though, in the latter case, the level of negativity depends on the type of optimization-based approach). Then, it is important to note that the analog reusable block must also speed up the synthesis process by providing the necessary input to the synthesis tools. Consequently, the development of each facet of the analog reusable block must be compliant with the specific synthesis tool it is devised to assist in the design reuse flow.

3.2 The design reuse flow

This section deals with how the hierarchical top-down bottom-up design flow is modified to become a design reuse flow, i.e., an analog design flow where reusable blocks, as defined above, can be seamlessly integrated. What is more, application of such a design flow is not limited to reusable blocks, but also to firstly-designed circuits (like star IPs) as long as the designer develops the corresponding entries to the synthesis tools used. In this sense, an automated synthesis approach that tries to solve the pitfalls of hierarchical approaches is proposed here.

The design reuse flow inherits the same structure of the top-down bottom-up flow described in Section 2.2 of Chapter 1 and illustrated in Fig. 14 on page 24. Such flow is improved thanks, first, to an automated synthesis approach and, second, to the properties of the reusable block that, as it is described below, helps minimizing iterations between design steps and provides also an already proven design entry. Fig. 5 shows the different steps of the design reuse flow[10], described below.

3.2.1 Adopted synthesis approaches

The synthesis approaches used in the design reuse flow are the following:

- For system, module, and cell sizing, the optimization-based engine called FRIDGE [Mede94] [Mede99] has been chosen. Although this engine is able to interact with any kind of performance evaluation approach, simulation is used simulation is used for the lower design entry time/effort required and the better accuracy levels provided. This choice is thoroughly justified in the following chapters. The implemented approach is a two-step one: in the first one, statistical optimization techniques are applied, while deterministic ones are

10. As in [Hamo03] and [Vand03] it is hereinafter assumed that the architecture of the system, and module-level blocks, as well as cell-level topologies have been decided, so architecture/topology selection is not carried out.

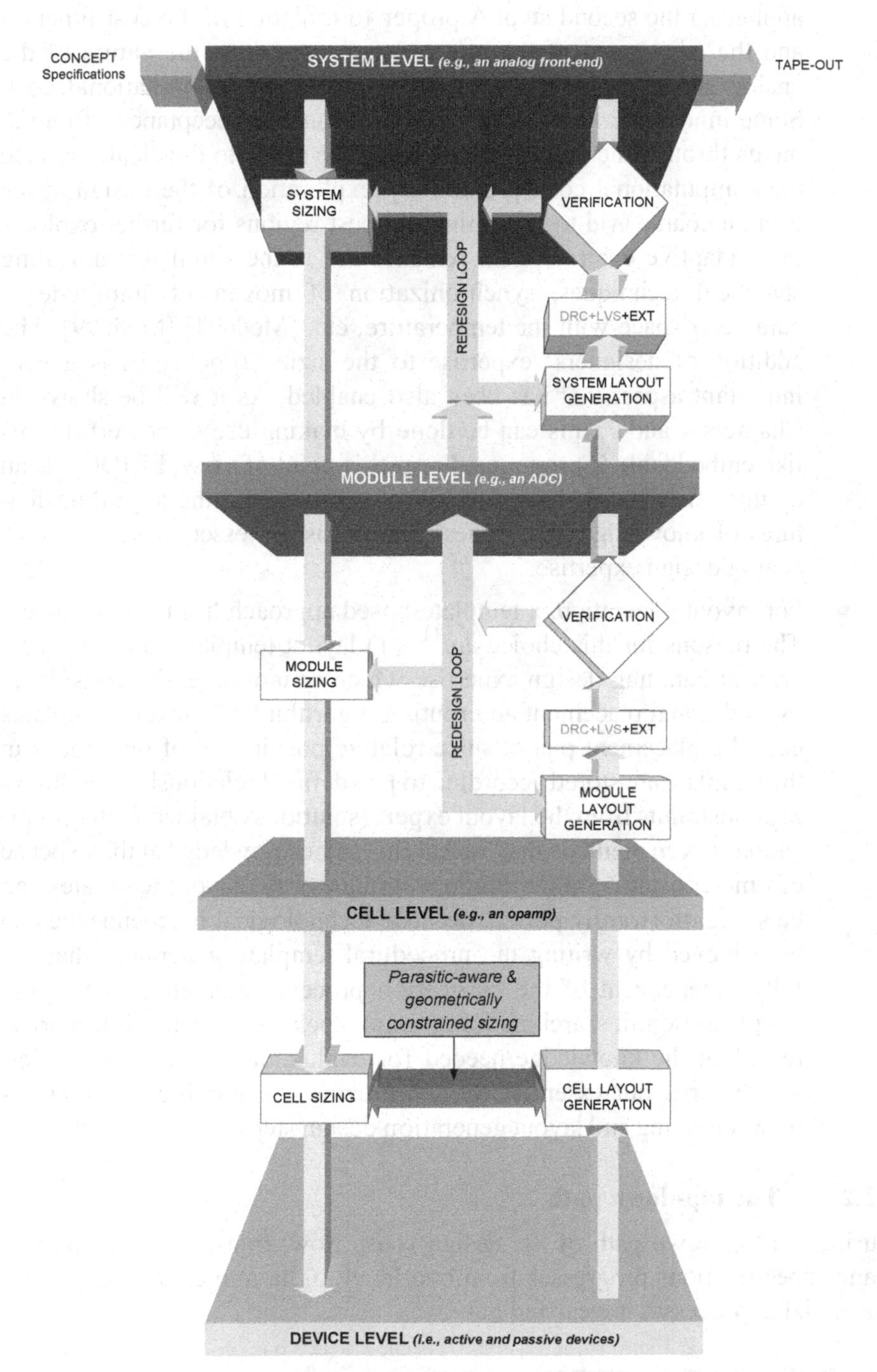

Figure 5. The design reuse flow.

applied in the second step. A proper formulation of the cost function and the adjustment of the movement generator to the nature of the analog sizing problem drastically decrease the computational cost. Some innovative features in the generation and acceptance of movements through the design parameter space allow to drastically reduce the computational cost: preliminary exploration of the design space using a coarse grid to determine the best regions for further exploration, adaptive control of the temperature in the simulated annealing statistical techniques, synchronization of movement amplitude in parameter space with the temperature, etc. [Mede94] [Mede99]. The addition of designers' expertise to the sizing procedures is a very important aspect that has been also enabled. As it will be shown in Chapters 4 and 7, this can be done by making use of powerful tools like embeddable C programs. From this point of view, FRIDGE is an optimization-based sizing approach incorporating the appealing features of knowledge-based ones. This turns out essential for reuse of analog design expertise.

- For layout generation, a template-based approach has been followed. The reasons for this choice are[11]: (1) layout templates are very efficient at handling design expertise which cannot be easily considered by traditional placement and routing algorithms; (2) layout templates ease the placement phase, since relative positioning of the blocks in the template are stored according to pre-defined relationships embodying constraints from the layout expert (solutions obtained with optimization-driven methods may reveal this same knowledge at the expense of time-consuming algorithmic techniques); (3) layout templates can be straightforwardly ported, because technological independence can be achieved by writing the procedural template generators that are fully independent of the fabrication process parameters; (4) layout templates permit searching for optimal block parameters while a priori revealing the knowledge needed for evaluation of layout parasitics, which turns out essential for minimizing the number of iterations between sizing and layout generation design steps.

3.2.2 The top-down path

During the top-down path of the design reuse flow, transmission of performance specifications progresses from one level to the immediate below one. Three sizing processes are carried out:

[11]. These reasons are further explained and documented in Chapter 5.

1. **System sizing**, where the system-level specifications are transmitted down to obtain performance specifications for every module-level component of the AMS system. This process is carried out by using the optimization-based approach described above. This optimization engine is combined either with equations relating module-level circuit performancs with system-level performance or with simulation of the system-level circuit with models of module-level components. Because of the complexity of the circuit at this level, behavioral models should be used to speed up the synthesis [Mede99].
2. **Module sizing**, where the performance specifications of each module-level block are mapped into performance specifications for each of its components (i.e., module, cell, and device-level blocks). This process is carried out by using the optimization-based approach described above. The performance evaluation is accomplished either by using equations relating the cell-level circuit performances with the module-level performance or by using simulation of the module-level circuit with models of the cell-level components. As in system sizing, due to the complexity of the circuit at this level, behavioral models should be used to speed up the synthesis.
3. **Cell sizing**, where each cell-level circuit is sized to obtain the value of device-level parameters (transistor width and length, resistor value, capacitor value, and so on) such that the performance specifications for the cell-level circuit are properly addressed. As in the levels above, the process is carried out by using the same optimization engine but now in combination with device-level simulation (i.e., low-level description models).

One main problem of traditional hierarchical top-down bottom-up analog design is to close the loop between circuit sizing and layout generation. When layout-induced errors degrade the circuit performance, circuit design changes are accomplished via circuit re-sizing [Giel00]. An important aspect of this problem is how to early evaluate layout-induced parasitics: in manual design, overestimation of parasitics results in wasted power and area, while underestimation of parasitics may lead to fatal performance degradation. Extraction of circuit parasitics within the circuit sizing process solves this problem. The extraction is performed by means of either layout generation or parasitic modeling. Heuristic-based or performance-driven approaches are currently too slow for layout generation within the circuit sizing process [Lamp99]. Procedural layout templates allow, on the other hand, fast generation of circuit layout since no time-consuming optimization algorithms are involved.

As described above, the optimization engine is able to incorporate design knowledge and use it during the sizing process. Once more, it is useful to include an optimization of the layout geometrical features (i.e., layout occupied area and its aspect ratio) directly during the circuit sizing. Thanks to the procedural nature of the adopted layout generation approach, it is possible to predict the exact shape of the circuit layout without actually moving to the layout phase. In addition, built-in circuit devices can be actually implemented in different ways (e.g., resistors and MOS transistors can be folded, the transistors in a differential input pair can be arranged in differing common-centroid arrangements, etc.), this knowledge being previously embedded in the circuit layout template. It is then possible to combine such layout knowledge with the optimization engine in order to find suitable implementations of the built-in devices that yield a layout with minimal occupied area and/or featuring a user-specified aspect ratio.

Including layout-induced parasitics in the circuit sizing process is known as **parasitic-aware** sizing; including layout geometrical information will be referred here as **geometrically constrained** sizing, since some design variables are constrained for the layout to meet certain user-defined geometric objectives. Note that both techniques are linked since different device implementation may give rise to several layout-induced effects (e.g., the diffusion capacitance of a transistor changes with the number of folds).

3.2.3 The bottom-up path

After cell-level sizing, layout is generated by instancing the corresponding template and the obtained sizing solution, which includes the implementation style of each device if parasitic-aware or geometrically constrained sizing has been carried out. No formal verification is required because the layout template is correct-by-construction. Verification of the extracted layout is not necessary either, since layout parasitics have been already considered. Sizing-layout-sizing spins are thus eliminated, thereby speeding up the overall flow.

Module layout is then generated by assembling the instanced layout of its cell-level components or by instancing the module layout template provided it has been incorporated as a reusable block itself. Only in the former case, formal design rule and layout vs. schematic checks of the module layout are performed. Extraction is necessary in both cases, since no parasitics were considered during module sizing[12]. Afterwards, the module performance is

12. Recall that at the module-sizing phase, cell-level sizing details are still unknown. From these details, the geometry of the routing wires interconnecting the cell-level blocks could be computed and, thereby, parasitics could be estimated. Therefore, unless there is a method to figure out the routing geometries directly from the cell-level performance specifications without actually going to the cell-level sizing phase, computing the layout-induced parasitics is not feasible.

verified by using simulation. Provided that the module-level circuit is not very complex, cell-level components can be replaced by their device-level description. Otherwise, each cell-level circuit can be replaced by a corresponding behavioral model, properly backannotated with the attained electrical performance (which includes, as said above, the performance degradation induced by the layout). A redesign loop to modify the module layout (i.e., layout elements other than cell layouts) or to repeat the module sizing is initiated in case that the module fails to meet the intended performance specifications.

Bottom-up verification of the system-level circuit follows the same methodology. First, the system layout is generated by assembling the module-level components or by instancing the system-level layout template provided it is available as reusable block. A formal verification precedes the performance verification where behavioral backannotated models of the module-level circuits (including performance degradation due to parasitics) can be used to reduce the simulation time.

3.2.4 The role of the analog reusable block

As depicted in Fig. 6, analog reusable blocks are used throughout the entire design flow. They accelerate the design process inasmuch as they provide a complete, knowledge-embedded input to the synthesis tools automating the design flow. The behavioral facet of cell-level reusable blocks is used at module sizing and module performance verification (the active facet of each reusable block is highlighted in Fig. 6). Their structural facet is used at cell sizing, whereas the layout facet is required at cell layout generation. If there are module-level reusable blocks, their behavioral facet is useful at sizing the immediately higher hierarchical level (i.e., the system or a higher module level), their structural facet can be required at module sizing and performance verification, and the layout facet is applied at the module layout generation. In the same way, the layout facet of system-level reusable blocks is used at system layout generation, while their structural facet can be required for system sizing and performance verification. Although not shown in Fig. 6, the behavioral facet of system-level reusable blocks can be used during chip verification and, if required, during initial partitioning of specifications for the complete mixed digital/analog system.

In the next three chapters, the design reuse flow is revisited to develop a convenient design for reusability methodology. The reuse-based framework developed has been implemented and put into practice with the retargeting, migration, and fabrication of an industrial-scale mixed-signal system. This application is described in detail in Chapter 6. Both **parasitic-aware** sizing and **geometrically constrained** sizing techniques are discussed in Chapter 7.

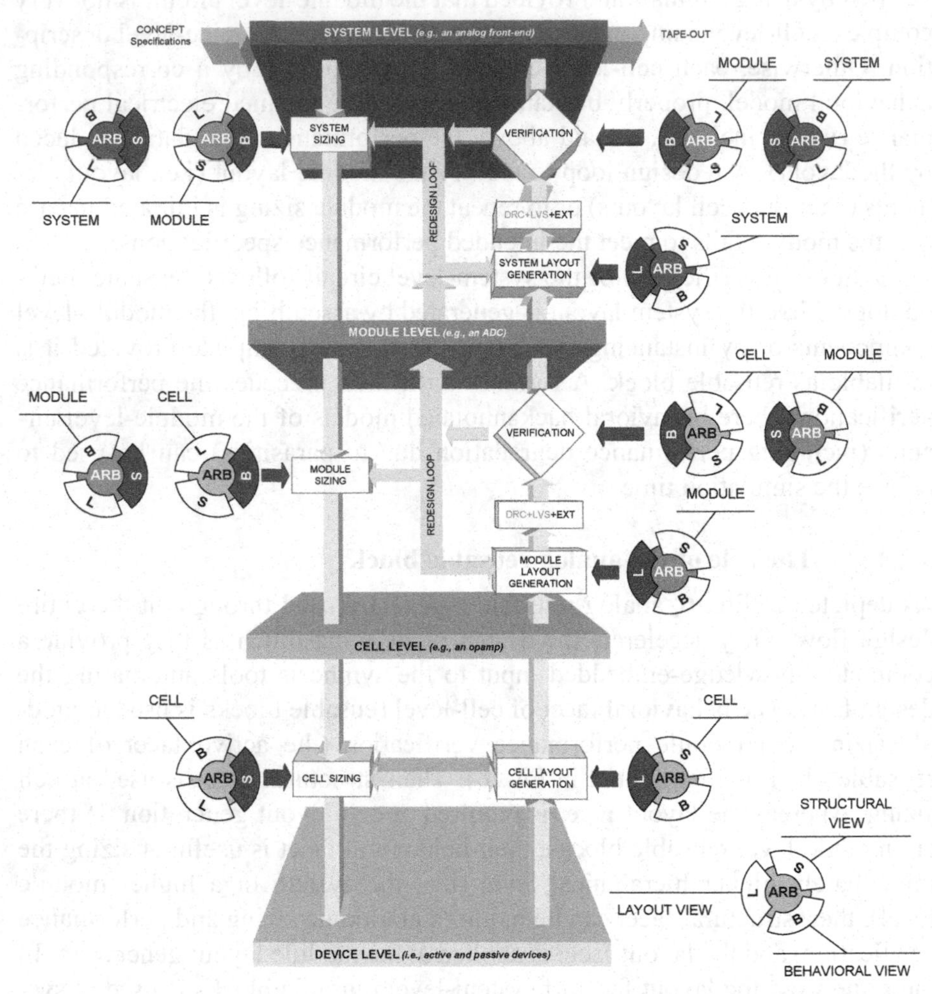

Figure 6. The role of analog reusable blocks in the design reuse flow.

3.3 The design for reusability methodology

As said above, the design for reusability methodology is closely related to the synthesis tools used in the design reuse flow. The analog reusable blocks are to be developed in compliance with the requirements imposed by these synthesis tools. For example, the structural view of the analog reusable block must comprise the necessary elements in conformity with the optimization-based sizing engine performing the cell sizing process. Therefore, design reuse and design for reusability are tightly connected.

The analog reusable block presented in Fig. 4 must be developed such that design retargeting and design migration are feasible. Here *feasible* means that a third-party user can perform both design operations with relatively little effort and faster than without reusable blocks[13]. This relevant feature can be achieved thanks to two essential, complementary principles, namely **intensive parameterization** and **design knowledge encapsulation**. These principles are intended to achieve the same features of flexibility and predictability of firm IPs.

Parameterization is the process by which the entire block description is turned completely independent of both the fabrication process and the specific value of the block performance features. That is, by means of parameterization, the analog block is progressively "released" from a specific performance and from an explicit set of process rules and constraints, becoming, thus, more and more independent. Changing the value of the parameters provides a new performance and expedites the design migration to a different fabrication process. In parameterizing a block there are, however, one obvious limitation: the achievable block performance after parameterization is limited to the achievable performance of the circuit architecture/topology. For instance, an opamp reusable block can address different values of its gain-bandwidth product by changing the value of its building devices (previously transformed into parameters of the reusable block); however, the topology limits the speed of the opamp, so not all values can be addressed.

Design knowledge encapsulation is the process that identifies the relevant design expertise for the block under development and embeds it into the appropriate facet of the analog reusable block. The design knowledge is fundamental to guide the synthesis tools throughout the design reuse flow.

The two principles described are said to be complementary because captured design knowledge is actually used to constrain the parameterization of the analog block. For instance, layout placement should not be fully parameterized because of the many degrees of freedom and the complexity of dealing with, among others, device mismatch, layout symmetries, or area requirements. Quite the contrary, placement should be guided by the same knowledge that a layout expert resorts to. Capturing and coding such knowledge is the way design for reusability constrains a "blind" block parameterization to smartly drive the synthesis process.

Another important factor that has been often ignored or underestimated is a thorough documentation of the reusable analog block. Unless this information is captured, future reuse of the analog block is severely limited because the design team would have to face serious time issues as they

---

13. This is true providing that the same synthesis tools are used.

struggle to recreate such information. Therefore, each facet of the analog reusable block must be accompanied with collateral deliverables containing a detailed description of the implemented parameters and the encapsulated design knowledge. Additional information on the applicability of the analog reusable block is also advisable.

4 SUMMARY

Circuit reuse is a design solution that can significantly accelerate product-to-market. Combined with the appropriate synthesis tools, an efficient management of the analog design complexity can be also achieved. The design gap between the chip density and designer's productivity can be closed by means of a reuse-based design framework, which integrates circuit reuse and design automation.

In this chapter, a reuse-based framework for AMS circuits has been introduced. A clear, structured definition of the analog reusable block has been provided. These blocks, composed of three separate views or facets (i.e., behavioral, structural, and layout), are devised to face the two scenarios of circuit reuse, namely design retargeting and design migration. A design for reusability methodology that systematizes the development of the analog reusable blocks has been outlined. Such methodology is founded on two basic, complementary principles: parameterization and design knowledge encapsulation. A design reuse flow has been also described. This flow not only allows the seamless incorporation of reusable blocks, but also tries to improve the hierarchical design flow described in Chapter 1.

Chapter 3

The Analog Reusable Block: Behavioral Facet

This chapter, the first in a series of three, leads into the description of the fundamental component of the design reuse methodology described in the previous chapter: the analog reusable block, which can be described by means of three differentiated facets: the behavioral, the structural, and the layout facets.

During system or module sizing phases of the design reuse flow described in the previous chapter, the performance specifications of a circuit comprising one or more reusable blocks are mapped into a set of performance specifications for each reusable block. This process, accomplished via a simulation-based optimization technique, requires a suitable behavioral description for each of these reusable blocks in order to speed up the circuit synthesis while maintaining an acceptable level of accuracy. Besides, such description turns out essential to reduce the overall design time during circuit verification.

This description constitutes the behavioral facet of the analog or mixed-signal reusable block. The goal of the present chapter is to describe how intensive facet parameterization and thorough design knowledge encapsulation (the principles of design for reusability) are attained at the behavioral level to successfully undertake both scenarios of design reuse, retargeting and migration. To set the background for the remainder of the chapter, the next introductory section explains the needs for a behavioral view and reviews its most general characteristics.

1 INTRODUCTION: WHY BEHAVIORAL DESCRIPTIONS?

In general, a circuit's behavioral description is a representation of the circuit's functionality by means of an abstract model. The abstraction level is an indication of the degree of detail specified about how a function is to be implemented. Abstraction is inversely related to the resolution of the detail. If there is much detail, or high resolution, the abstraction is said to be low. Moreover, the abstraction level forms a hierarchy. A design at a given abstraction level is described in terms of a set of constituent items and their inter-relationships, which in turn can be decomposed into their constituent parts at a lower level of abstraction [Vsia01]. So far, the lowest level of abstraction is achieved by using device-level descriptions of each of the circuit devices (like, for instance, SPICE-like transistor models). In contrast, behavioral models try to capture as much circuit functionality as possible with far less implementation details than the device-level description of the circuit.

The derived benefit is that behavioral models significantly speed up the simulation time when compared with simulation using device-level models. The unavoidable price to pay is, however, a reduced accuracy of the simulation results, with accuracy being the magnitude of the matching between the intended circuit performance (i.e., the real performance of the circuit being modeled) and the results predicted by its behavioral model [Dias92]. This suggests a definite trade-off between accuracy and the abstraction levels of the behavioral model. Thus, it is expected that the more accurate the model is, the lower the abstraction level (for instance, device-level models exhibits the lowest abstraction level and the best accuracy level).

Whereas accuracy is not all too critical for digital design, it turns out a key concern for analog and mixed-signal (AMS) design. The reason is that any analog or mixed-signal circuit model not only has to emulate the functionality the circuit is to perform (i.e., the first-order performance), but also its second-order performance. This second-order performance is typically plagued with analog-intrinsic non-idealities, their modeling resulting essential to discover design problems as early as possible.

Therefore, behavioral descriptions of AMS circuits bring in simulation speed at the cost of a reduction in accuracy when compared to device-level descriptions. Nevertheless, trading speed with accuracy in such a way may be useful in the following stages of the hierarchical AMS design flow (see Fig. 1):

- During the top-down design flow, appropriate architectures of the system building blocks (modules) are firstly selected by intensive **exploration** of the system-level design space given the system-level specifications. Exploration can also be used to select the correct topology of the module building blocks (cells) given the set of specifications of the module[1]. Behavioral modeling can successfully assist

[1] Exploration is also known as architecture or topology selection.

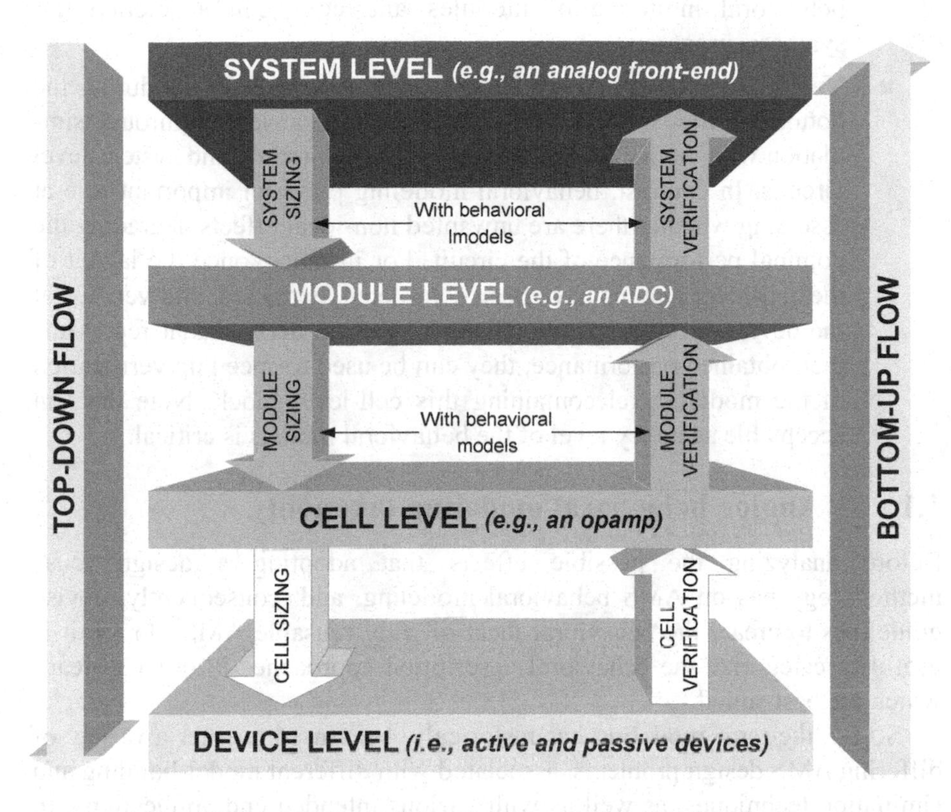

Figure 1. Use of behavioral description in the hierarchical design flow.

the designer to explore such large design spaces that would otherwise take a prohibitively large amount of time and resources if explored using device-level circuit descriptions.

- Also during the top-down flow, it is necessary to propagate the system-level specifications down to the rest of levels. In other words, hierarchical design implies that specifications for the module-level blocks are to be obtained from the system-level specifications, specifications for the cell-level blocks from the module-level blocks, and specifications for the device-level blocks from the cell-level blocks[2]. This process is known as **electrical synthesis**[3]. For both system-to-module, module-to-module, and module-to-cell specification mapping,

[2] In the latter case, typical device-level specifications are the transistor widths and lengths, and the resistance or capacitance values of the passive devices.

[3] Other given names are sizing and specification mapping or transmission.

behavioral modeling of modules and cells can accelerate the synthesis process.

- **Verification** of the circuit performance at several levels during the bottom-up flow may take excessively long if carried out through simulation with device-level descriptions of the module and system-level circuits. In contrast, behavioral modeling plays an important role at assessing whether there are unwanted non-ideal effects degrading the nominal performance of the circuit. For instance, once the layout of the cell-level circuits have been completed, extracted, and verified at the device level, a behavioral model can be derived that represents their obtained performance; they can be used to speed up verification of the module-level containing this cell-level block. Note that, an acceptable accuracy level of the behavioral models is critical.

1.1 Analog behavioral modeling taxonomy

Before analyzing the possible effects that adopting a design reuse methodology has on AMS behavioral modeling, and, consequently, devise guidelines to create the behavioral facet of truly reusable AMS blocks, it is useful to categorize the behavioral description approaches in order to learn which are best suited.

So far, the term **modeling** has historically been used to refer a variety of differing AMS design problems, associated with different model-building and simulation techniques, as well as with various intended end applications. In the same way, concepts such as **macro**, **high-level**, or **functional** have been commonly used as equivalent adjectives of **behavioral modeling.** In this book, the term *behavioral* is used in reference to any description approach other than the one done at the device-level. Fig.2 shows a taxonomy of the different approaches that clarifies the purposes of analog behavioral modeling and the rest of concepts mentioned above [Liu02]. The three separate classes in this taxonomy, namely **use**, **dimensionality**, and **style**, are represented with separate axes explained below.

- **Use**. In this axis, a distinction is made between behavioral models primarily built for top-down design, that is, to carry out both synthesis and exploration, and behavioral models used for verification in bottom-up design flows. These two purposes are not mutually exclusive as the same behavioral model, if properly built, can be used both at top-down and bottom-up design phases. Then, the difference really appears at the other two axes.

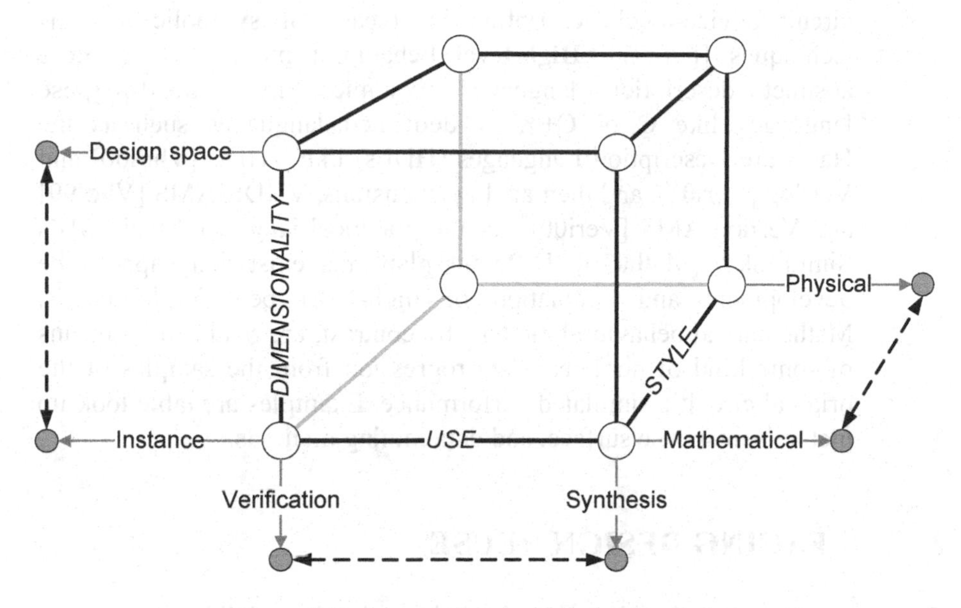

Figure 2. Taxonomy of analog behavioral modeling.

- **Dimensionality**. This class distinguishes between behavioral models that represent the behavior of a single circuit, or **instance,** from models that represent a range of circuits, or **design space**. Whereas instance behavioral models emulate a particular defined set of electrical characteristics of a single circuit, and, thus, can be used for verification, design space behavioral models are deliberately parameterized to cover a range of electrical characteristics of one or more circuits (obviously with equivalent electrical behavior, such as different operational amplifier topologies). Therefore, they are suitable not only for verification but for exploration and synthesis.
- **Style**. This axis refers to the nature of the circuit's behavioral description. At one end, there are behavioral models that represent the circuit's behavior by means of an underlying, explicit **physical** model. At the other end, **mathematical** models intend to "fit" some of the circuit's "true" responses without any underlying physical model. An example of the physical style is the **macromodel**, a simplified version of the circuit that reduces the simulation time thanks to the reduced number of nodes and devices. Physical models can be further classified in terms of the description language used. Thus, macromodels use device-level description languages (e.g., SPICE-like syntax). The same physical description can be attained by explicitly using **s** or **z**-domain transfer functions to model some of the circuit's responses. These functions can as well be generated from a

circuit device-level description by means of symbolic analysis techniques [Fern97]. **High-level** behavioral models use a more abstract description language. Examples are general-purpose languages like C or C++, or dedicated languages such as the Hardware Description Languages (HDLs) like VHDL [Vhdl00] and Verilog [Veri01], and their analog extensions, VHDL-AMS [Vhdl99] and Verilog-AMS [Veri00][4], or the graphical language MATLAB's SimuLink® [Math02b]. Different platforms exist that support the development and simulation of high-level behavioral models. Mathematical behavioral models, by contrast, are obtained by means of some kind of nonlinear data regression from the samples of the original circuit's simulated performance. Examples are table look-up methods, response surface, and data mining methods.

2 FACING DESIGN REUSE

The previous section has introduced the behavioral modeling concept, its many applications in modern hierarchical design flows, and a three-class taxonomy of behavioral modeling approaches. This section examines the relationship between behavioral modeling and the design reuse methodology developed. More specifically, it analyzes the characteristics that AMS behavioral models must have in order to become the behavioral facet of the AMS reusable blocks in the reuse-based framework.

2.1 The design reuse flow: top-down electrical synthesis

As depicted in Fig. 3, the top-down design reuse flow proceeds as follows:

1. The system-level specifications are transmitted down to obtain performance specifications for every module-level component, as well as for any separated cell or device-level component of the analog or mixed-signal system. This process is known as **system-level sizing**.
2. For each module-level component, its performance specifications are mapped into performance specifications for each one of its module-level components[5], its cell-level components, as well as sizes and

4. An intermediate approach between general-purpose and dedicated languages is found in SystemC [Arno00] and its analog extension SystemC-AMS [Vach03]. SystemC provides hardware-oriented constructs as a class library implemented in standard C++.

5. Recall from Chapter 1 that we have considered that there can be more than one module level in the hierarchy.

values for any device-level components the module might have. This process is known as **module-level sizing**.

3. Each cell-level circuit is sized to obtain the value of device-level parameters (transistor width and length, resistor value, capacitor value, etc.) such that the performance specifications for the cell-level circuit are properly addressed. A model describing each device component of the cell-level circuit is required. This process is known as **cell-level sizing**.

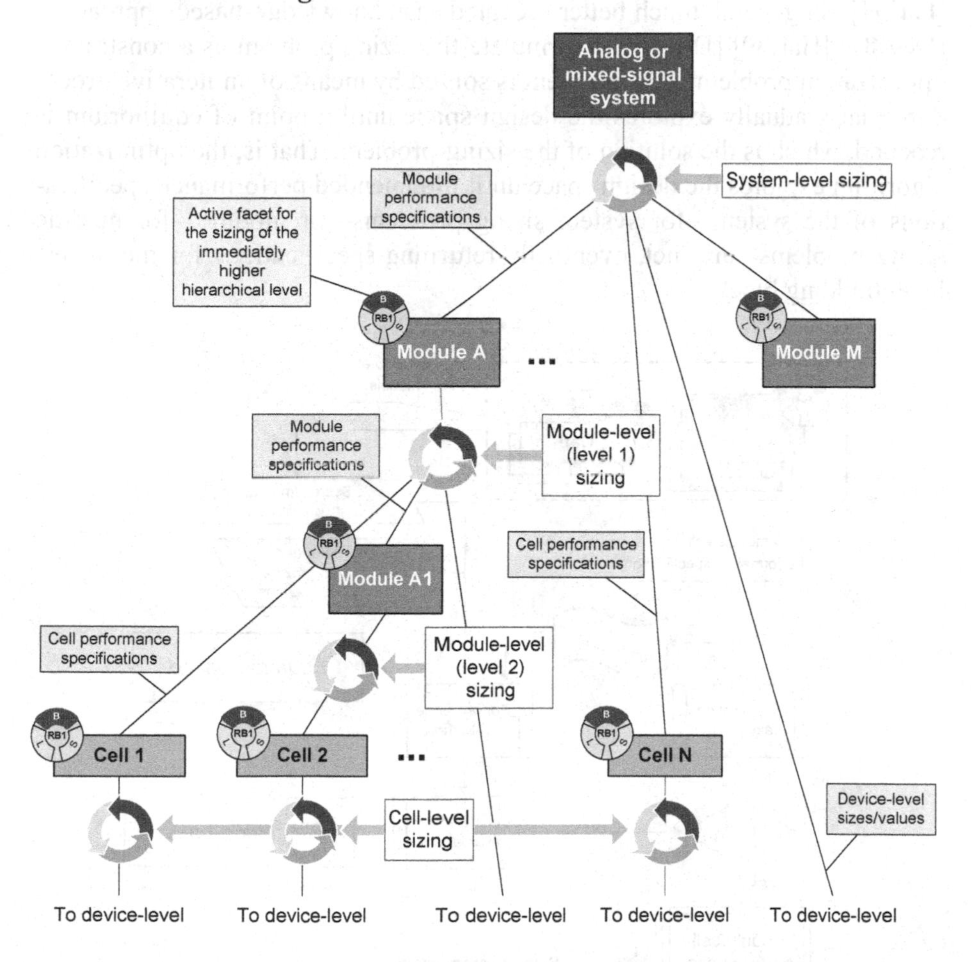

Figure 3. System and module sizing of the top-down phase of the design reuse flow.

As described in Section 1, the behavioral model is used at system and module-level sizing phases where the circuit is too complex (in terms of number of components, nodes, or the nature of the simulations required to evaluate the performance of the circuit). For a system-level circuit, speeding up the electrical synthesis process can be achieved by means of behavioral models of its module and cell-level components. For a module-level circuit, it also implies using a behavioral model for each of its components: cells or other lower-level modules.

As outlined in Chapter 2, both system and module sizing involve the optimization-based process depicted in Fig. 4. Optimization-based approaches [Giel90] [Koh90] [Mede94] [Mede99] [Nye88] [Ocho96] [Phel00a] [Deb02] [Fari04] –in general, much better accepted than knowledge-based approaches [Degr87] [Harj89] [Dess01a]– formulate the sizing problem as a constrained optimization problem. This problem is solved by means of an iterative procedure that gradually explores the design space until a point of equilibrium is reached, which is the solution of the sizing problem. That is, the optimization algorithm explores the design space until the intended performance specifications of the system –for system sizing problems– or module –for module sizing problems– are met, eventually returning specifications for the lower-level building blocks.

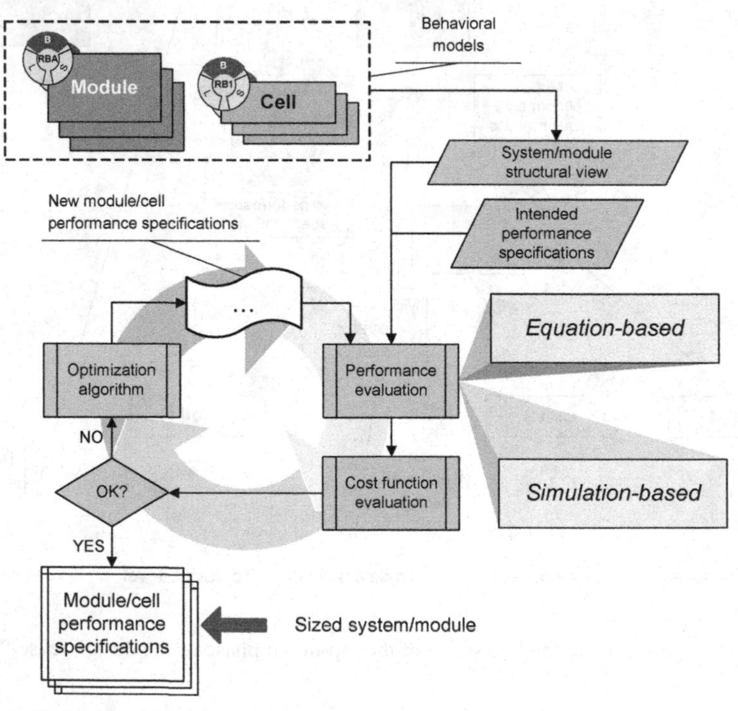

Figure 4. The optimization-based process: system and module sizing.

At each visited point of the design space, the sizing process has to rely on a fast yet accurate technique to estimate the quality[6] of such visited point. Such an estimate can be carried out by means of either equation-based techniques [Koh90] [Giel90] or simulation-based techniques [Nye88] [Mede94] [Mede99] [Nye88] [Ocho96] [Phel00a]. In the equation-based technique, design equations expressing the performance of the circuit to size (system or module) as a function of the performance of all its components (modules, cells, and devices) are used. In the simulation-based technique, the system or module-level circuit is simulated by replacing the device-level descriptions of its module and cell-level building blocks with adequate behavioral models. This is the technique selected to carry out the top-down design flow.

For a system or module-level circuit with a collection of performances features $\Pi = \{\pi_1, .., \pi_M\}$ and specifications $S_i(\pi_i, \pi_i^\circ)$, where π_i° is the targeted value or directive of the specification[7] and $i = 1, \ldots, M$, the sizing problem consists in mapping these specifications into a target value or directive for each performance feature of the N_c building components (other than device-level components, for which a size or value is returned). This is illustrated in Fig.5. The solution of the sizing problem is composed of N_c sets of module and/or cell-level performance specifications, $s_{k,1}(\rho_{k,1}, \rho_{k,1}^\circ)$, where $k = 1, \ldots, N_c$ and $l = 1, \ldots, N_k$, with N_k being the number of performance features of module/cell k. The goal of the optimization is thus to find the target values or directives $\rho_{k,l}^\circ$, whose fulfillment implies that the system or module-level performance specifications are confidently met.

As it is also pointed out in Fig. 5, the sizing process involves a behavioral model B_k for each of the k module or cell-level components, which reproduces its electrical behavior with performance features $\rho_{k,l}$. During the sizing process, the optimization algorithm samples each of these performance features on a previously defined variation range $[\rho_{k,l_S}, \rho_{k,l_E}]$[8]. The complete set of all the performance variation ranges of all module and cell-level behavioral models defines (together with the variation range of any designable character-

---

6. The term "quality" refers to the degree of compliance of the visited point with the system or module specifications, i.e., the optimization goals. In the optimization procedure used in this book, such a degree is quantified through an adequate cost function [Mede99].

7. There are many different ways of defining a performance specification. For example, if it is required that $\pi_i \geq \pi_i^\circ$, then its corresponding specification can be defined as $S_i = \pi_i - \pi_i^\circ \geq 0$. A similar specification is defined for $\pi_i \leq \pi_i^\circ$, so having $S_i = \pi_i - \pi_i^\circ \leq 0$. This type of specifications involving inequalities are known as *design restrictions*. A specification is called a *design objective* when it involves a *directive* (e.g., to minimize or maximize certain performance feature) rather than a numerical value. For instance, $S(\pi_i, \text{minimize})$ is a design objective meaning that the performance feature π_i is to be minimized during the sizing process.

8. These variations ranges are sampled following a specified grid.

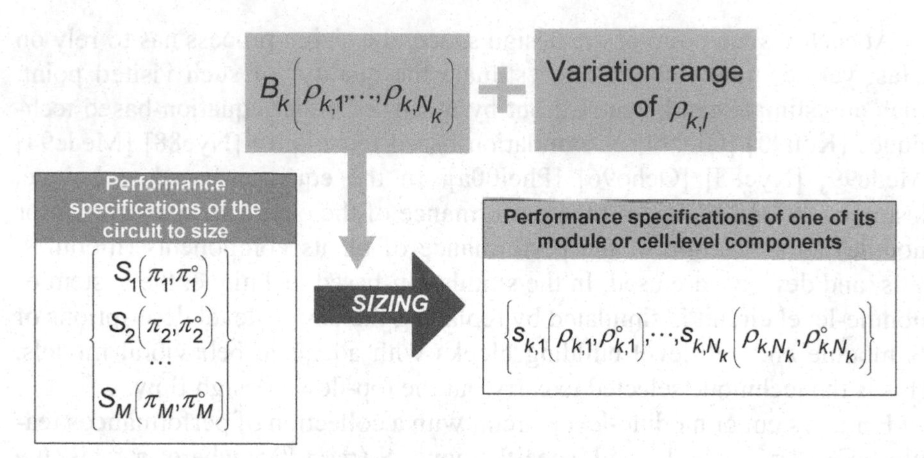

- S_i is the i-th performance specification of the circuit to size (system or module), and i=1,...,M, with M being the number of specifications.
- $s_{k,l}$ is the l-th performance specification of the k-th component (module or cell):
  - k=1,...,N_c, with N_c being the number of components.
  - l=1,...,N_k, with N_k being the number of performance features of component k.

Figure 5. Illustration of system/module sizing processes.

istic of a device-level component that might be present) the available design space for the system and module-level sizing problems.

2.2 The design reuse flow: bottom-up verification

During module and system verification, the reusable behavioral facet can be also used to expedite the required simulations whenever the circuit's complexity is such that a full device-level simulation results prohibitive in terms of computational resources and/or time. This bottom-up verification process is illustrated in Fig.6. Both for module and system-level circuits, verification is carried out by using a behavioral model for each cell or module-level component. These behavioral models are obtained by adapting the corresponding reusable behavioral facet to the electrical behavior that has resulted from the verification of the component itself. For instance, the reusable behavioral facet of *cell 2* (see Fig. 6) is adapted to reproduce the performance yielded by cell-level verification of *cell 2*. The behavioral model so derived of *cell 2* is used in the module-level verification of *module A1*; this verification is then used to adapt the reusable facet of *module A1* and derive its behavioral model, necessary to the verification of *module A* (which, by the way, also uses behavioral models of cell 1, and process-specific models of its device-level components).

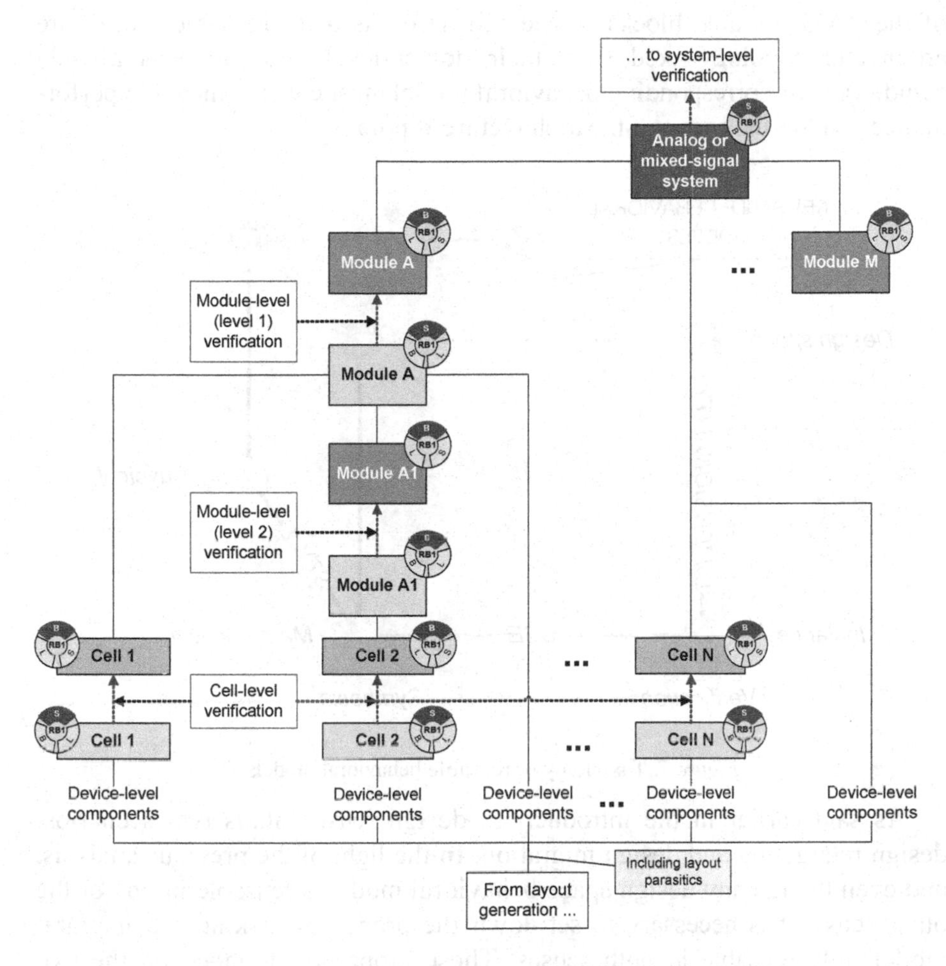

Figure 6. Illustration of the bottom-up verification processes.

2.3 Characteristics of the behavioral facet of the AMS reusable block

A valuable conclusion concerning the taxonomy of AMS behavioral modeling shown in Fig. 2 can be derived from the precedent analysis: the behavioral models used in the design reuse methodology have to cover a sufficiently wide design space rather than representing a single instance of the circuit architecture/topology being modeled. This conclusion is illustrated in Fig. 7: whatever the style (from physical to mathematical) or the use (synthesis or verification), the dimensionality of the reusable block has to be such that a range of circuit performances is covered. The conclusion, however, must be slightly qualified for the behavioral models representing the behavioral facet of the AMS reusable blocks considered here. As these reusable blocks are architecture/topology-fixed (i.e., their device-level description is already decided), their corresponding behavioral model must cover achievable performances within the scope of the architecture/topology.

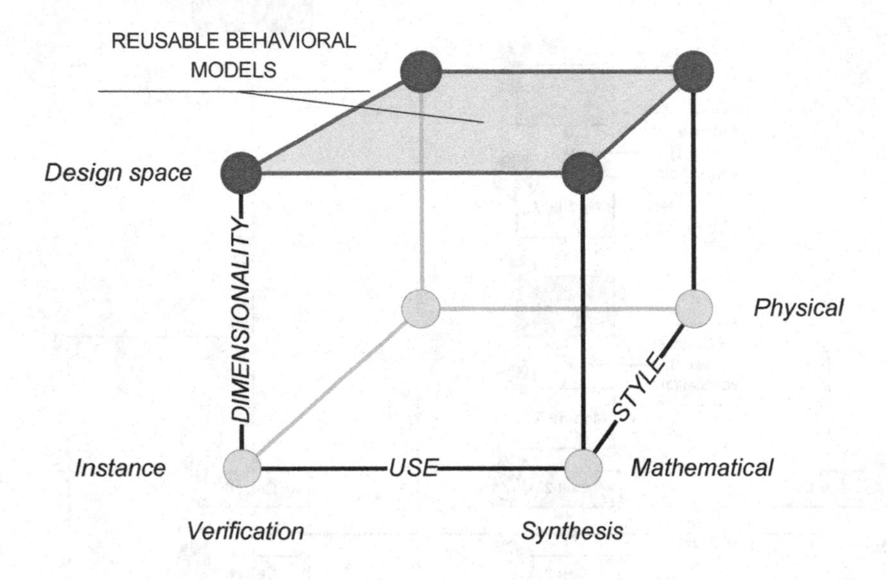

Figure 7. Taxonomy of reusable behavioral models.

As said earlier in the introduction, design reuse entails two scenarios: design retargeting and design migration. In the light of the previous analysis, and even though any design space behavioral model is reusable in one or the other sense, it is necessary to set down the properties making a behavioral model truly reusable in both senses. These properties, founded on the two chief, complementary reusability principles –intensive facet parameterization and design knowledge encapsulation– are the following:

- **Parameterization**. As illustrated in Fig. 7, the reusable behavioral model has to be such that it is easy to emulate the electrical behavior at many different points of the block's design space. Covering such a design space facilitates, as explained above, retargeting of the higher-level circuit containing the reusable block. To have design space behavioral models instead of simply instance behavioral models, it is required that the behavioral model has a set of input parameters representing its performance features. In this sense, it is important to note that the behavioral and structural facets of a reusable block are linked by the these input parameters. Consider, by way of illustration, a cell-level reusable block. Its input parameters are both design variables during the module-level sizing phase (where the behavioral facet of the cell-level reusable block is used), and design goals during the later cell sizing phase (where the structural facet of the cell-level reusable block is required)[9]. A similar link exists during bottom-up verification: the outcome of cell verification, which needs its structural facet, is the required input for the behavioral facet to reproduce the correct cell performance during module verification.

- **Modularity**. To allow block reuse in different retargeting operations, the parameters of the behavioral model have to be implemented separately, in such a way that it is possible that one or more parameters can be excluded (from the module sizing process and, hence, from the cell-level sizing as well) if the retargeting procedure requires so.
- **Design knowledge encapsulation**. Capturing valuable design expertise into the behavioral facet is essential to enhance its degree of reusability. In this sense, properly choosing the modeling style, the different modularity characteristics, and the type of parameterization, are factors that altogether stem from the accumulated experience regarding the design of the block whose behavioral model is under development. The modeling style is important to improve top-down synthesis and bottom-up verification through a much more efficient behavioral simulation that suits the nature and characteristics of the circuit block (an illustrative example is the behavioral modeling of discrete-time Sigma-Delta modulators, a mixed-signal block whose characteristics suggest a modeling style appropriate to event-driven behavioral simulation [Dong01] [Cast03]). Selecting modularity is crucial to define which performance features are to be modeled and managed during top-down specification mapping or bottom-up verification. The type of parameterization is the way that these performance features are modeled, which certainly calls for a deep understanding of the block functioning; these features are transformed into the parameters of the behavioral model. In relation to this, a variation range or a default value for each parameter has to be defined and stored within the reusable block database to ease block reuse. This variation range is used to define a design space for the system or module-sizing problem[10].

---

9. Strictly speaking, a particular value of each input parameter, which belongs to the solution of the module-level sizing phase, is required during cell-level sizing.

10. Note that a proper definition of each variation range cannot always be easily derived as it strongly depends on the circuit's architecture/topology and on the fabrication technology. Quite the contrary, it may require a great deal of accumulated design experience.

- **Technology independence**. In order to procure a seamless migration of the reusable behavioral model to different fabrication processes, any explicit reference to a particular process in the behavioral description has to be totally and utterly avoided. If it cannot be done, the reference should be incorporated in a generic format (e.g., as an additional variable), so it updates with every different process.

The remainder of this chapter covers the practical implementation of the reusable behavioral model properties. As illustration vehicle, behavioral modeling examples of an AMS industrial-scale circuit are used.

3 CASE STUDY: A QUADRATURE DA TRANSMIT INTERFACE

This case addresses the macromodel behavioral description style by analyzing a realistic design problem in which analog reusable blocks are used. To provide a common background, an industrial-scale system is firstly analyzed, its levels of hierarchy examined, and the top-down design flow outlined. Afterwards, the behavioral-facet of the analog reusable blocks used is described in detail.

3.1 System description

Quadrature transmit interfaces are seen in a wide range of wireless communication standards such as GSM, PMR, PHS, CDMA, DECT or W-CDMA [Rapa96]. To lower fabrication costs, these systems should be integrated on a mainstream digital CMOS technology. Essentially, such modems provide the interface between complex digital signals coming from the baseband digital processors and complex analog signals in quadrature phase (channels I and Q) for the RF transceiver. An implementation of quadrature transmit interfaces is the IQ digital-to-analog (IQ DA) system [Fran99a] whose block diagram is shown in Fig. 8. In this system, each channel is formed by a digital processing unit for signal shaping and interpolation, and a DA converter (DAC) and an analog back-end –the shaded area at the end of each channel. An additional calibration unit is employed to adjust the unavoidable offsets and mismatches between both channels[11]. From now on, the case study is focused on the

[11] The reader is referred to Chapter 6 for a comprehensive description of the system and detailed reuse examples.

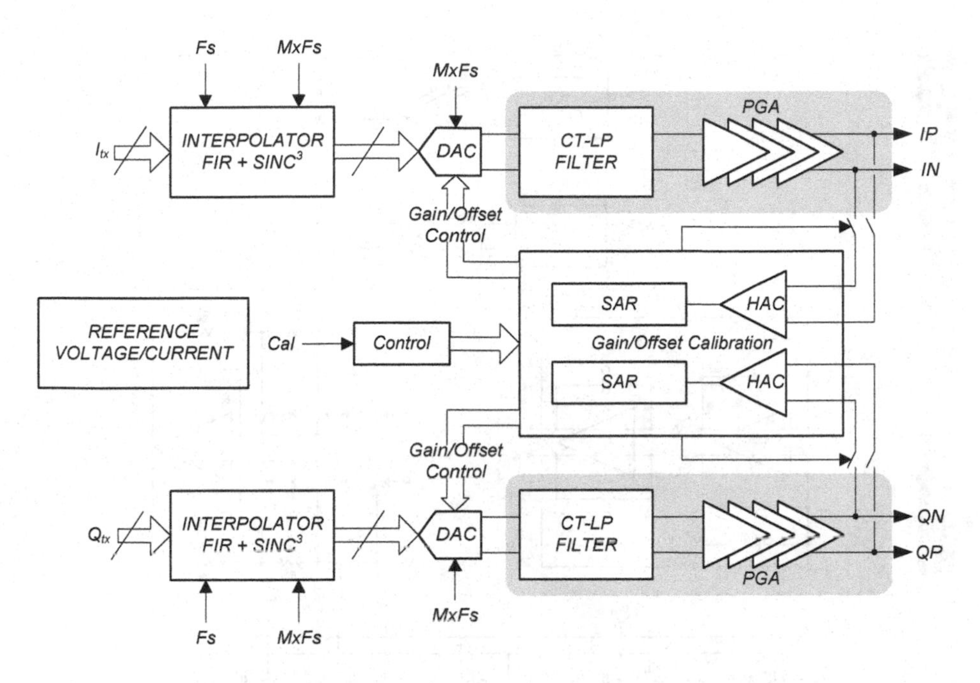

Figure 8. Block diagram of the IQ DA transmit interface.

analog back-end section of the IQ DA transmit interface, considering it as the analog sub-system.

The hierarchical levels of such analog back-end are shown in Fig. 9. At the module level, it consists of a continuous-time second-order low-pass filter (CT-LP filter) and a programmable-gain amplifier (PGA). The purpose of the CT-LP filter is to attenuate the image components of the baseband spectrum at multiples of the clock frequency, to smooth the output signals generated by the preceding segmented current-steering DA converter, as well as to provide the required current-to-voltage conversion. The purpose of the PGA, on the other hand, is to provide a digitally-controlled amplification of the differential output signal delivered by the CT-LP filter, to enforce the image rejection function of the CT-LP filter, and to support the buffering of its output differential signal to the external load of the chip. Both modules are composed of passive elements (resistors and capacitors) and active elements (opamps).

Designing the analog back-end to cover a wide range of applications or across different fabrication processes implies being able to retarget its building components, so avoiding the start of a new design from scratch. By using a design reuse methodology and properly designed reusable blocks, such retargeting process can be speeded up. Let us consider the module-sizing phase of the top-down design reuse process, and the analog back-end performance specifications for two wireless applications, GSM and DECT, shown

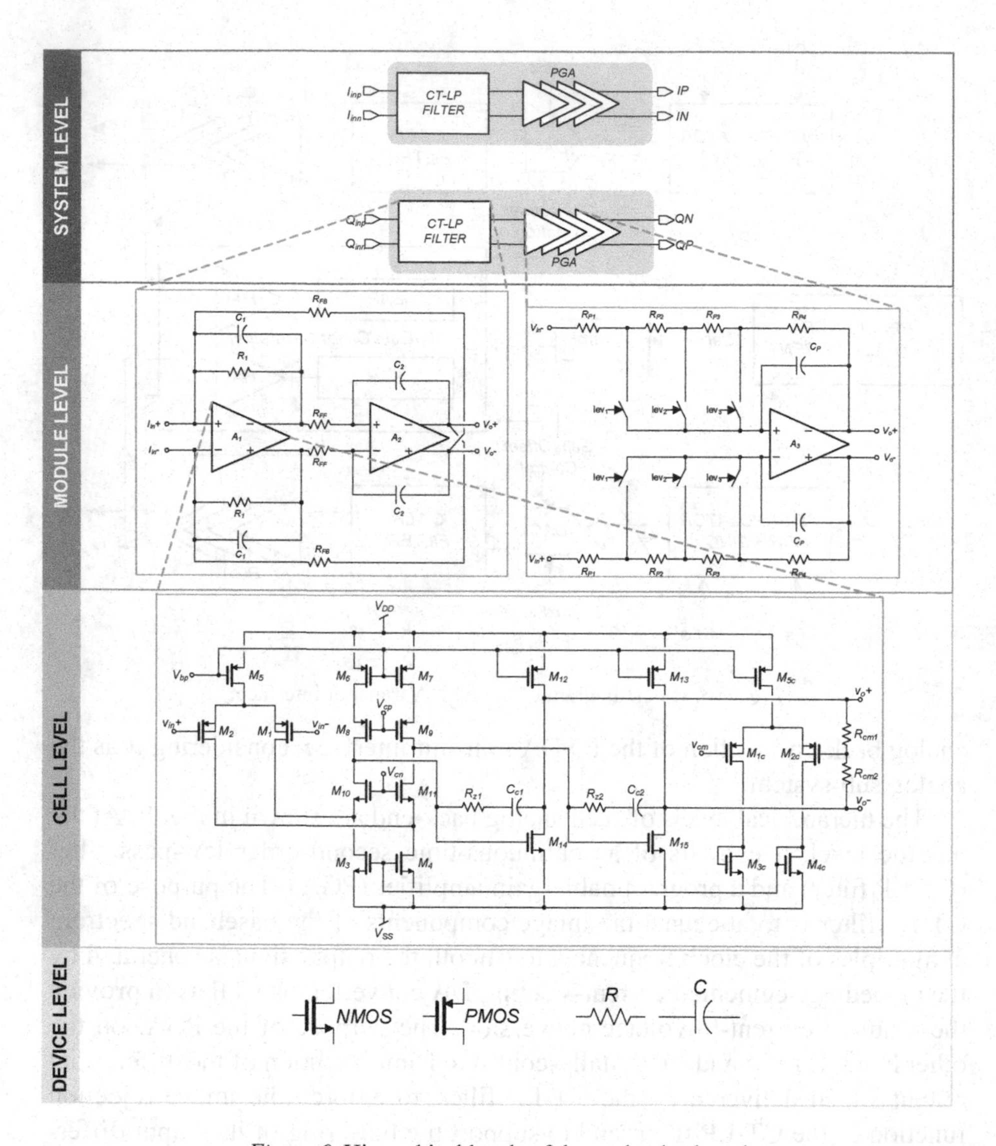

Figure 9. Hierarchical levels of the analog back-end.

in Table 1. As said in Section 2.1, sizing the modules of the analog back-end (i.e., the CT-LP filter and the PGA) involves the mapping of the module-level specifications, (previously obtained at sub-system sizing, in which the sub-system performance specifications in Table 1 are mapped into performance specifications for the CT-LP filter and the PGA modules), into performance specifications of their cell-level components. That is, the module sizing process eventually returns the value of the passive elements as well as the performance specifications for each of the opamps. In this process, reusable

Table 1. Sub-system performance specifications.

	Performance Feature	*Unit*	*GSM spec*	*DECT spec*
Electrical	Oversampling factor, M	-	12	8
	Sampling frequency, FS	MHz	13	36.864
	Signal bandwidth, SB	kHz	100	700
	Minimum attenuation of the analog back-end @FS	dB	49.8	52.6
	Maximum I/Q gain mismatch	dB	± 0.35	± 0.35
	Maximum group delay deviation (DC to SB)	ns	8.3	9.5
	Maximum I/Q group delay mismatch (DC to SB)	ns	33.5	6.5
	Maximum I/Q phase mismatch @SB	°	1.3	1.75
Optimization	Power	mW	minimize	minimize
	Area	μm^2	minimize	minimize

behavioral descriptions for each of the active components of the analog back-end modules are required. The macromodel style is used to implement the behavioral facet of the reusable opamps. These models are explained below.

3.2 Reusable macromodels

As said in Section 2.3, the design variables of the macromodels are high-level specifications such as DC gain, gain-bandwidth, or phase margin, which are then transferred down by the design reuse flow to size their corresponding devices. In this section, the macromodel descriptions of the opamps involved in the analog back-end are developed such that major non-idealities of real opamp implementations are conveniently represented.

All three opamps in the analog back-end are implemented as a fully-differential two-stage topology in order to increase the open-loop gain and dynamic range. As shown in Fig. 9, which depicts the first amplifier of the CT-LP filter, such topology consists of a folded-cascode differential stage followed by two compensated common-source amplifiers, one per differential output, and a current-steering common-mode feedback (CMFB) circuit. The different opamps have been macromodeled with the fully-differential amplifier model whose simplified view as a block diagram is depicted in Fig.10. This macromodel is composed of three stages –input, intermediate, and output–, each modeling different non-ideal effects.

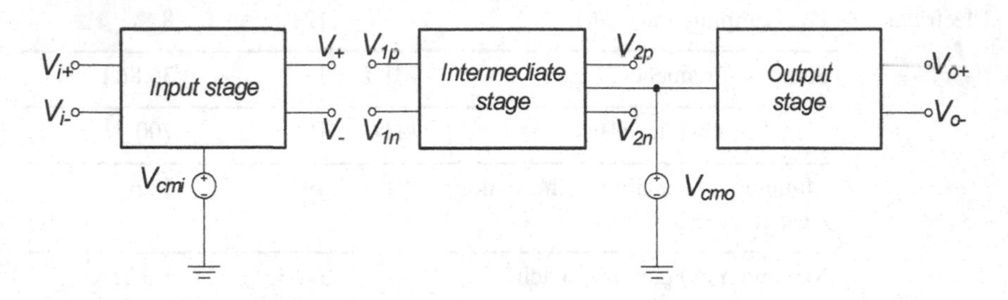

Figure 10. Simplified view of the opamp macromodel.

These non-ideal effects are, as previously stated, the input parameters of the opamp macromodel, shown in Table 2. The non-idealities considered here not only address frequency response limitations but also large signal deviations that take into account the finite dynamic range of real opamps and the existence of harmonic distortion mechanisms. Moreover, these are the performance features for which a target value is obtained at the end of the module-sizing process, such that the analog back-end meets the sub-system performance specifications, and which are then considered as goals of the cell-level sizing process (where the structural facet of each reusable opamp is used).

The opamp's reusable behavioral model has been developed by following a complete portrait of the different signal paths in fully-differential opamps, including those arising from transistor mismatch and variations on the power supplies [VanP90] [Duque93]. This increasingly complex representation approach allows enhancing the modularity of the behavioral view of the reusable block, thus providing an efficient way to "turn on/off" the different second-order phenomena, as retargeting or migrating the opamps over various reuse scenarios requires so.

Figure 11 shows the composition of each stage of the opamp's reusable macromodel. The role of each stage in modeling the set of opamp non-idealities is as follows:

1. **Input stage**. This stage models the first three input parameters in Table 2, i.e., the input offset voltage, the common-mode input capacitances, and the differential input capacitance.
2. **Intermediate stage**. The non-idealities modeled here are: the differential-input differential-output low-frequency gain, dynamic effects, slew-rate, nonlinear input transconductance, as well as *CMRR*, *PSRR*

Table 2. Macromodel input parameters.

Macromodel parameter	*Default*	*Unit*	*Description*
v_{os}	0.0	V	Input offset voltage
c_{ic}	0.50e-12	F	Capacitance from input terminal to AC ground
c_{id}	0.20e-12	F	Differential input capacitance
add_db	70.0	dB	Differential-input differential-output low-frequency voltage gain
fp_1	1e3	Hz	1st pole position
fp_2	1e6	Hz	2nd pole position
sr	35.0	V/μs	Slew-rate
acc_db	-40	dB	Common-mode input to common-mode output voltage gain
CMRR[a]	60	dB	Low-frequency common-mode rejection ratio
$PSRR^+$	80	dB	Low-frequency positive power supply rejection ratio
$PSRR^-$	80	dB	Low-frequency negative power supply rejection ratio
fz_{dd}	2.0e+04	Hz	First zero of the positive power supply transfer characteristic
fz_{ss}	2.0e+04	Hz	First zero of the negative power supply transfer characteristic
fz_{cc}	2.0e+05	Hz	First zero of the common-mode transfer characteristic
r_{out}	1.00e+03	Ω	Resistance from output terminal to common-mode output voltage
os_{pos}	1.0	V	Positive output swing from common-mode output voltage
os_{neg}	1.0	V	Negative output swing from common-mode output voltage

a. It is a well-known fact that CMRR and PSRR in fully-differential amplifiers are due dominated by mismatches introduced by imperfect fabrication. Therefore, the corresponding default values should be actually defined as worst-case values.

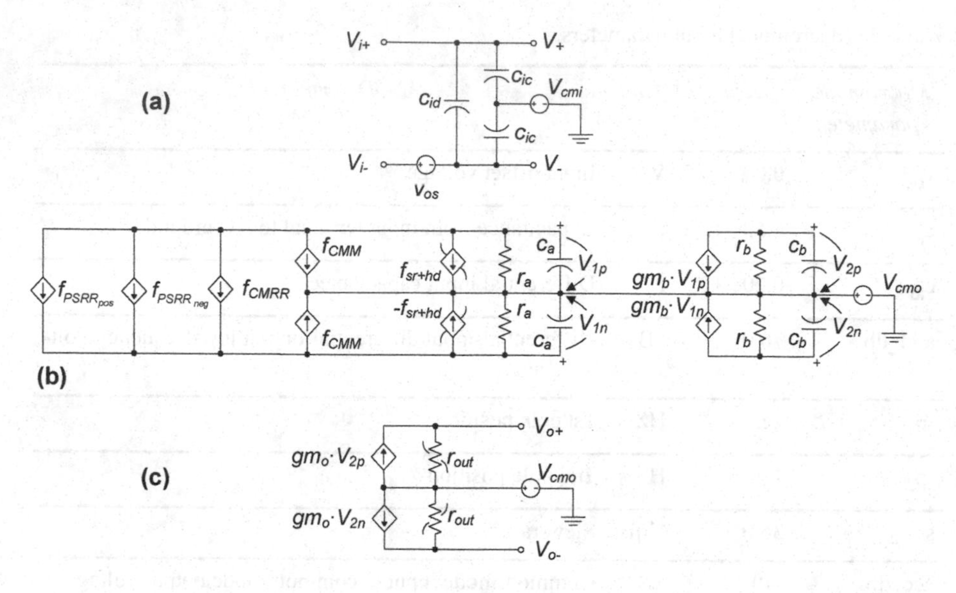

Figure 11. (a) Input, (b) intermediate, and (c) output stages of the opamp's reusable macromodel.

effects, and the impact of the non-ideal common-mode transfer function (i.e., common-mode input to common-mode output).

3. **Output stage**. The output stage models the positive and negative saturation voltages as well as the output resistance.

The macromodel, as it is shown above, is composed of circuit elements such as transconductances, resistances, and capacitances, which are to be assigned a value prior to each performance evaluation. For the input and output stages, the value of the macromodel elements are directly given in Table 2. The output saturation voltage is implemented with the nonlinear resistor depicted in Fig. 12.

To derive the elements of the intermediate stage in order to correctly model the different non-idealities, a set of design equations are used. For the dynamic effects and the slew-rate, the corresponding design equations are shown in Table 3. Note that two additional design equations can be used provided that the input parameters are the unity-gain frequency, ft, and the phase margin, pm, instead of fp_1 and fp_2. Note also that it is necessary to specify the value of intermediate variables in order to fully calculate the value of the macromodel elements[12], which further complicates the model characteriza-

12. In contrast to the input parameters in Table 2, the value of intermediate variables are not necessarily transmitted down to size the opamp.

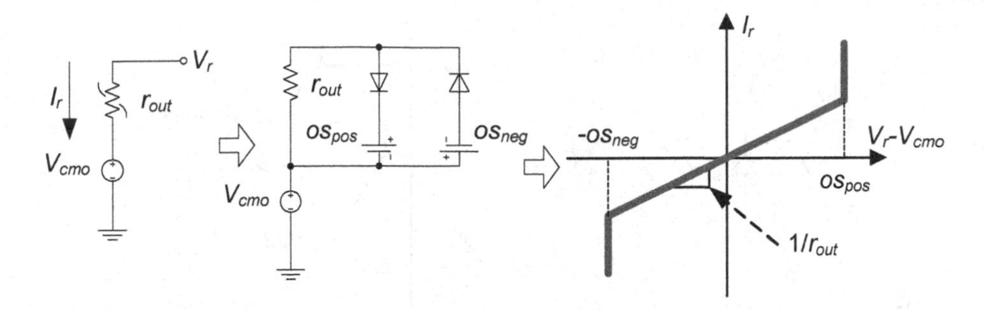

Figure 12. Modeling the output voltage saturation with a nonlinear resistor.

Table 3. Design equations to model dynamic and slew-rate effects.

Equation	*Intermediate variables*
$add = gm_a \cdot gm_b \cdot r_a \cdot r_b$ [a]	r_a, r_b, gm_b
$\omega p_1 = 2 \cdot \pi \cdot fp_1 = 1/(r_a \cdot c_a)$	
$\omega p_2 = 2 \cdot \pi \cdot fp_2 = 1/(r_b \cdot c_b)$	
$gm_o = 1/r_{out}$	
$\omega p_2 = [(add \cdot \tan(pm) - 1) \cdot \omega t]/[add + \tan(pm)]$ [b]	
$\omega p_1 = (\omega t \cdot \sqrt{\omega t^2 + \omega p_2^2})/(\sqrt{(add^2 - 1) \cdot \omega p_2^2 - \omega t^2})$	
$ib = sr \cdot c_a$	

a. $add = 10^{((add\_db)/20)}$

b. $\omega t = 2 \cdot \pi \cdot ft$

tion. As it is explained later, the drawback of using such additional variables can be overcome by using a different macromodeling approach, based on the ability of many description languages to directly specify transfer functions.

The slew-rate is modeled by adequately modifying the transconductance gm_a, as shown in Fig. 13(a). The value of the maximum current available to charge the equivalent capacitance of the first-frequency stage, is calculated through the input parameter sr and the dominant (first) pole frequency ωp_1.

Including nonlinearities in this transconductance to model harmonic distortion is carried out by means of transforming each of the nonlinear voltage-controlled

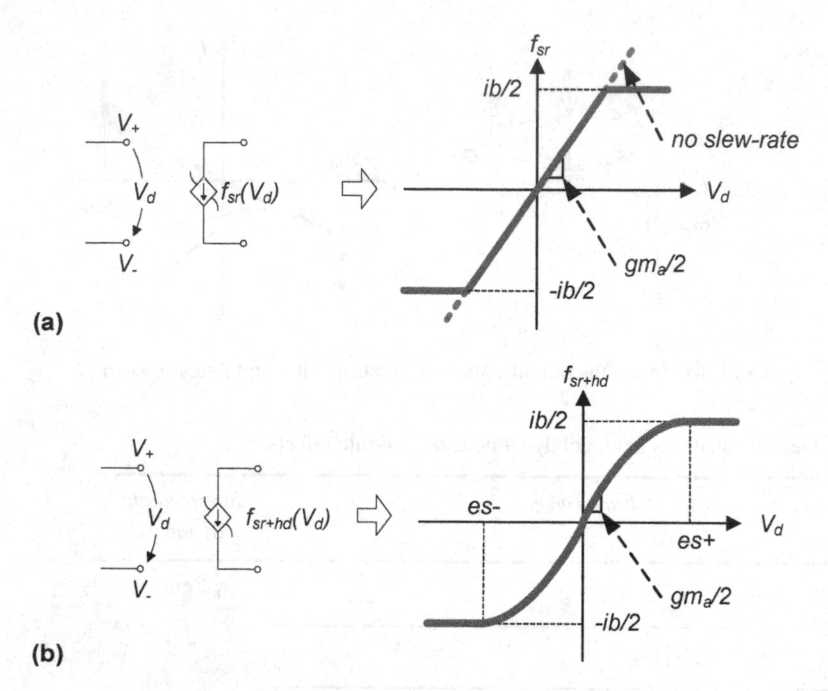

Figure 13. (a) Modeling slew-rate effects, (b), plus large-signal deviations non-linearities.

current sources, gm_a, as depicted in Fig.13(b). Between saturation limits es+ and –es–, the transconductance gm_a is a nonlinear function, as expressed in the following equation:

$$f_{sr+hd}(V_d) = \frac{gm_a \cdot V_d}{2} \cdot \sqrt{1 - \left(\frac{gm_a \cdot V_d}{2 \cdot ib}\right)^2} \tag{1}$$

with es+ and es– being:

$$es+ = -es- = \frac{2 \cdot ib}{\sqrt{2} \cdot gm_a} \tag{2}$$

In this equation, gm_a and ib are calculated through the corresponding set of design equations in Table 3.

The intermediate stage is progressively augmented with appropriate elements to model a finite common-mode rejection ratio (*CMRR*), power supply rejection ratios (*PSRR*), and a non-ideal common-mode transfer function. To model a finite *CMRR*, the voltage-controlled current source f_{CMRR}, whose characteristic is shown in Fig.14(a), is used. To model a positive *PSRR* (i.e., the rejection ratio with respect variations in V_{DD}), a negative *PSRR* (i.e., the rejection ratio with respect variations in V_{SS}), or the common-mode transfer function effects, the same configuration in Fig. 14(b) with adequately derived

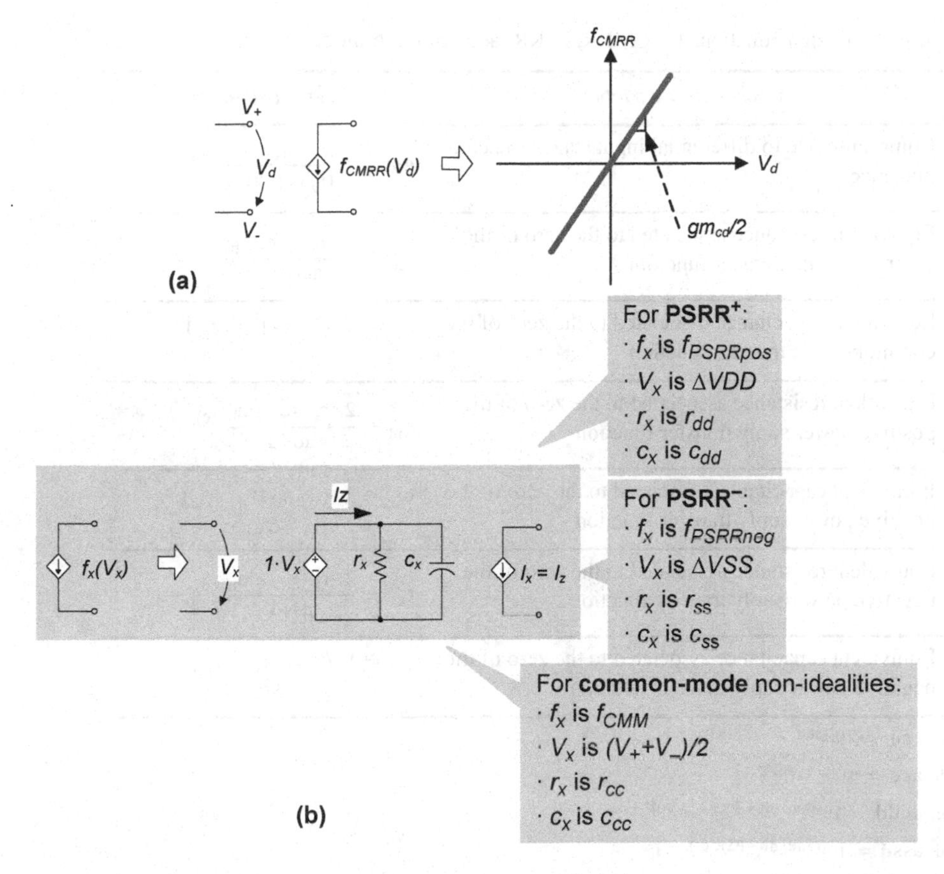

Figure 14. Elements modeling the CMRR, PSRR, and common-mode transfer function characteristics.

macromodel elements, is used. The design equations in Table 4 are used to derive the elements of the configurations in Fig. 14.

At this point, and coming back to the general concept of reusable behavioral model, it is interesting to comment on two relevant aspects of reusable behavioral macromodels. The first one is that, both during top-down electrical synthesis and bottom-up verification, the behavioral model must be able to accurately reproduce the electrical behavior of the circuit with specified values of its input performance features. This aspect is illustrated in Fig. 15. Considering the reusable behavioral model, B_k, and a set of N_k input performance features with values $\{\rho_{k,1}, \ldots, \rho_{k,N_k}\}$, (coming either from a selection made by the optimization tool or from a simulation for verification of the corresponding circuit), it is expected that the emulated performance features, $\{\rho'_{k,1}, \ldots, \rho'_{k,N_k}\}$, equal its input performance features $\{\rho_{k,1}, \ldots, \rho_{k,N_k}\}$.

Table 4. Design equations for CMRR, PSRR, and common-mode effects.

Macromodel variable	*Design equation*
Common-mode to differential input stage transconductance	$gm_{cd} = \frac{acd}{(r_a \cdot r_b \cdot gm_b)}$ (a)
Equivalent resistance associated to the zero of the common-mode transfer function	$r_{cc} = \frac{r_a \cdot r_b \cdot gm_b}{acc}$ (b)
Equivalent capacitance associated to the zero of the common-mode transfer function	$c_{cc} = 1/(2 \cdot \pi \cdot fz_{cc} \cdot r_{cc})$
Equivalent resistance associated to the zero of the positive power suply transfer function	$r_{dd} = \frac{2 \cdot r_a \cdot r_b \cdot gm_b}{addd}$ (c)
Equivalent capacitance associated to the zero of the positive power suply transfer function	$c_{dd} = 1/(2 \cdot \pi \cdot fz_{dd} \cdot r_{cc})$
Equivalent resistance associated to the zero of the negative power suply transfer function	$r_{ss} = \frac{2 \cdot r_a \cdot r_b \cdot gm_b}{assd}$ (d)
Equivalent capacitance associated to the zero of the negative power suply transfer function	$c_{ss} = 1/(2 \cdot \pi \cdot fz_{ss} \cdot r_{ss})$

a. $acd = 10^{[(add\_db - CMRR)/20]}$

b. $acc = 10^{(acc\_db/20)}$

c. $addd = 10^{[(add\_db - PSRR^{+})/20]}$

d. $assd = 10^{[(add\_db - PSRR^{-})/20]}$

During top-down synthesis, a possible mismatch between $\{\rho_{k,1}, \ldots, \rho_{k,N_k}\}$ and $\{\rho'_{k,1}, \ldots, \rho'_{k,N_k}\}$ is not critical since it just implies a shift in the design space equal to the difference between $\{\rho'_{k,1}, \ldots, \rho'_{k,N_k}\}$ and $\{\rho_{k,1}, \ldots, \rho_{k,N_k}\}$; the problem can be easily overcome by transmitting down $\{\rho'_{k,1}, \ldots, \rho'_{k,N_k}\}$ instead of $\{\rho_{k,1}, \ldots, \rho_{k,N_k}\}$. By contrast, the mismatch between $\{\rho_{k,1}, \ldots, \rho_{k,N_k}\}$ and $\{\rho'_{k,1}, \ldots, \rho'_{k,N_k}\}$ is certainly decisive during bottom-up verification, for it can lead to completely wrong verification results (e.g., rejecting a successful design or vice versa). In this sense, it could be reasonable to use one behavioral model for top-down synthesis, and a different, more accurate behavioral model for bottom-up verification; instance behavioral modeling approaches such as look-up tables, in general very accurate for the instance at hand, can be used in the design reuse flow as long as a method is available with which automatically build the table from the bottom-up verification results of the reusable block itself.

The possible mismatch between these two sets of performance features depend on how the behavioral model is built. For instance, although in the

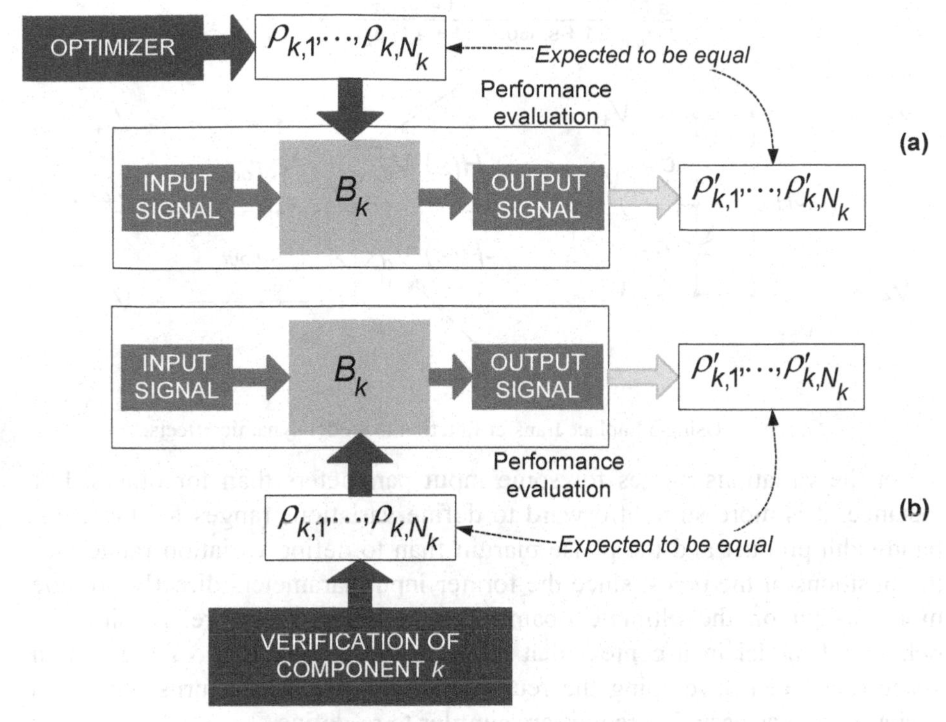

Figure 15. (a) Top-down electrical synthesis and, (b), bottom-up verification with cell-level behavioral models.

case of the fully-differential opamp behavioral model described above, no error is committed in translating the input parameters (e.g., f_t and pm) into adequate values of the macromodel elements, for more complex opamp macromodels (including, for instance, a zero and an additional high-frequency pole through adequate compensation elements between stages in Fig. 12(b)), an inherent error may be, however, committed. This error depends on the particular equations used, and may eventually cause a mismatch between input parameters (i.e., positions of the poles and the zero) and the actual emulated dynamic performance. In this case, the problem can be alleviated by directly resorting to Laplace transfer function, as illustrated in Fig. 16, since the positions of the poles can be specified directly[13], where ωz is the frequency of the zero and ωp_3 is the frequency of the additional pole.

The second relevant aspect mentioned above concerns the use of certain input parameters rather than others. Recalling that the input parameters and their variation range define the explorable design space on which sizing (i.e., optimization) takes place, it is necessary to remark that it can be much easier

[13] This also alleviates the problem of specifying the value of intermediate variables beforehand.

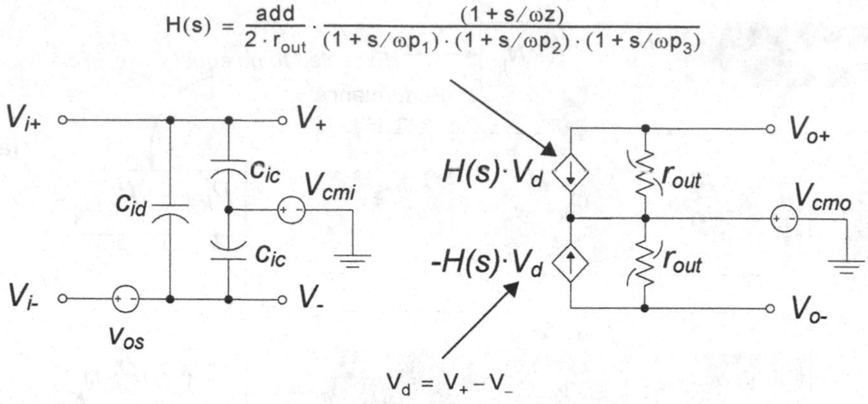

$$H(s) = \frac{add}{2 \cdot r_{out}} \cdot \frac{(1 + s/\omega z)}{(1 + s/\omega p_1) \cdot (1 + s/\omega p_2) \cdot (1 + s/\omega p_3)}$$

V_{i+}, V_+, C_{ic}, V_{cmi}, C_{id}, C_{ic}, V_{i-}, V_-, V_{os}, $H(s) \cdot V_d$, $-H(s) \cdot V_d$, r_{out}, r_{out}, V_{o+}, V_{cmo}, V_{o-}

$V_d = V_+ - V_-$

Figure 16. Using a Laplace transfer function to model dynamic effects.

to set the variations ranges for some input parameters than for others. For instance, it is more straightforward to define variations ranges for the gain-bandwidth product and the phase margin than to define variation ranges for the positions of the poles, since the former input parameters directly provide more insight on the ultimate opamp performance. Therefore, reusing the behavioral model in a context that requires the re-definition of the design space (and even developing the reusable behavioral facet), turns out much easier as long as variation ranges are simpler to re-define.

4 SUMMARY

This chapter has described the behavioral facet of the AMS reusable blocks needed for the design reuse methodology presented in this book. This facet serves to obtain the performance specifications of a cell or module-level block from those of the block containing it and lying immediately up in the hierarchy to meet the previously derived performance specifications. Besides, the behavioral facet can be advantageously used during bottom-up verifications. The use of behavioral modeling in the design reuse methodology has been examined and the properties of the reusable behavioral models (i.e., the behavioral facet of the AMS reusable blocks) have been developed and illustrated with a circuit example.

In the process of transforming the performance requirements of a particular circuit block into requirements for all its components, as well as in assessing the circuit block's performance during the bottom-up flow, not only the behavioral facet of each module-level and cell-level component turns out essential, but also a separate facet of the circuit block itself. This reusable facet is called the structural facet, and it is the goal of the following chapter.

Chapter 4

The Analog Reusable Block: Structural Facet

This chapter describes the structural facet of the analog reusable block. Whereas its behavioral facet is used to map the performance specifications of the circuit immediately above in the hierarchy into performance specifications for the reusable block itself, the structural facet facilitates the mapping of these performance specifications into an appropriate sizing of all the reusable block's components (in terms of specifications, or device characteristics, sizes and values). In this latter sizing process, in turn, the behavioral facet of each component of the reusable block (obviously except for any device-level component, for which there is no behavioral facet) helps expediting the translation of specifications down the hierarchy. Besides, the structural facet is also used, as explained in Chapter 2, to assess the performance of the reusable block during the bottom-up flow.

1 INTRODUCTION

A structural description of a circuit block consists in a number of functional blocks (module and cells) and devices, (such as MOS transistors, resistors, and capacitors), which are interconnected to compose the circuit's **netlist**. This netlist defines two characteristics concerning the building component: how they are interconnected and which are the their particular characteristics.

The interconnect description is commonly referred as the circuit's **architecture/topology**. It basically includes the circuit **nodes**, at which the **terminals** of each component (e.g., the inputs/outputs of a module-level circuit block, or the drain, source, and gate of a MOS transistor) are attached, and the circuit **pins**, the nodes at which the circuit's input, output, supply, and bias signals (currents and voltages) are defined.

On the one hand, the resulting circuit performance emerges altogether from the circuit's architecture/topology, from the **component models**, and from the specific value of their characteristics. Such models are quantitative descriptions of the various phenomena that are responsible for the behavior of the component, as a function of its externally applied voltages or currents and, possibly, the fabrication technology. At the device level, the electrical characteristics depend upon the device type; typical examples are the MOS transistor width and length, the capacitance value of a passive capacitor, and the resistance value of a passive resistor; the device models are process-specific (e.g., SPICE-like transistor models). For hierarchically higher components (i.e., cell, module, and system levels), the component model refers to its behavioral model; the characteristics of the component are the parameters of the behavioral model (see Chapter 3). The whole set of device characteristics is called the circuit **sizing**. These three factors –architecture/ topology, models, and sizing– are circuit-specific. On the other hand, the circuit performance also depends upon the **peripheral setup**. This setup comprises the input signals, the output loading conditions, the supply voltages, and the biasing currents and voltages. For the same circuit architecture/ topology, models, and sizing, differing peripheral setups may render differing circuit performances.

Feasibility of performance retargeting –the change in performance while retaining the fabrication process– implies that the component characteristics are to be freely changed in order to attain various circuit performances. In this sense, the structural view of the reusable block should comprise an architecture/topology-fixed block whose component characteristics are fully parameterized. Similarly, feasibility of process migration involves correct updating of the component models to the new fabrication process (e.g., updating the MOS transistor models to the new technology). Again, the structural view has to be based, at the very least, on a fixed architecture/topology whose component models can be easily adapted to new processes. That is, both the component characteristics and the component models should be parameters of the structural view.

Given the description above, it is straightforward to conclude that most commonly used formats of circuit netlists could be effortlessly reused. Consider, for instance, a SPICE-like circuit netlist of a cell-level circuit: both device characteristics and models are user-defined settings easily customizable. It seems that it would be just necessary to re-execute the automated sizing process on said netlist with new specifications in the goal technology. Yet these circuit descriptions alone are not fully reusable. The reason why becomes evident when carefully considering the analog design flow. Recalling Chapter 2, **sizing** was defined as any constructive process that maps **performance specifications** into **design parameters**. These design parameters can be device sizes, performance specifications of module-level or cell-level components, and peripheral setup parameters, such as biasing currents or voltages. Automated circuit sizing (i.e., sizing supported by an automated sizing EDA tool) implies not only that the parameters of the reusable block can be unrestrictedly changed, but also that the sizing process itself is guided by a certain amount of accumulated, circuit-specific **design knowledge**. This knowledge is fundamental to manage efficiently the complexity of analog circuit design, as explained in Chapter 2. Therefore, parameterization does not suffice. It is also necessary that the reusable block includes adequate design knowledge and that this knowledge can be easily reused by a design team or by the sizing tool. Which design knowledge must be *encapsulated* within the structural view of the reusable block is one of the issues addressed in the present chapter.

From a practical point of view, the software database representing the structural view of the analog reusable block, illustrated in Fig.1, may be given in several different formats, from plain text files to schematic capture databases. When considering the technology portability of this database, practical problems arise that render the reuse process unnecessarily slow. Due to the lack of a standardization of device-level circuit descriptions among fabrication foundries, solving this problem may result rather intricate. This chapter also discusses the problem in more detail. As a practical implementation example, a method for the migration of the structural view database in the Cadence® commercial tool suit *Design Framework II* (*DFII*) is outlined.

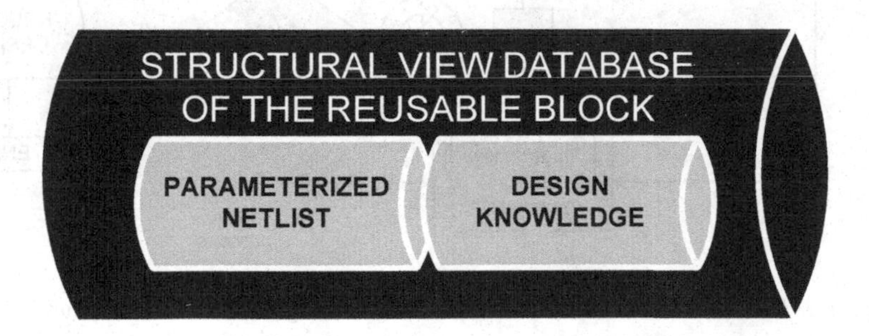

Figure 1. The elements of the analog reusable block's structural view.

To sum up, this chapter covers the realization of structural views by addressing the following issues:

- Design knowledge encapsulation (Section 2).
- Database software reuse (Section 3).

Needless to say, the database format should fit up with the requirements of the tool performing the sizing process. Furthermore, it is important to note that the type of sizing approach followed may dictate the type of knowledge that must be coded into the structural view of the reusable block. The optimization-based sizing approach adopted in the design reuse methodology described in this book has already been outlined in Chapter 2. The next section briefs its most important features.

1.1 Adopted sizing approach

Optimization-based approaches translate the circuit-sizing problem into a constrained optimization problem. This problem is solved by means of the iterative procedure in Fig. 2. Starting from an initial position $\mathbf{x}_0$, the design space is gradually explored until a point of equilibrium is reached, which is the solution of the sizing problem. This procedure benefits from the use of a **cost function** that quantifies the degree of compliance of the design with the targeted performance.

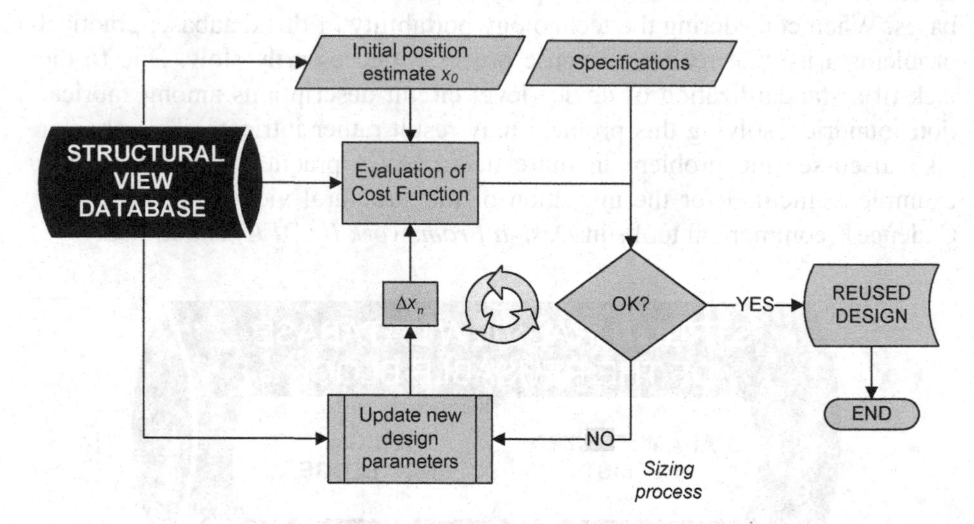

Figure 2. The optimization-based sizing process for design reuse of the structural view.

The cornerstones of optimization-based sizing are: (1) an adequate formulation of the cost function; (2) a fast yet accurate method to evaluate the cost function; and (3) an efficient technique to generate the next movement over the design space. The cost function is formulated to quantify the compliance of the attained performance at each iteration. This is accomplished by comparing the attained value of the performance features against the performance specifications of the goal design. The performance specifications are here considered in a wide sense, including performance restrictions on the circuit performance –those specifications involving inequalities like $\text{DC-gain} > 70\text{dB}$ or $\text{bandwidth} > 50\text{MHz}$– and design objectives –whose intention is to maximize or minimize some performance feature like power consumption or area occupation– Mathematically, the fulfillment of these groups of performance specifications can be formulated as a multi-constrained optimization problem [Mede94]. With respect to the evaluation of the cost function, we have opted, as explained in Chapter 2, for an optimization-based sizing approach using simulation [Mede94] [Mede99]. In manual design, the selection of the magnitude and orientation of the movement $\Delta \mathbf{x}_n$ (see Fig. 2) is based on certain indicatives of the circuit performance (obtained by simulation) and, more importantly, on the accumulated design knowledge about the circuit being sized. It is for this reason that the addition of designers' expertise has been enabled in the sizing tool. Such expertise takes the form of arbitrary **constraints** or relationships between the design parameters, and the performance restrictions and/or design objectives. From this point of view, the adopted sizing tool is an optimization-based system incorporating the appealing features of knowledge-based ones. Note that such constraints are fully independent of the tool core. Therefore, they can be easily expanded, deleted, or modified, making the approach fully open.

2 DESIGN KNOWLEDGE ENCAPSULATION

Two conclusions can be drawn from the previous section:

1. The structural view of the reusable block has to comprehend not only a parameterized netlist, but also a complete design knowledge database.
2. This knowledge database must be self-consistent with the sizing process used for the design reuse methodology.

As described above, a simulation-based optimization tool realizes the sizing task for either a performance retargeting or a process migration. The ensuing description of the required design knowledge elements for a structural view is, therefore, suited to such a sizing approach.

The required elements of the reusable block's structural view are shown in Fig. 3. As mentioned in earlier paragraphs, these elements can be organized into two groups: the parameterized netlist and the design knowledge database. The former contains the circuit architecture/topology with all the component characteristics transformed into parameters of the view. The latter is composed of all the elements that, by capturing the designer's intent, make the sizing process more efficient.

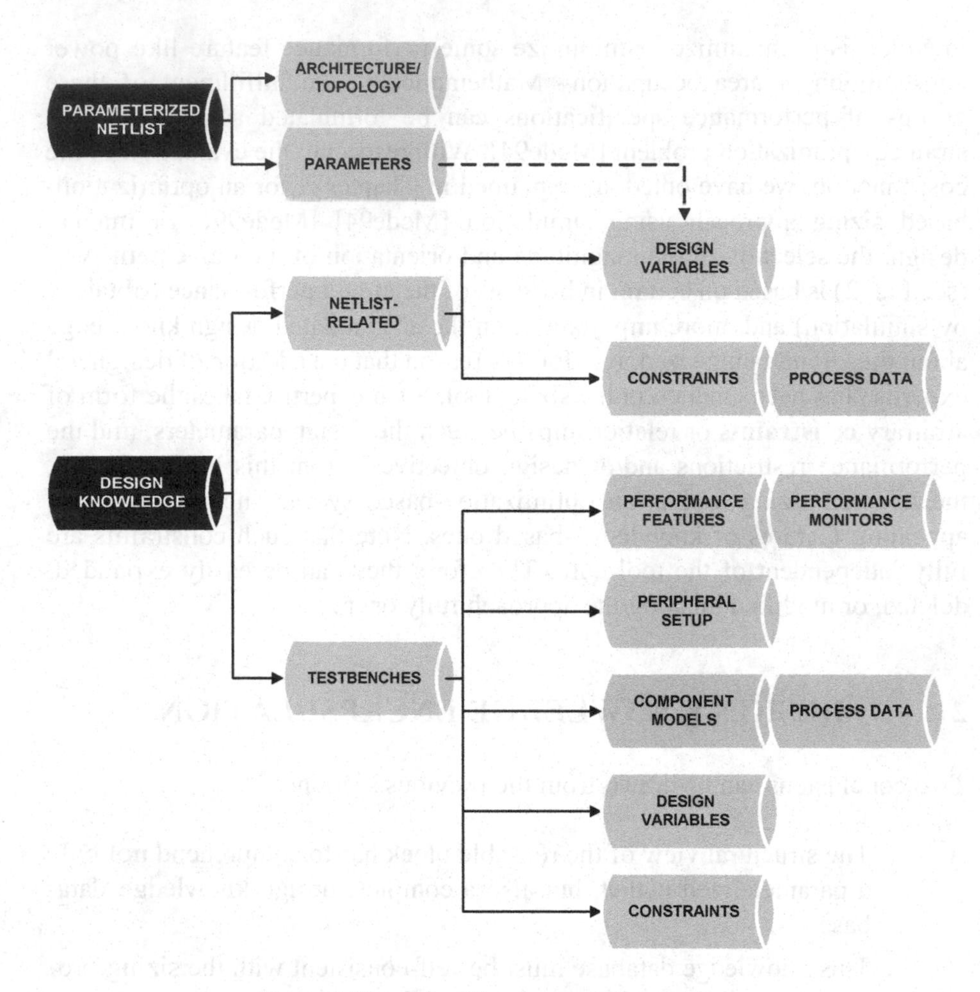

Figure 3. Elements of the structural view database.

For the sake of clarity, a cell-level circuit block is employed to illustrate the description of the elements of the structural view. A module or system-level example would comprise equivalent elements, with the addition of behavioral models of some of their cell and module building blocks.

Then, let us consider the circuit block shown in Fig. 4. It is a fully-differential amplifier –common-mode feedback circuitry (CMFB) included– with two compensated folded-cascode paths. The parameterized netlist database for this circuit block is composed of the topology, defining the connectivity of the device-level components, and a collection of parameters for each device.

Obviously, the component parameters depend upon the model being used for each type of device: the more elaborated the model the more number of parameters might be required (e.g., statistical models of resistors require that

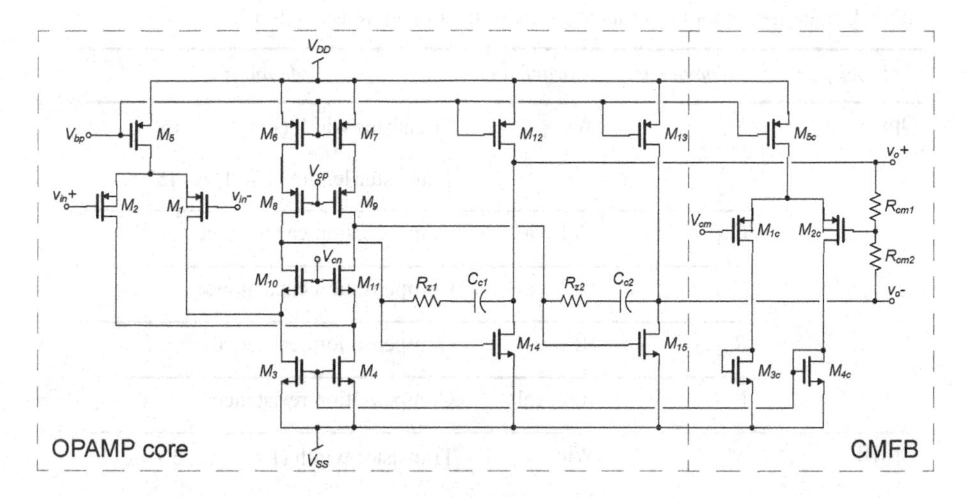

Figure 4. Fully-differential two-stages opamp and its CMFB circuitry, used as illustration vehicle.

both width and length of the resistive strip be specified). The models, in turn, depend upon the fabrication process and the type of simulation to perform during the sizing process (e.g., statistical simulations). Nevertheless, fundamental device parameters (e.g., MOS transistor width and length, resistor value, and capacitor value) remain the same for different fabrication processes. It is also important to note that whatever the models used, all the netlist parameters are also parameters of the reusable block's layout facet, explained in the following chapter.

Table 1 shows the list of parameters for the opamp example. This basic set of 46 parameters (34 for the core and 12 for the CMFB circuitry) suffices to make retargeting the circuit block feasible by just considering each parameter as a variable of the sizing process. As it will be explained below, further parameters can be added that deal with several aspects of analog design (e.g., layout parasitics).

Such a large parameter set entails, however, an extremely lengthy exploration of the design space, as every visited point **x** has 49 degrees of freedom[1] (i.e., $\mathbf{x} \in \Re^{49}$). Encapsulation of designer's expertise reduces, as it will be illustrated below, the complexity of the design space exploration by reducing the degrees of freedom of the design problem, just as expert designers deal with the subtle characteristics of analog design. Nevertheless, design knowledge is not solely used to expedite design space search. Actually, design knowledge plays the role of an "absent" experienced

1. This parameter set includes as well the biasing voltages V_{bp}, V_{cp}, and V_{cn}.

Table 1. Parameters for the structural view of the opamp reusable block.

Block part	*Component*	*Parameters*	*Meaning*
Opamp core	M_i	Wi	Transistor width (i = 1, ..., 15)
		Li	Transistor length (i = 1, ..., 15)
	C_{c1}	Cc1_val	Compensation capacitance
	C_{c2}	Cc2_val	Compensation capacitance
	R_{z1}	Rz1_val	Compensation resistance
	R_{z2}	Rz2_val	Compensation resistance
CMBF	M_{ic}	Wic	Transistor width (i = 1, ..., 5)
		Lic	Transistor length (i = 1, ..., 5)
	R_{cm1}	Rcm1_val	Common-mode detection resistance
	R_{cm2}	Rcm2_val	Common-mode detection resistance

designer, guiding the sizing process toward sizing solutions as satisfactory and robust as the ones achieved by such a designer. It is worth noting that the particular sizing approach adopted allows the incorporation of design knowledge in a rather wide sense: in effect, the reduction of the design space's cardinality and the correct guidance of the sizing process can be attained by several means, which, as shown below, range from simple relationships between design variables, defining suitable design space contours, to complete, elaborate design plans[2] (i.e., any degree of freedom is removed), every one of these means steaming from accumulated design expertise.

The following is a detailed description of the elements upon which the knowledge database of the structural view is elaborated.

2.1 Netlist-related elements

The design knowledge database is firstly constructed by specifying which (and which not), from the complete set of parameters, are to be considered as

[2] Obviously enough, a complete design plan would eliminate the need for an optimization process; nevertheless, the fact that such a design plan can be included as encapsulated knowledge within the reusable block demonstrates the extent to which the followed synthesis approach is ready for design knowledge reuse.

design variables of the sizing process, such that a reduction of a possibly huge design space is achieved. This reduction is attained by:

1. Adding a relationship between design variables (**independent variables**) and the rest of the netlist parameters (**dependent variables**). These relationships are called **constraints**, because they restrict independent variables to a value determined by one or more dependent variables; the scope of constraints ranges from simple analytical relationships to complex design heuristics or algorithms relating independent and dependent design variables.
2. Confining the variation range of each design variable within limits such that the resulting defined design space is wide enough for block reuse feasibility, but small enough to avoid lengthy explorations.

Both resources, directly drawn from the accumulated design experience for the circuit block, are detailed and illustrated below.

2.1.1 Design variables

The selection of the design variables is made, to a large extent, upon the designer's experience and it is closely related to the constraints that he/she has defined. The purpose of the structural view is to capture this experience. For instance, both dependent variables and design (independent) variables for the opamp example in Fig. 4 are shown in Table 2. Once the design variables have been picked out, it is necessary to define their scope or variation range. In doing so, it is necessary to take into account the following issues:

- The lower and the upper limits have to comply with process design rules, mainly those referred to minimum and maximum values. For example, the width and length of the transistor must be over their corresponding process minimum values.

- The lower and/or upper limits of a particular design variable have to comply with available design considerations. These considerations, based on valuable design knowledge, are in the form of relationships between design variables and process parameters. The design variable is still independent save for the fact that its minimum and/or maximum values are constrained. Coming back to the opamp example in Fig.4, let us consider the nulling resistors R_{z1} and R_{z2}. As it is well known, the transfer function of the opamp exhibits a zero at $\omega z \approx 1/[(1/g_m) - R_z] \cdot C_c$, with g_m being the transconductance of M_{14} and M_{15}. For $R_z \approx 1/g_m$, such zero moves to infinity, a design

Table 2. Design variables for the structural view of the opamp reusable block.

Dependent variables	*Design variable*	*Range*	*Unit*
W2	W1	$[MINW^{a}, MINW \cdot F_{W1}]$	µm
W4	W3	$[MINW, MINW \cdot F_{W3}]$	µm
W6, W7, W14, W15, W3c, W4c	W5	$[MINW, MINW \cdot F_{W5}]$	µm
W9	W8	$[MINW, MINW \cdot F_{W8}]$	µm
W11	W10	$[MINW, MINW \cdot F_{W10}]$	µm
W13, W14, W15	W12	$[MINW, MINW \cdot F_{W12}]$	µm
L2	L1	$[MINL^{b}, MINL \cdot F_{L1}]$	µm
L3, L4, L6, L7, L12, L13, L5c	L5	$[MINL, MINL \cdot F_{L5}]$	µm
L9	L8	$[MINL, MINL \cdot F_{L8}]$	µm
L11	L10	$[MINL, MINL \cdot F_{L10}]$	µm
L15, W14, W15	L14	$[MINL, MINL \cdot F_{L14}]$	µm
Cc2_val	Cc1_val	$[MINC^{c}, MINC \cdot F_{Cc1\_val}]$	pF
Rz2_val	Rz1_val	$[MINRz, MINRz \cdot F_{Rz1\_val}]$	Ω
W2c	W1c	$[MINW, MINW \cdot F_{W1c}]$	µm
W3c, W4c	W5c	$[MINW, MINW \cdot F_{W5c}]$	µm
L2c	L1c	$[MINL, MINL \cdot F_{L1c}]$	µm
W3c, W4c, L4c	L3c	$[MINL, MINL \cdot F_{L3c}]$	µm
Rcm2_val	Rcm1_val	$[MINRcm^{d}, MINRcm \cdot F_{Rcm1\_val}]$	Ω

a. MINW is the process-defined minimum MOS transistor width.

b. MINL is the process-defined minimum MOS transistor length.

c. MINC is the minimum process-achievable capacitance.

d. MINRcm is the maximum between the minimum process-achievable resistance and the advisable value derived from the designer experience. In particular, this resistance must be much greater than the opamp output impedance ($r_{oM12} \| r_{oM14}$) so as to avoid lowering the open-loop gain.

option which turns R_{z1} and R_{z2} into dependant variables. Furthermore, making the resistor R_z greater than $1/g_m$ moves the zero into the half-plane, which can be used to provide positive phase shift at high frequencies and improve the phase margin of a feedback circuit that uses this opamp. This design knowledge can be stored by setting the lower limit of R_{z1} and R_{z2}, $MINR_z$, to the maximum value between the minimum process-achievable resistor and $1/g_m$, which depends on the dimensions of M_{14} (M_{15}), on the current flowing through M_{14} (M_{15}), and on the process parameters μ_N (mobility of NMOS transistors) and C_{ox} (oxide capacitance).

- Last, the grid according to which each design variable changes must fit the type of the variable (e.g., integer, floating-point number) and the variable nature. For example, if the design variable corresponds to a layout dimension, such as the MOS transistor width and length, its grid has to equal the process' layout grid[3].

The opamp example is a cell-level reusable block, i.e., it only has device-level components. Then, most of its design variables are design parameters of these devices (e.g., W and L of MOS transistors). For a system or module-level reusable block, its components are also other cell and module-level reusable blocks. The design variables are to be selected, in the same way, from among the design parameters of these components as well. As it will be explained later in this chapter, these cell and module-level components are included in the structural view by using their behavioral facet, i.e., their behavioral model (see Chapter 3). Hence, the design variables are picked out from the parameters of each behavioral model. Note that variation ranges of these parameters are already defined in the reusable behavioral facet. Then, the structural facet of a circuit comprising cell or module-level building components, simply "inherits" the variation ranges defined in the behavioral facet and, if necessary, adapts them to fit the requirements of the sizing process.

Table 2 also displays the variation ranges of the design variables for the opamp block in Fig. 4. Note that the upper limits of the variable ranges are defined as the lower limit value multiplied by a certain factor F_{DV}, where DV is the design variable. This factor is defined upon heuristic design considerations. For example, the factor for the transistor width may range from several tens to several hundreds, whereas the MOS transistor length design variables are typically set at a very much lower value ($< 10\ \mu m$) both to ease

---

3. Although it might not be strictly necessary to assign on-grid values, it is always preferable to do so in order for the sized circuit to be as similar as possible to the eventually generated circuit layout, for which all sizes must be "on grid".

the design space exploration and to minimize design objectives such as area or power consumption. Note as well that, although the variable variation ranges are stored by default in the reusable block database, they can be also varied in case the optimization process fails to obtain a suitable circuit sizing.

As said above, design variables are selected from the set of netlist parameters of the circuit block. This does not mean that all the remaining parameters are kept fixed, but that some of them depend upon the design variables. The rest of netlist parameters must only change to deal with a design migration to a different fabrication process[4].

2.1.2 Constraints

Constraints relate the circuit's dependent variables and the circuit's independent variables (design variables). For instance, they include, but are not limited to, analytical expressions. A great deal of the accumulated design experience can be stored and thereinafter reused by means of these constraints. The chief goal at including constraints is to make the optimization process more efficient as many "invalid" points of the design space are avoided. Therefore, the fulfillment of these constraints is essential for a successful optimization. To illustrate this point, consider the sizing of transistors M_{3c} and M_{4c} of the CMFB circuitry of the opamp in Fig. 4. Proper biasing is needed to make the currents in current sources M_3 and M_4 in the opamp core larger than $I_{M5}/2$ in the balanced point. Thus

$$I_{M3} = I_{M10} + \frac{I_{M5}}{2} \tag{1}$$

$$I_{M4} = I_{M11} + \frac{I_{M5}}{2} \tag{2}$$

Since $I_{M3} = (S_{M3}/S_{M3c}) \cdot I_{M3c}$ and $I_{M10} = (S_{M6}/S_{M5}) \cdot I_{M5}$ (where S_{Mx} denotes the aspect ratio W_{Mx}/L_{Mx}), an expression for the aspect ratio of transistors M_{3c} and M_{4c} can be derived, resulting that

$$S_{M3c} = \frac{2 \cdot I_{M3c} \cdot S_{M3}}{2(S_{M6}/S_{M5}) \cdot I_{M5} + I_{M5}} \tag{3}$$

$$S_{M4c} = \frac{2 \cdot I_{M4c} \cdot S_{M4}}{2(S_{M7}/S_{M5}) \cdot I_{M5} + I_{M5}} \tag{4}$$

---

4. The parameters of power-down devices serve as an example. In low-power design, it is common to put these devices in the biasing circuitry to implement a power-down operation mode. The value of these parameters are normally irrelevant to the circuit performance, so they are fixed during the sizing process as long as retargeting is solely involved. When porting the reusable block, however, sizing of power-down devices should be carefully considered so minimum width and length process design rules are still fulfilled.

As the current through M_{3c} and M_{4c} is

$$I_{M3c} = \frac{I_{M5c}}{2} = \frac{S_{M5c}}{2 \cdot S_{M5}} \cdot I_{M5} \tag{5}$$

$$I_{M4c} = \frac{I_{M5c}}{2} = \frac{S_{M5c}}{2 \cdot S_{M5}} \cdot I_{M5} \tag{6}$$

the final correct sizing for M_{3c} and M_{4c} is given by the following expressions:

$$S_{M3c} = \frac{S_{M5c} \cdot S_{M3}}{2 \cdot S_{M6} + S_{M5}} \tag{7}$$

$$S_{M4c} = \frac{S_{M5c} \cdot S_{M4}}{2 \cdot S_{M7} + S_{M5}} \tag{8}$$

If such relationships are not guaranteed per construction, that is, if the aspects of M_{3c} and M_{4c} are not constrained during each iteration of thc optimization process, most iterations will lead to wrong results because current sources M_3 and M_4 do not provide enough current. Although design space exploration algorithms can be powerful enough to bring the value of S_{M3c} and S_{M4c} close to those expressed in Eq. (7) and Eq. (8), respectively, it will however require a larger number of iterations that can be saved by just including this relationships as constraints. Two additional constraints for the opamp circuit can be also derived. The first one ensures that the cascode transistors M_{10} and M_{11} still operate in the saturation region for unbalanced operation of the input differential pair. The constraint is expressed with the following equation:

$$S_{M6} = S_{M7} = S_{M5} \cdot (0.5 + \kappa) \tag{9}$$

where $\kappa \cdot I_{M5}$ is the minimum current flowing in M_{10} and M_{11} to ensure saturation operation. Typically, this current is around 10% of I_{M5} (so $\kappa = 0.1$). The second constraint can be used to maintain a constant current density flowing through the output transistors M_{14} and M_{15}. By doing so, their overdrive voltage, $V_{ov} = V_{GS} - V_{TH}$, is constrained to vary over a range such that points of the design space featuring better output swing are favored. The constraint is captured with the following equation:

$$W_{M14} = W_{M15} = \frac{1}{V_{ov}^{\ 2} \cdot (\beta/2)} \cdot \frac{W_{M12}}{W_{M5}} \cdot I_{M5} \cdot L_{M14} \tag{10}$$

with $\beta = \mu_N \cdot C_{ox}$.

Another type of constraint that is used to reduce considerably the number of design variables simply consists in making two or more netlist parameters equal. This becomes very useful in circuits like the one used here for illustration purposes, with two differential signal paths whose devices should perfectly match.

As it is easily concluded from the above illustrative example, capturing the designer's intent into constraints proves highly valuable to reduce the otherwise huge design space. It is also useful to enhance the reusability property of the analog circuit block, which furthers third-party reuse because the required knowledge to retarget (or migrate) the circuit block needs not to be derived again.

Table 3 summarizes the collection of netlist-related constraints defined for the opamp circuit in Fig. 4. These constraints relate the dependent design variables (first column) with the netlist design variables (second column) by means of a proper relationship (third column).

It is important to remark that these approximate equations do not affect accuracy of the performance evaluation done during the optimization process since they are just used to assign the value of the dependent variables. In the same way, generating constraints to store design knowledge requires an insignificant effort when compared to the effort required to generate equations to perform equation-based optimization or to create a customized design plan to realize a knowledge-base circuit sizing. This advantage stems from the fact that by using optimization and simulation, it is no more necessary to generate design equations to solve each and every one of the degrees of freedom (i.e., each variable of the synthesis problem), as it has to be done when resorting to equation-based optimization or to a design plan. Constraints, relating only dependent variables to design variables rather than to the circuit performance features (e.g., the DC gain, the phase margin, the output swing, etc.), reduce but a fraction of these degrees of freedom, hereby improving the efficiency of both the optimization process and the solution. Moreover, whereas constraints usually require (though they are not restricted to) simple relationships, most of them derived from Kirchhoff's current and voltage laws, design plans and equations entail involved calculations that may take as much as several months of development.

Although all dependent variables are set to a value that results from the application of a constraint, it might happen that the resulting value is violating a process design rule (e.g., minimum MOS transistor width) or is out of its grid (e.g., if the dependent variable is an integer number and the constraint relationship returns a floating-point number). Therefore, it may also be necessary that, as part of the stored design knowledge, the lower and upper limits as well as a grid value are defined for all or some dependent variables (except, of course, for those variables whose associated constraint makes them match a design variable). The sizing engine used in the design reuse flow presented here automatically rejects those optimization iterations, which result in any

Table 3. Netlist-related constraints for the structural view of the opamp reusable block.

Dependent variables	*Design variables*	*Constraint*	*Range & grid*
W2	W1	=	Not necessary
W4	W3	=	Not necessary
W6, W7	W5	Eq(9)	*minimum value*: MINW *grid*: layout grid
W9	W8	=	Not necessary
W11	W10	=	Not necessary
W13	W12	=	Not necessary
W14, W15	W5, W12, L14, V_{ov}	Eq(10)	*minimum value*: MINW *grid*: layout grid
L2	L1	=	Not necessary
L3, L4, L6, L7, L12, L13, L5c	L5	=	Not necessary
L9	L8	=	Not necessary
L11	L10	=	Not necessary
L15	L14	=	Not necessary
Cc2_val	Cc1_val	=	Not necessary
Rz2_val	Rz1_val	=	Not necessary
W2c	W1c	=	Not necessary
W3c, W4c	L3c, W5, W3, L5	Eqs.(7) and (8)	*minimum* value: MINW *grid*: layout grid
L2c	L1c	=	Not necessary
L4c	L3c	=	Not necessary
Rcm2_val	Rcm1_val	=	Not necessary

dependent variable having a value out of range (and the performance of corresponding netlist performance is not evaluated). In addition, if a grid has been assigned to a dependent variable, the resulting value is rounded to the nearest integer multiple of the grid. Examples of this type of knowledge are shown in the fourth column of Table 3.

Finally, it is interesting to comment on two important applications of constraints that reveal the advantages of capturing the designer's intent and store it in the structural view of the reusable block. Both applications concern the capture of layout design knowledge and its inclusion in the cell-level sizing process. As it will be explained in the following chapter, the reusable block at the layout level is created by embedding relevant analog layout expertise. This expertise is coded to obtain a layout *template*, in which the layout implementation style and the position of each circuit component (e.g., whether a transistor is folded or not) are both known without actually going down to layout generation. The first application concerns the inclusion of layout knowledge in the sizing process to obtain a solution for which the area and shape of the eventually implemented layout are optimized. This sizing procedure has been referred as geometrically constrained sizing. Consider, for instance, that the MOS transistors in the structural view of the opamp reusable block example (see Fig. 4) have, in addition to the width and length parameters, a *multiplicity* parameter controlling the number of fingers. Like the width and length parameters, the multiplicity parameter is implemented as a design variable. If the task of finding which values of the multiplicity variables optimize the layout area and/or shape is completely left to the sizing tool, the exploration of the design space, now wider due to the new variables, will possibly take much longer. Thus, if constraints are defined between the width and length variables and the multiplicity variables (turning the latter ones into dependent variables), such that for a given width and length value there is a multiplicity value that optimizes the layout area and shape features[5], the design space is reduced and so the duration of the sizing process.

Knowing how the circuit layout is implemented leads us to the second application: the inclusion of layout parasitics in the circuit sizing process, also known as parasitic-aware sizing. Rather than to shrink the design space, constraints can be added in this case to help reducing the number of sizing-layout spins by providing a circuit sizing that is already robust against layout-induced parasitics. As these parasitics typically represent a capacitance or resistance between every two nodes of the circuit netlist[6], these can be previously added to the circuit netlist. Their values can be computed at each iteration of the optimization process from the device sizes and the specific layout knowledge stored in the layout template (e.g., the implementation style

5. The task of finding this value is transferred to an independent algorithm that interacts with the sizing tool. This algorithm is explained in Chapter 7.

6. Among the parasitic capacitances, the most important ones are the device parasitics, such as the MOS transistor diffusion capacitances, and the interconnect parasitics.

of a group of MOS transistors). Evaluation of the circuit performance is carried out by also considering parasitics into account, so the obtained circuit sizing is made robust against the layout-induced parasitic effects. For example, if M_{12} (see opamp in Fig. 4) is folded, a relationship can be added that relates the value of its diffusion area and perimeter (to calculate the diffusion parasitic capacitances) with the value of its width, length, and number of fingers. Such a relationship has been derived and can be found in Appendix A on page 354[7]. In short, these constraints store the knowledge derived from the layout designer's intent for the reusable block, and relates the value of the layout parasitics to the value of the design variables.

2.2 Testbench setups

As explained in Section 1, the sizing process involves the evaluation of a cost function. This cost function computes the degree of compliance of the circuit performance with a set of performance restrictions and design objectives. Since the adopted sizing approach is based on simulation, the cost function is evaluated by performing one or more simulations of the circuit electrical behavior. To expedite design reuse, there are certain relevant questions that should be already answered: what kind of simulation to run?; which operating conditions?; how to analyze the simulation results?; how can the performance specifications be managed to obtain better optimization results? By storing the answers to these and other relevant questions (actually valuable design knowledge), the design process can be carried out even when the initial designer is unavailable and, therefore, the end user of the reusable block can be spared of answering them, thus reducing the overall design time. A **testbench setup** is the database where such knowledge is captured. It comprises the required elements that allow a partial evaluation of the cost function. The complete evaluation is given by the whole set of testbench setups. As explained later in this section, testbench setups are also used during bottom-up verification.

---

[7] Some MOS transistor models already incorporate the diffusion capacitances, which are computed from the diffusion area and perimeter (parameters of the transistor model). Some simulators, like HSPICE® [Hspi04], calculate the diffusion area and perimeter of a single transistor as a function of its width and length. HSPICE® provides also the (limited) capability of calculating them when the diffusion node is shared by one or more transistors (i.e., for transistor folding) by means of the parameters *ACM* and *GEO* separately applied to each of the unitary transistors in the stack. Unfortunately, this is not appropriate for optimization-based circuit sizing as the number of fingers (i.e., unitary transistors) changes from one iteration to the next and the description of the stack changes as well (e.g., removing or adding unitary transistors along with values for *ACM* and *GEO*). This limitation is partly due to the fact that most circuit simulators do not provide a resource to describe the layout implementation of each device in the circuit netlist.

Each testbench setup is related to a specific type of simulation, e.g., DC, AC, transient, Monte Carlo, worst-case process parameters, etc., and comprises the following elements:

- **Performance feature elements**. Each testbench setup must include the definition of a number of performance features as well as adequate procedures to obtain the value of these features from the simulation results, the value of the design variables (i.e., from the visited point of the design space), or both.
- **Peripheral setup elements**. These setup elements configure the circuit surroundings to perform the simulation defining the testbench setup. Then, elements such as the circuit block configuration (e.g., open or closed loop), the circuit stimuli (input and biasing sources), the loading conditions, and the temperature and supply conditions, among others, must be defined for each testbench setup.
- **Component model and process data elements**. These elements define which model is used for each component as well as other process-specific data that might be required to perform the simulations.

Besides, each testbench setup may require the addition of specific design variables, dependent variables, and constraints. Note also that the elements described above can be shared by two or more testbench setups. For example, the same peripheral setup can be used at several testbench setups. The following subsections describe each group of elements in detail.

2.2.1 Performance feature elements

The knowledge database of the reusable block's structural view contains a number of performance features that are optimized taking into account the associated specifications (restrictions or design objectives). Each performance specification (e.g., $\text{DC gain} \geq 50\text{dB}$ or minimize power) is then defined as the combination of a **performance feature** (e.g., DC gain, power) plus a **target value** (e.g., $\geq 50\text{dB}$) or **directive** (e.g., minimize). The whole set of performance specifications for which the reusable circuit is sized are distributed amongst the testbench setups, depending upon the type of simulation and peripheral setup required to obtain the actual value of each performance feature. By way of illustration, Fig. 5 shows some of the performance features commonly used in the synthesis and analysis of opamp circuits.

The performance features that can be defined in the structural view stem from two different sources. On the one hand, there are performance features that are transmitted down in the hierarchy from one previous level L_{i-1} to

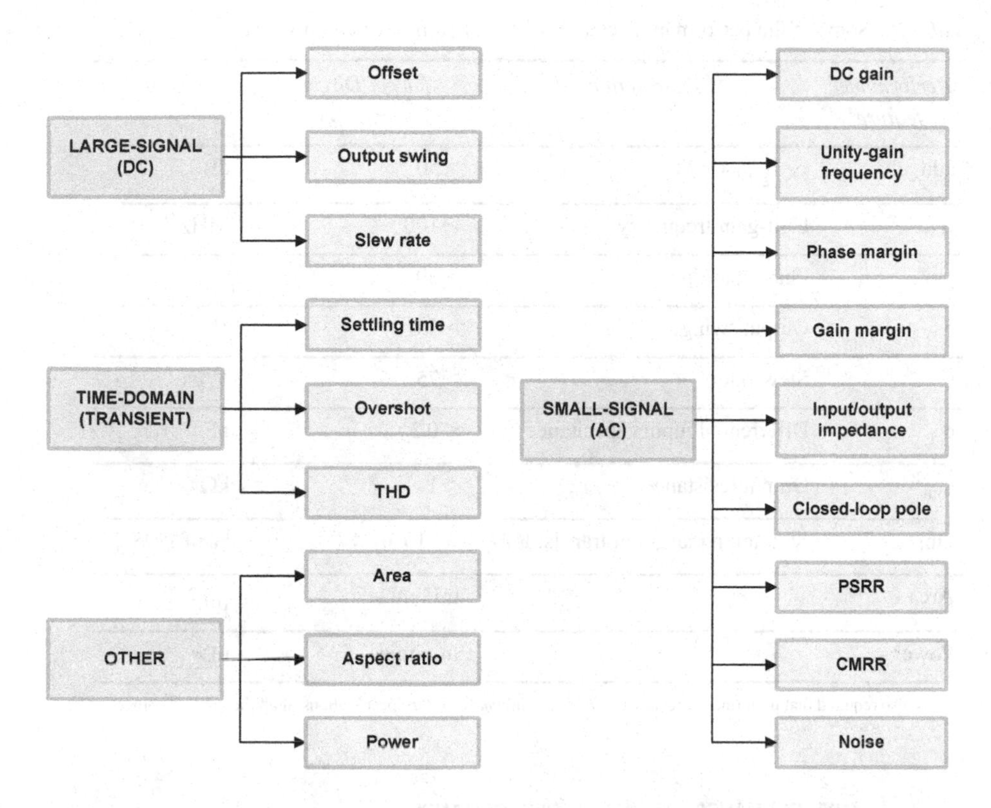

Figure 5. Examples of performance features commonly used in opamp circuit design.

the one where the structural view is used in the sizing process, L_i. Therefore, these performance features have been design variables in the sizing process of level L_{i-1} of the design flow and they are consequently used as performance features in the sizing process of level L_i. By way of illustration, Table 4 lists some of the performance features of the opamp example; add_db, ft, and pm, for instance, from these, belong to the first type of performance features. On the other hand, there are performance features that, though not typically managed during the top-down transmission of specifications, can be explicitly defined to add valuable design knowledge by means of which the electrical behavior of the circuit (whose structural view is being defined) and their components is noticeably enhanced. For instance, it is possible to check the saturation margin of certain MOS transistors in the opamp circuit to ensure correct operation. In this case, the performance feature is the saturation margin dm_i in Table 4.

Each performance feature is accompanied by a **performance monitor**, a procedure by which the actual value of the performance feature can be obtained, as Fig. 6 illustrates. The monitor acts upon the simulation data (e.g.,

Table 4. Some of the performance features of the opamp's structural view.

Performance feature	*Description*	*Target/Directive example*	*Unit*
add_db	DC gain[a]	> 50	dB
ft	Unit-gain frequency	> 100	MHz
pm	Phase margin	> 60	°
os	Output swing	> 3.6	V
sr	Slew-rate	> 75	V/μs
c_{id}	Differential input capacitance	< 0.3	pF
r_{out}	Output resistance	< 1	kΩ
dm_i	Saturation margin of transistor M_i	> 110	% of VDS_{sat}
Area		minimize	μm^2
Power		minimize	mW

a. It is also required that performance features *add_db*, *ft*, and *pm* fulfil the specifications in all the process corners.

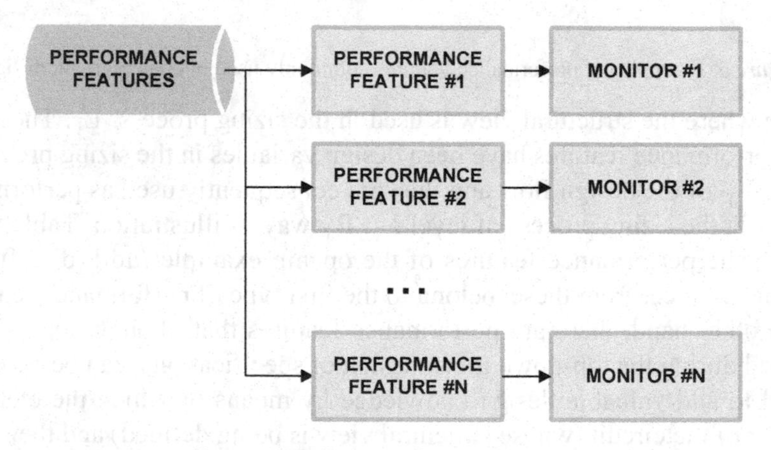

Figure 6. Performance feature elements.

the time evolution of the voltage on a circuit node) or upon the value of the design variables for the ongoing point of the design space. The complexity of the procedure on which the performance monitor is based ranges from that of a simple analytical expression (e.g., to calculate the slew-rate) to that of signal processing algorithms (e.g., to calculate the Total Harmonic Distortion, *THD*).

Table 5 shows some performance monitors of the opamp circuit in Fig. 4[8].

Table 5. Some performance feature elements of the structural view of the opamp reusable block.

Simulation type	*Performance feature elements*		
	Performance feature	*Monitor*	*Monitor Description*
AC	DC gain	add_db	Find the magnitude of the output voltage at zero frequency
AC	Unit-gain frequency	ft	Find the frequency at which the magnitude of the output voltage is zero for the first time
AC	Phase margin	pm	Find the phase of the output voltage when its magnitude is zero
AC	Power consumption	power	Average the consumed power
Operating point	Saturation margin of transistor M_i	dmi	Compute $\left[\frac{V_{DS}}{V_{GS}-V_{TH}}\right]\cdot 100$ for the transistor *Mi*

2.2.2 Peripheral setup elements

To obtain the value of each performance feature, a simulation of the circuit performance is required. This value is used by the sizing tool to evaluate the degree of compliance with the target value via the cost function. A suitable setup of elements, required to carry out such a simulation, must be stored in the database of the structural view. By means of this setup, the simulation data can be obtained so the performance monitors can be applied and the resulting value of the performance feature passed to the cost function evaluator. This setup is called a **peripheral setup**, as all elements are external to the circuit netlist itself. Several performance features may entail the same simulation, and, hence, the same peripheral setup. The elements of any peripheral setup are:

8. The actual implementation of the performance monitors is subordinated to the capabilities of the sizing tool, the simulator used, and to the facilities of the design environment. For instance, if the sizing is carried out by using the HSPICE® simulator, the .MEAS and the .PARAM sentences, together with voltage or current-controlled ideal sources, can be combined to implement a wide range of performance monitors [Hspi04]. In the Cadence® *DFII* design environment, to give another example, performance monitors can be implemented following a dedicated SKILL syntax [Skill04].

- **Stimuli elements**. These are the elements providing a electrical stimulus to the circuit block at its input/output pins. Examples are AC, DC, and transient (TR) voltage and current sources.
- **Biasing elements**. These elements provide a DC current deliberately made to flow, or DC voltage deliberately applied, between two points for the purpose of controlling a circuit.
- **Driving elements**. These elements specify the power-consuming components (typically capacitances and resistances) that are connected to a circuit. They represent the loads that the circuit must drive.
- **Circuit configuration**. It consists on the particular external wiring of the circuit's input and output pins to drive the circuit into different operation modes (e.g., open and close loops).
- **Initial conditions**. These elements comprise any possible initial voltage or current value at time t = 0 in transient analyses.
- **Supply** and **clock**.
- **Temperature**.
- **Analysis setup**. These elements define the type and characteristics of the simulation carried out at the testbench setup. Each of the three types of simulation (AC, DC, and TR) is defined by a number of parameters (e.g., frequency sweeping on an AC analysis, or starting and ending time value on TR simulations) that have all to be stored. Additionally, it is necessary to specify whether the simulation is single-point (producing a single result, or a single set of output data) or a multi-point (performing a single-point analysis sweep for each value in an outer loop sweep, e.g., Monte Carlo and corner analyses).

By way of illustration, a peripheral setup of the opamp reusable block in Fig. 4 is schematically depicted in Fig. 7. This setup is used to obtain the nominal value of the opamp performance features shown in Table 5 via an AC analysis. The stimulus source is the AC source vac, while sources VCM, CN, CP, and IB with diode-connected transistor M_{bp} are the biasing elements. RL and CL are the driving elements. The opamp configuration is open loop. Sources VDD, and VSS provide the supply voltages, and the temperature as well as the analysis characteristics are shown in the same figure.

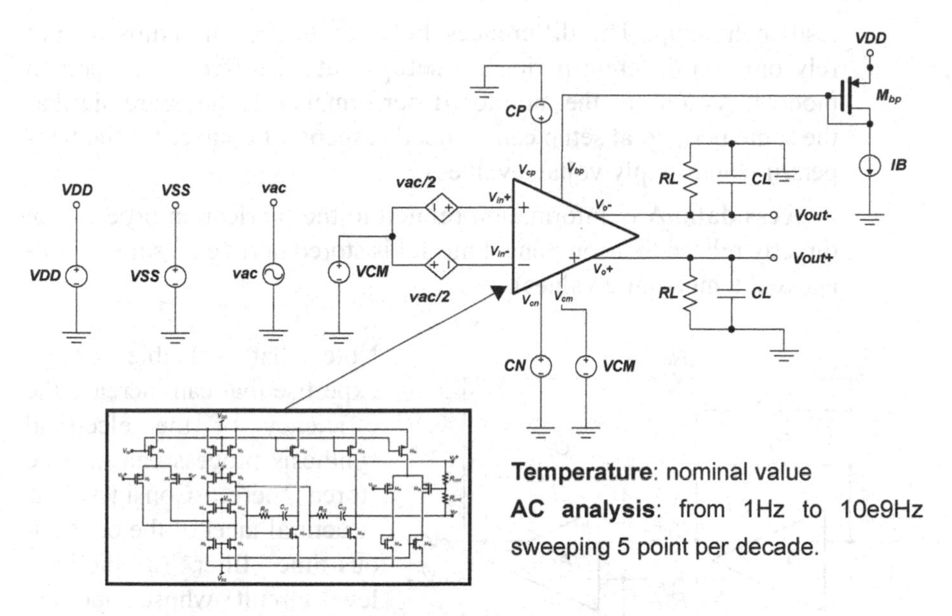

Figure 7. Peripheral elements of a testbench setup for the cell-level reusable block example.

2.2.3 Component model and process data elements

Not only is each testbench setup defined by the type of simulation but also by specific process data and a description model for each component of the reusable block. Thus, every testbench setup comprises the following elements:

- **Component models**. In order to perform the simulation of the reusable block, both for sizing and verification, each component of the reusable block must be accompanied by a corresponding model describing its behavior. For a device-level component, the model typically employed is the foundry-provided technological model (e.g., SPICE BSIM3v3 model). For cell or module-level reusable components, their behavioral facet is used here. Besides, a number of the components (typically device-level components) may require different models depending upon the type of simulation defined in the testbench setup (e.g., typical, worst-case speed, worst-case power, or Monte Carlo MOS transistor models). That is why components models were not specifically considered as part of the netlist-related database. Returning to the opamp example, to obtain the worst-case values of the performance features add_db, ft, and pm (see Table 4) it is necessary to perform separate AC simulations, each with a different set of component models and values of temperature and supply. Each set represents a corner of the process, each one requiring a separate testbench setup. The differences between testbench setups do not rely only on different peripheral setups but on different component models. As long as the monitored performance features are similar, the same peripheral setup can be used (except, of course, for the temperature and supply voltage values).

- **Process data**. Any information related to the fabrication process not directly related to a component model is stored here (e.g., supply voltage and temperature values).

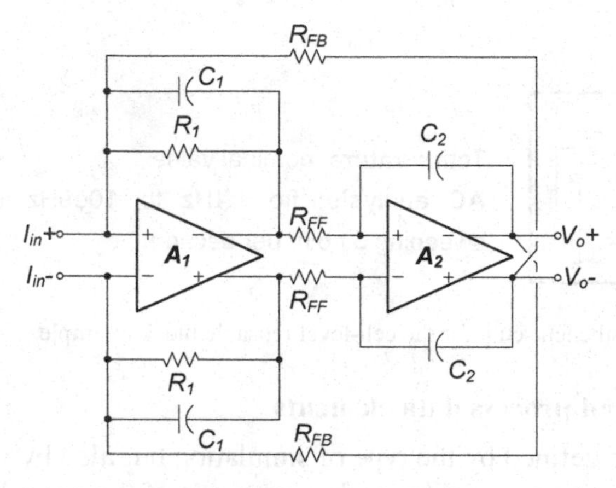

*Figure 8.*A continuous-time filter.

Note that valuable design expertise that can increase the efficiency of the electrical synthesis process can also be stored here. Consider the structural facet of the continuous-time filter (a module-level circuit) whose topology is shown in Fig. 8. It comprises a feedback/feedforward network made of passive devices and two active components, opamps A_1 and A_2 (cell-level circuits). Process-specific models are used to describe the device-level components; reusable behavioral models are used for the opamps. The design variables are the set of device characteristics and the opamp performance specifications. What is essential to understand is that some of their performance characteristics should be attained at the expense of the passive network rather than of the opamps' dynamic performance. Clear examples are the filter's attenuation characteristics. Consider, by contrast, that the filter is synthesized such that an acceptable level of attenuation is achieved thanks to the position of the poles and zeros of one or more opamps rather than to the actual values of the passive devices in the feedback/feedforward network. Consider also that the required opamp performance specifications stemming from such a synthesis experiment are $\rho_{k,1}{}^{\circ}$ (with ρ being the 1-th performance feature of the opamp A_k, with $k = 1, 2$)[9]. For many of the opamps' performance features, this means either that $\rho_{k,1} \geq \rho_{k,1}{}^{\circ}$ or $\rho_{k,1} \leq \rho_{k,1}{}^{\circ}$[10] (e.g., add_db $\geq$ 50 dB or

[9] Later on the design flow, sizing the opamps (i.e., finding an appropriate size or value of all their device-level components) is carried out such that all these performance specifications are to be fulfilled.

[10] Using $\rho_{k,1} = \rho_{k,1}{}^{\circ}$ may otherwise over-constrain the synthesis process in most practical cases.

rop ≤ 70 kΩ). This entails that for all the scenarios where the obtained opamp performance features are above (or, alternatively, below) their required values (i.e., $\rho_{k,1} > \rho_{k,1}^{\circ}$ or $\rho_{k,1} < \rho_{k,1}^{\circ}$) the positions of the opamps' poles and zeros may be different from those yielding an acceptable attenuation level, which, eventually, may translate into a different and possibly unacceptable attenuation level of the filter. As said above, this problem arises because an essential fact has been left out, namely that some of the filter's performance features rely on the passive network as well as on ideal amplification. That is, the filter should be synthesized such that, first, the intended performance specifications are addressed when considering ideal opamps (i.e., infinite gain and bandwidth product, zero output impedance, and so on) so the their dynamics do not contribute to the intended filter performance, and, second, the non-idealities of the opamps do not prevent from attaining those performance specifications.

A method to carry out this synthesis process is illustrated with the filter example in Fig. 9. Two testbench setups are defined whose only difference consists in the component models used for the active elements: reusable behavioral models of ideal opamps in the first, and reusable behavioral models of real opamps in the second in which, as explained in the previous Chapter 3, non-ideal effects are modeled. The sizing process of the filter is then carried by simulating both setups at each iteration of the optimization process while addressing the same design goals for the filter (e.g., the same attenuation level). In this way, whatever the eventual values of the opamps' performance features, and obviously providing they fulfill the required opamp performance specifications (e.g., when $\rho_{k,1} = \rho_{k,1}{}^{*} > \rho_{k,1}^{\circ}$ if its required that $\rho_{k,1} \geq \rho_{k,1}^{\circ}$), correct performance of the filter is guaranteed, since it is determined by the passive network and ideal amplification (ensured by the achievement of design goals with testbench setup #1) and deviations stemming from the opamp non-ideal performances are sufficiently small (ensured by the achievement of design goals with testbench setup #2).

2.2.4 Design variables, dependent variables, and constraints

New elements included in the peripheral setup may require the addition of new design variables. For example, the value of the biasing current source *IB* in Fig.7 is a new design variable optimized by the sizing tool which will allow further design of the opamp biasing stage (considered here as a separate circuit block)[11]. Considerations on the variable range and grid explained in Section 2.1.1 also apply here.

---

[11] Some new design variables in the testbench setup may conflict with some of the netlist-related design variables. For instance, in defining the opamp current biasing either IB or V_{bp} is redundant. In such a situation, one of the design variables should be removed.

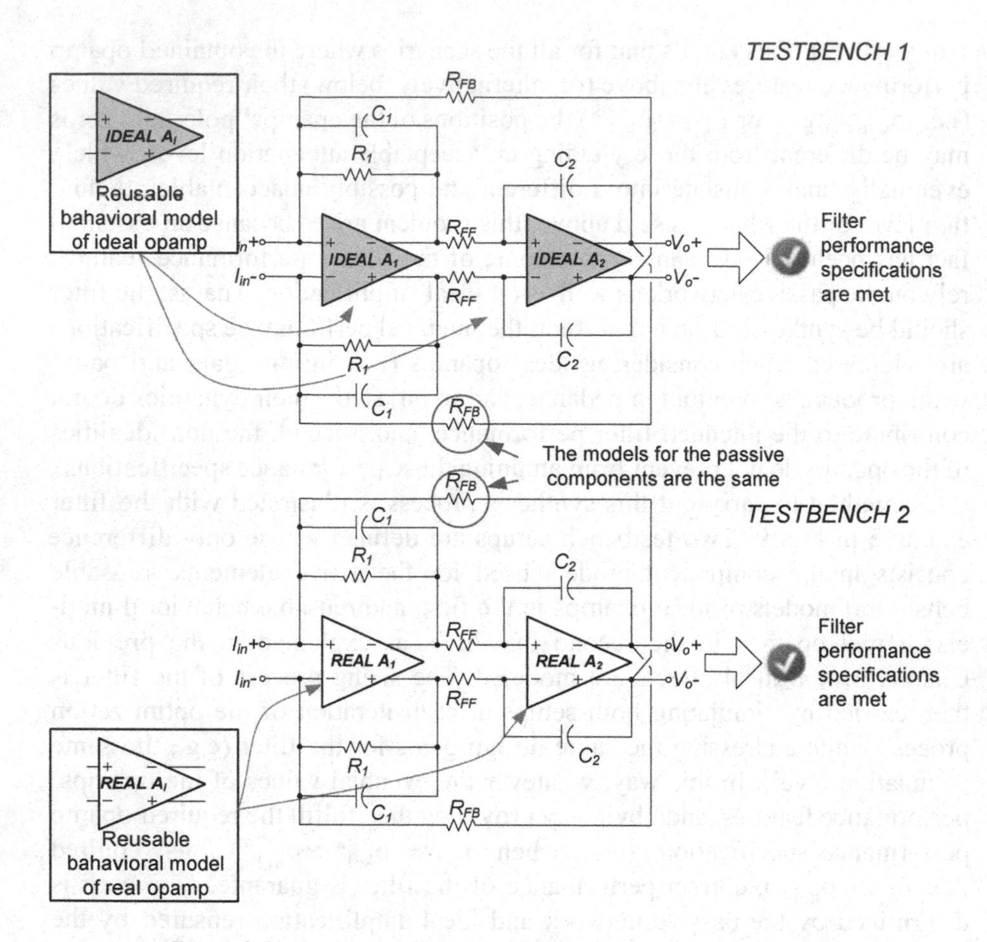

Figure 9. Improving the filter sizing process with ideal opamp models.

In the same way, the incorporation of dependent variables and their corresponding constraints may be necessary. Continuing with the opamp example, the width, and length of transistor M_{bp} are dependent variables whose constraint makes them equal the width and length of transistor M_5 in the opamp core. In like manner, the considerations on the value range and grid of the dependent variable described in Section 2.1.2 are still valid.

Figure 10 summarizes the elements of the testbench database. Suppose that the circuit sizing is carried out with N performance specifications. The corresponding set of N performance features, PF_j, with $j = 1, \ldots, N$, is broken up into M subsets, each with n_i performance features such that:

$$\sum_{i=1}^{M} n_i = N \tag{11}$$

The criterion followed to perform such a division is that evaluating each performance feature in the subset i requires the same peripheral setup and process data. With the addition of a performance monitor, PM_j, for each performance feature, PF_j, the testbench setup B_i is completed, and so is the testbench knowledge database of the reusable block's structural view.

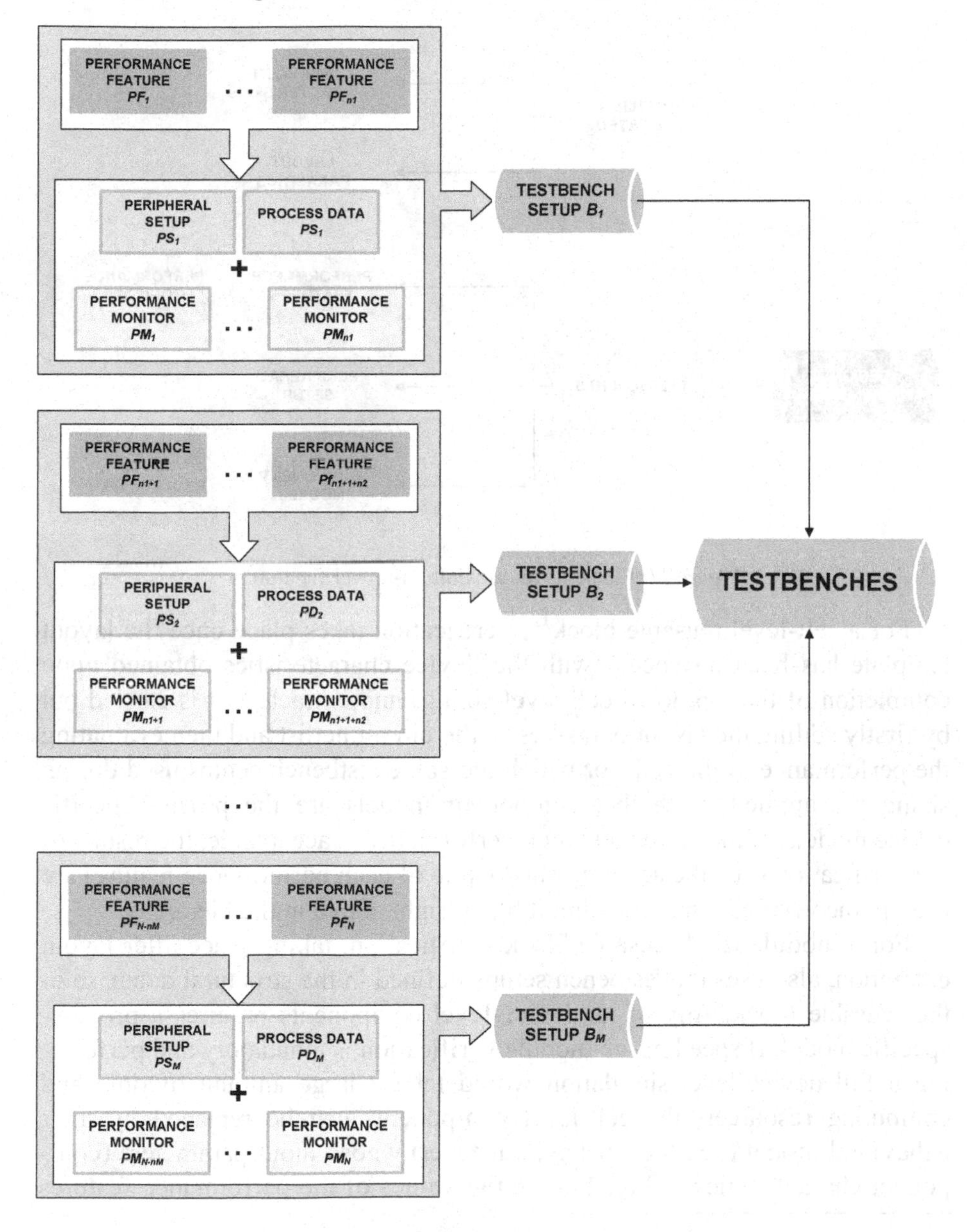

Figure 10. Elements of the testbench database.

As described earlier, the structural facet of a reusable block serves also to carry out bottom-up verification. The elements of the structural database required for the complete evaluation of the reusable block performance are shown in Fig. 11. For the sake of clarity, let us take two scenarios to explain how verification using this structural database is carried out.

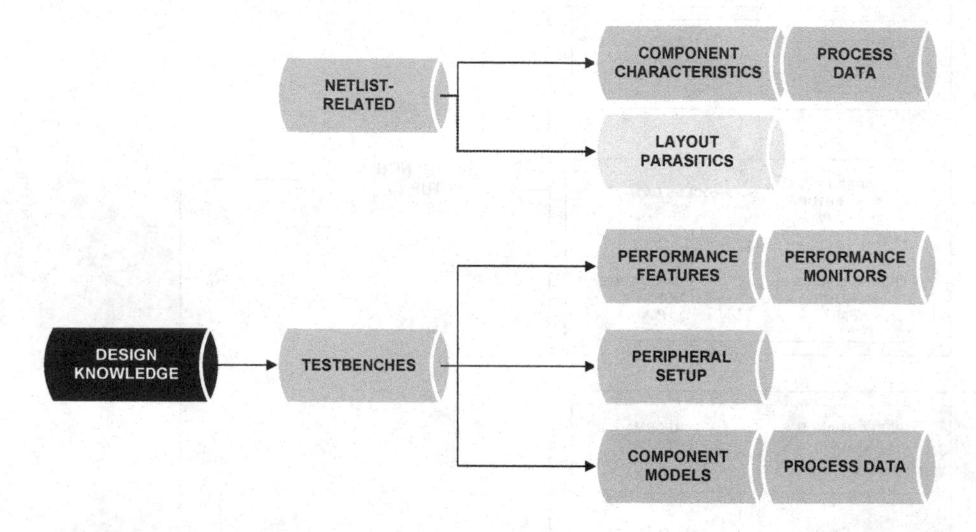

Figure 11. Elements for the bottom-up verification.

For a cell-level reusable block[12], verification takes place once the layout template has been instanced (with the device characteristics obtained upon completion of the top-down cell-level sizing) and extracted. It is carried out by firstly adding the layout parasitics to the circuit netlist and then evaluating the performance of the cell, for which the same testbench setups used during sizing are applied. Note that component models are the process-specific device models. If the cell-level block performance is acceptable, the results of the verification (i.e., the accomplished value of each performance feature) are used in the verification of the circuit block right above in the hierarchy.

For a module-level reusable block, verification, taking place after layout extraction, also uses the testbench setups defined in the structural database of the reusable block. Any of its device-level components requires a process-specific model. If speeding-up module verification is mandatory and performing a full device-level simulation would take a large amount of time and computing resources, the cell-level components can be replaced by their behavioral model (i.e., their behavioral facet) whose input parameters (component characteristics in Fig. 11) are the values of the performance features

[12] This explanation assumes that no parasitic-aware sizing (see Chapter 2) has been carried out.

obtained after cell verification; in the same way, any module-level component can be replaced by its behavioral model (i.e., its behavioral facet) whose input parameters (component characteristics) are the values of the performance features obtained after its verification phase. Layout parasitics included are only those found on the interconnections between the components of the reusable blocks (the impacts of parasitics "within" each component are already accounted for in the values of each component's performance features).

3 PRACTICAL ASPECTS OF STRUCTURAL VIEW REUSE

Previous sections have described the characteristics of a structural view used with the design reuse methodology explained in this book. Its two chief underlying aspects are the parameterized netlist and the design knowledge databases.

Reusing such a circuit block for a change in the performance specifications (called design retargeting) consists in carrying out the sizing process for a new set of performance target values. In this case, circuit reuse just consists in defining the new target values of the performance specifications as well as providing the component models and process data required.

Reusing this circuit block for a change of the fabrication process (called design migration) implies first correctly porting the databases to the targeted fabrication process and, possibly, re-executing the sizing process to amend the performance deviations caused by changes of the process characteristics (e.g., supply voltages and device models). The latter problem is basically a retargeting problem, which, as mentioned above, has already been treated in preceding sections. The complexity of porting the databases largely depends on the framework the databases are built upon, and, obviously, on the complexity of the reusable block structural view (mainly in terms of its number of components). This section deals with several practical aspects of the database porting process within the Cadence® *DFII* design environment[13].

Conceptually speaking, porting the databases of the reusable block's structural view from one process to another involves correctly pointing to the new component models (as long as these models are process-dependent) and process parameter values[14]. As already noted, this seemingly simple task may

[13] The design reuse methodology on which this book is focused has been put into practice in this design environment, thereof the motivation for the ensuing discussion.

[14] For the sake of simplicity, we will consider that porting the databases is done between different fabrication processes but on the same design environment, so the circuit description format (ASCII format, schematic capture format, etc.) remains the same and no further precautions must be taken in that sense.

actually pose serious problems depending upon the design environment or platform on which all the design reuse flow is being performed. Consider, for example, that cell-level sizing is undertaken by using the sizing tool in collaboration with an electrical simulator whose inputs are plain ASCII files (e.g., HSPICE®). The database for a structural view –parameterized netlist and design knowledge– are thus implemented as ASCII files (see Fig.12). Porting the databases just consists in updating the component models and process parameters, which, for text files, can be done by using simple parsing scripts (e.g., using PERL [Perl93]).

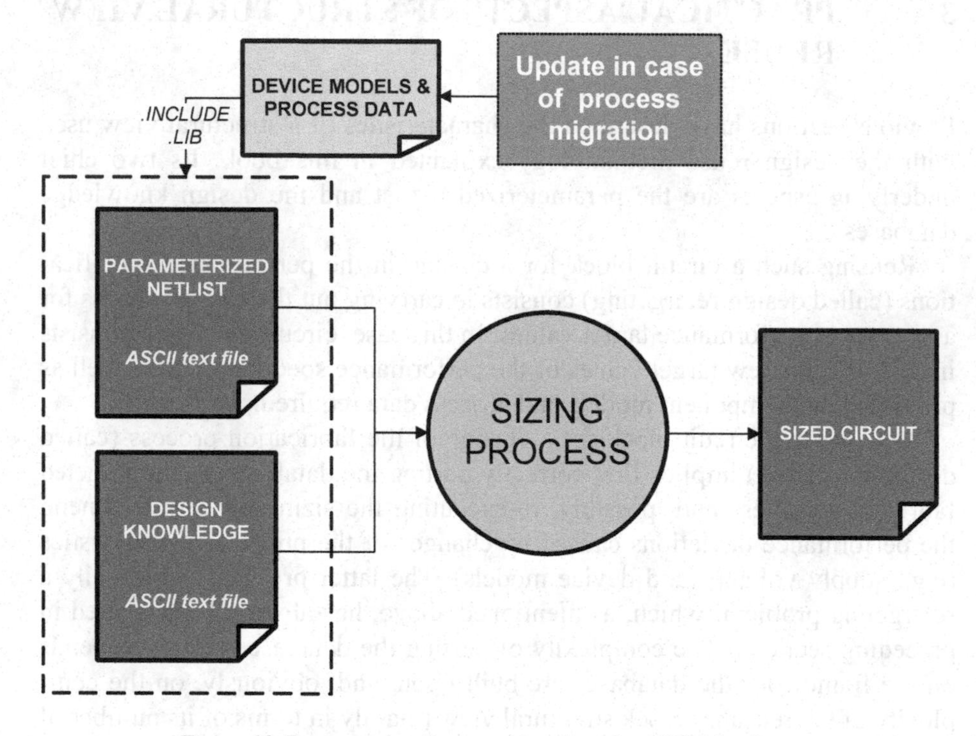

Figure 12. Process migration considerations for ASCII databases.

The following considerations must be taken into account:

1. Porting from one initial process to a goal process is only feasible as long as both processes have equivalent device-level primitives (e.g., MOS transistors, resistors, and capacitors) in terms of functionality, number of terminals, and fundamental parameters.
2. All numerical reference to the fabrication process should be avoided. Instead, symbolic references are used for both the device-level models and other process data. The numerical values are retrieved by

pointing to predefined technology files where the values of both device model parameters and other process data are stored (see Fig.12). For example, if the simulator linked to the sizing tool is HSPICE®, the .INCLUDE and .LIB sentences can be used.

In the Cadence® *DFII* environment, the parameterized netlist as well as some elements of the peripheral setup (stimuli, biasing, and supply sources) and the performance monitors, are created by using the Composer schematic capture tool [Comp04]. There is a schematic view for the circuit netlist and as many schematic views as testbenches with different peripheral setups have been defined. These views are composed of component symbols and interconnect wiring (see Fig. 13). The real problems arise at porting these schematics to several, different processes (which in the context of the *DFII* environment are called *kits*) as long as each foundry requires using its own, possibly different, set of device-level primitives (e.g., MOS transistors, resistors, capacitors)[15]. Consider that the schematics for a reusable block are ported to the new process kit (which is another way to say "import the schematic cells and attach them to the new process technology file"[16]) and the "old" devices replaced with the new ones. Due to the lack of primitive device standardization noted above, this schematic migration process poses the following problems:

- **Connectivity** problems. Differences in the graphical symbols of the primitive devices (see Fig. 13) due to differing size and terminal positions[17] may damage the circuit connectivity. For example, devices that were connected in the initial process kit, become unconnected in the goal process kit.
- **Device description** problems. Each type of device is described with a special set of parameters (called *Component Description Format®* [Comp05], or CDF, parameters). These CDF parameters not only specify which are the device parameters (e.g., MOS transistor width, length, and number of fingers) but also provide essential information to other design tools (e.g., for netlisting, simulation, and LVS). Porting the schematics to a new process kit entails dealing with large differences in the CDF parameters, especially between kits from

---

15. Porting cell or module-level behavioral models should not pose a serious problem as long as they follow the technology-independence guidelines explained in Chapter 3.

16. Importing the schematics can be done by copying them into the new work directory or by using the *Edif300®* translation tool [Edif04]. In the latter case, device symbol replacement can be done simultaneously.

17. Considerations 1 and 2 above still apply here.

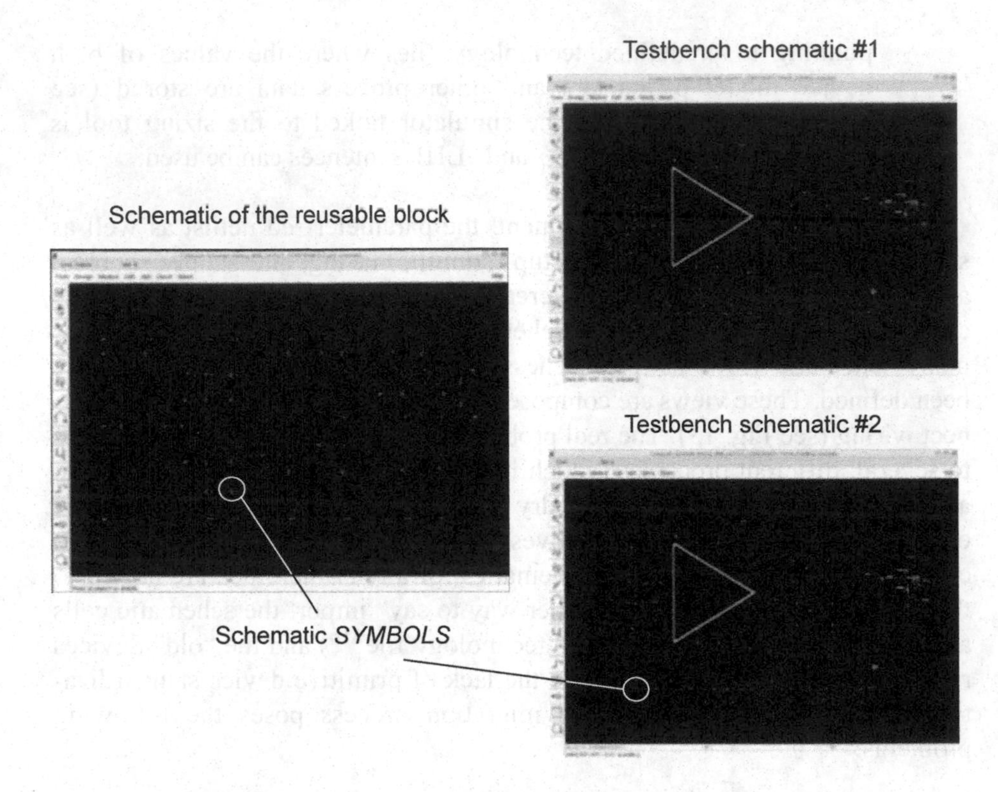

Figure 13. Schematic views of some elements of a reusable block database.

different foundries[18]. Furthermore, simply updating the CDF parameters of each primitive device type does not preserve the parameter values of each circuit device (unless these parameters are specified by the same CDF parameters in both start and goal technologies).

To overcome these difficulties, the four-step method illustrated in Fig. 14 can be used. This method is based on composing the schematics using dummy primitive devices, which are used throughout the several process kits.

The dummy devices are process-correct since they mimic the process devices by using their same set of CDF parameters. Once the schematics have been composed for the first (and only) time in any initial technology *TECH A*, with these dummy devices (which actually are equivalents of the primitive devices in that technology) and their parameter values inputted, the method can be applied to port the schematics to a goal technology *TECH B*. The method outline is as follows:

---

18. It worsens when the CDF parameters use *callback* procedures (SKILL procedures for propagating changes in one parameter's value to the rest of parameters). Besides, many of these callbacks use process data to carry out the required calculations.

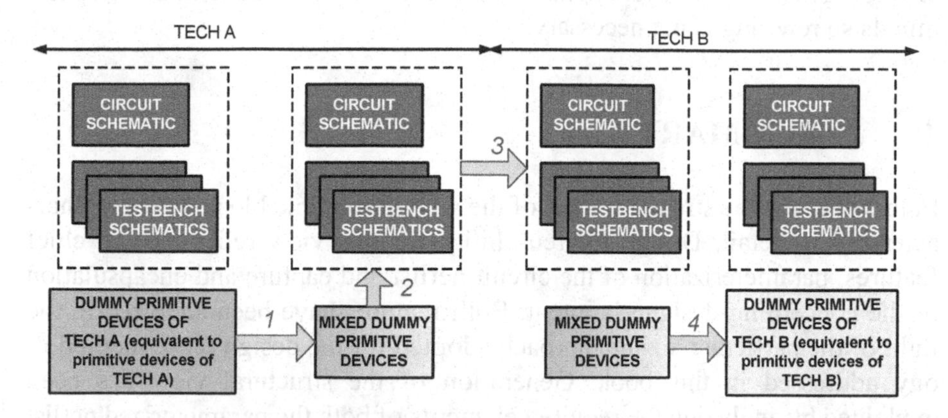

Figure 14. Method to port schematics in *DFII* from Cadence®.

1. The CDF parameters of the dummy devices are complemented with all those CDF parameters of the devices of *TECH B* that are to be preserved during the porting process and whose equivalent *TECH A* is differently defined. For example, if the width parameter of the MOS transistor primitive device is called *width* in *TECH A* and *W* in *TECH B* (which implies a difference in the CDF parameter), this latter parameter must be added to the CDF parameters of the dummy MOS transistor primitive device.
2. Still in *TECH A*, the value of the parameters for which a CDF parameter has been added is set to the value of the equivalent parameter in *TECH A*. For instance, if a MOS transistor has *width* equal to 10μm, the so far non-valued *W* parameter (the added parameter) is set to 10μm as well. By doing so, the values of the device parameters are preserved in the migration process as it is explained below.
3. The schematics are imported to *TECH B*.
4. All the CDF parameters of the dummy devices are replaced by those of the devices of *TECH B*. This process preserves the value of the parameters whose CDF description shares the same parameters. That is why the CDF parameters are mixed in steps 1 and 2.

The advantages of this method are that no graphic problems arise as the symbols used (those of the dummy devices) are the same and no CDF-related problems result as the dummy devices are made equivalent of the process' own devices. The limitation is that the method can only be applied when each type of primitive device in the initial technology has an equivalent in the goal technology (in terms of functionality) and they have the same number of terminals so rewiring is not necessary.

4 SUMMARY

In this chapter, the structural view of the analog reusable block has been characterized in detail. Design for reusability for this view relies on two chief features, parameterization of the circuit netlist and capture and encapsulation of the concerning designer's intent. Both features have been analyzed in the light of the particular sizing approach adopted for the design reuse methodology addressed in this book. Generation of the structural view has been explained by analyzing the required elements of both the parameterized netlist and the design knowledge database. The practical aspects of structural view migration in a well-known design environment have been discussed. Guidelines have been given that ease such a process migration of the structural view database and a method has been outlined that support such a process migration.

Chapter 5

The Analog Reusable Block: Layout Facet

In previous chapters, the reusable analog block, the cornerstone of the design reuse methodology presented in this book, has been characterized at both its behavioral and the structural facets. This chapter deals with the remaining one: the layout facet. In the course of the chapter, a method to implement this layout-reusable analog block is described in detail.

1 INTRODUCTION

Recalling what was defined in Chapter 2, design reuse is the use of previous design knowledge, experiences, and databases in the implementation of a different design, perhaps in a different fabrication process. This reuse process has to be carried out efficiently and rapidly. From the layout point of view, design reuse implies two different scenarios:

a. Reuse of the circuit layout database[1] for changes in the circuit performance specifications. This concept of design reuse has one limitation: the specifications changes must be such that the new specifications can be addressed by using the same circuit architecture/topology. This does not mean, however, that the required changes translate into minor adjustments at the layout level at all [Giel00]. Quite the opposite, specifications changes, though within the circuit's achievable behavior, may translate into drastic modifications of the circuit device sizes[2] and biasing conditions,

1. A circuit layout has different forms of representation: a GDSII file, a coded procedural layout database, or simply an input file to a layout generation tool.

2. For instance, the width and length of a MOS transistor, the resistance of a passive resistor, and the capacitance of a passive capacitor.

and, thereby, in the circuit layout. Whatever the layout generation approach, it has to consider, therefore, how to accommodate these specification changes. In this scenario, layout reuse is called **layout retargeting** as the circuit target performance is modified and the previous working circuit architecture/topology is reused.

b. Reuse of the circuit layout database for a change of the fabrication technology. In this case, layout reuse is known as **layout migration**, as the layout database is moved from the technology it was designed for, to a different goal technology, perhaps from a different foundry. In a sense, layout retargeting can be seen as a component of layout migration, as changing the fabrication process, while trying to obtain the same circuit behavior, would likely require to adapt the layout to new device sizes as well (actually what layout retargeting aims at). Nevertheless, layout migration will be considered here, for the sake of simplicity, as a stand-alone aspect of analog and mixed-signal layout reuse, meaning only the adaptation process of the layout database (i.e., database migration), and not of the circuit device sizes, to another fabrication process, with different design rules and mask layers.

So far, design reuse is not new. Every designer, at the beginning of the circuit design process, tries to reuse, among the already working designs, a good solution to what he/she is looking for. Most of the times, unfortunately, reusing this solution may be rather involved and may take possibly a large amount of effort and time. To make matters worse, reusing a circuit layout manually, for either a change in the device sizes or a change of the fabrication process, could become a quite laborious and slow task. Actually, the great specificity of analog and mixed-signal designs is the main factor that makes direct layout reuse unfeasible. Therefore, a suitable design for reusability methodology is required that implements the reusability property at the layout descriptions of analog and mixed-signal circuit blocks.

To attain such a design-for-layout-reusability methodology, it is first essential to understand how the reusable block has to cope with the two design reuse scenarios, namely, layout retargeting and layout migration. This will allow us to know which are the requirements that layout reuse imposes on the methodology.

2 LAYOUT RETARGETING

Layout retargeting is performed when any of the circuit devices and/or any of the biasing conditions need to be modified to address the changes in the circuit performance specifications, and, as a result, the circuit layout must be updated.

When the layout is modified to accommodate these new device sizes and biasing conditions, two different aspects have to be considered. First, the layout has to remain compliant with the process design rules. This aspect will be treated below in Section 3. Second, several characteristics of analog layout quality may result spoiled, so they have to be carefully treated and maintained. An analysis of these characteristics is provided below.

2.1 Device mismatch

All devices in an IC occupy the same piece of silicon, therefore suffering from the same manufacturing imperfections. There are devices which are specifically constructed to keep a known constant ratio between them (for instance, between the values of two capacitors [OLca91] [McNu94] or between the current factor β of the MOS transistors in a differential pair [Pelg89] [Laks86] [Bast96b]) and they are thus called matched devices. There are many different sources for device mismatch, but all of them can cause a type of mismatch that can be categorized into any of these two categories [Laks86]:

- **Random mismatch** is caused by microscopic fluctuations of several circuit parameters such as dimensions, doping, and oxide thicknesses, all of them affecting final component values. They scale with the device dimensions and, although they cannot be completely avoided, appropriate rules can be followed in order to minimize their influence.
- **Systematic mismatch** stems from a host of causes such as edge effects, wafer gradients, or source-drain asymmetry. These errors can limit the obtainable accuracy of the IC because they do not scale with device dimensions. In other words, systematic mismatch is caused by mechanisms that affect all of the samples of the wafer die in the same manner. The goal of the layout designer is then to identify and eliminate the source of systematic errors with the information provided by the silicon foundry.

The mismatch between two components is usually expressed as a deviation of the measured ratio from the intended device ratio. If the intended values are X_1 and X_2, and the measured values are x_1 and x_2, then the mismatch δ is:

$$\delta = \frac{(x_2/x_1) - (X_2/X_1)}{(X_2/X_1)} = \frac{X_1 x_2}{X_2 x_1} - 1 \tag{1}$$

By measuring the deviations of a large number of devices it is possible to derive an average mismatch m_δ and a standard deviation of the mismatches σ_δ, using the following equations [Hast01]:

$$m_\delta = \frac{1}{N}\sum_{i=1}^{N} \delta_i \tag{2}$$

$$\sigma_\delta = \sqrt{\frac{1}{N-1}\sum_{i=1}^{N} (\delta_i - m_\delta)^2} \tag{3}$$

The mean value in Eq. (2) represents the systematic mismatch while the standard deviation in Eq. (3) quantifies the random mismatch, caused, as said above, by statistical fluctuations in the material used in the fabrication process or in the process itself.

Six major layout geometric factors can affect the matching of identical devices [Rute02] [Hast01]:

- **Size** and **shape**: MOS mismatch has been experimentally measured for a number of processes (see, for instance, [Bast96b]). Results reveal that both the current factor (β) mismatch and the threshold voltage (V_{th}) mismatch vary inversely with the square root of the active gate area [Pelg89]. Therefore, large MOS transistors match more precisely than small ones. In addition, long-channel transistors match better than small-channel ones because longer channels reduce line width variations and channel-length modulation. Creating devices with identical shape improves matching since the geometry-induced errors affect them in a similar way.
- **Symmetry** and **separation**: process parameters are affected by process gradients (i.e., variations of a process parameter, such as the oxide thickness, over the wafer). These gradients induce differences between matched devices. This gradient-induced mismatch can be minimized by reducing the distance between the geometric center of the devices, and even reducing it to zero. Sensitivity to gradient-induced errors can also be alleviated by placing matched devices in close proximity. The structures known as *common-centroid* arrays can cancel out entirely the effects of long-range variations as long as

these are linear functions of distance. In common-centroid arrays, a number of unitary elements are distributed in one-dimensional (1-D) or two-dimensional (2-D) arrays, as shown in Fig. 1. A common-centroid array should fulfill the following rules: *coincidence* (the centroids of the matched devices should exactly coincide), *symmetry* (the array should be symmetric with respect to both the horizontal and vertical axes), *dispersion* (the fingers should be distributed throughout the array as uniform as possible), and *compactness* (ideally, the array should be nearly square) [Hast01].

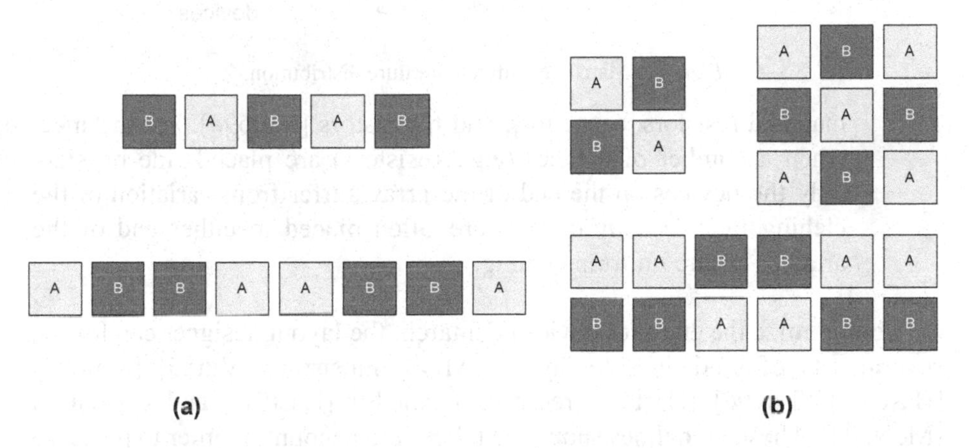

Figure 1. Examples of common-centroid arrays: (a) 1-D; (b) 2-D.

- **Orientation**: pressure on silicon results in an anisotropic stress distribution [OLea91]. If a set of devices require precise matching and these are laid out with different orientations, they would experience different degrees of stress. This may yield variations on the mobility of MOS transistors, or a change in the resistivity of resistor layers. For instance, stress-induced mobility variations can induce a current mismatch in MOS transistors of as much as 5%. Therefore, matched devices should be drawn with the same orientation, and should not be rotated with respect to one another. The orientation of matched devices is also important with respect to power dissipation. For elements with temperature dependence, gradients of the temperature distribution in the chip can cause a change in the element value. The solution is thus to lay out the matched devices symmetrically with respect to the main power dissipation elements, as illustrated in Fig. 2.
- **Boundary**: although drawn structures of the layout are regarded as two-dimensional, these structures are, in fact, three-dimensional. Variations in etching process rate are very important to design of

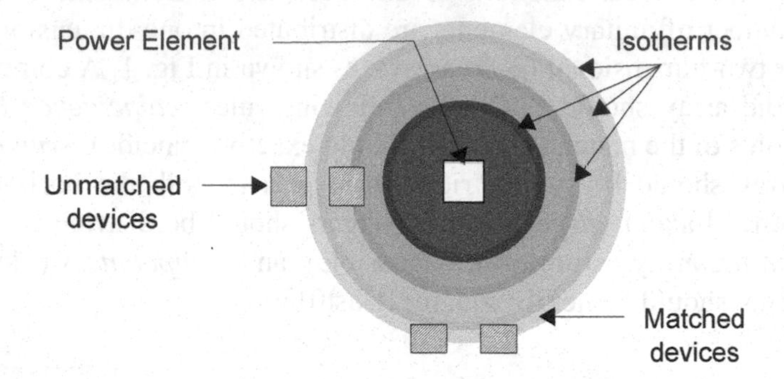

Figure 2. Matching with temperature distribution.

matched resistors, capacitors, and transistors [Malo94]. For instance, when a number of devices (e.g., resistors) are placed side by side, only the devices on the end of the array suffer from variation of the etching rate. *Dummy* devices are often placed to either end of the array to ensure uniform etching.

To minimize the effect of device mismatch, the layout designer can follow several sets of guidelines to lay out MOS transistors [Vitt85] [Malo94] [Bast96a] [Tsiv96] [Hast01], resistors [Lane89] [Hast01], and capacitors [McNu94]. These guidelines should be taken into account in order to preserve the quality of the analog layout when retargeting it to new circuit device sizes and biasing conditions.

2.2 Loading effects

The physical nature of materials used in the fabrication process introduces capacitive and resistive effects[3]. These effects are referred as resistive or capacitive *parasitic*s. RC parasitic elements associated with the inter-device wiring can seriously degrade the circuit performance because of their resulting loading effect, and they depend on the type of material used for the wiring as well as on the geometry of the wire. For instance, capacitive parasitics are proportional to the wiring area, and resistive parasitics proportional to the wiring length, and inversely proportional to the wiring width. These dependencies suggest a reduction of the wiring length by bringing the devices closer to each other, adjusting the wiring widths, or by using lower resistive (capacitive) mask layers [Prie01]. There are also RC parasitics associated with the geometry of the devices themselves, e.g., the

[3] At sufficiently high frequencies, inductive effects arise as well [Rute02].

capacitive effects derived to the MOS diffusion area. These effects are remarkably sensitive to detailed rectangle-level layout. These parasitic elements can be controlled through appropriate selection of minimum area devices, by using *folding* geometries for devices with large width-to-length ratios, by using parallel side-by-side abutment of shared source or drain terminals (called *device merging*), etc.

The non-zero resistance of the source or drain terminals may cause a voltage drop. This effect can be also avoided by minimizing the aspect ratio of the diffusion area, by merging diffusions, and by using low-resistivity layers to contact diffusion.

Another important aftermath of parasitics elements concerns analog and mixed-signal circuits with fully differential topologies. The success of these structures lies in taking the difference between the signals on two signal paths; if there is any interference common to both paths, it cancels out in the difference. Therefore, the layout should be such as to ensure that the interference is, indeed, common. This implies that not only the two signal paths have to be matched, but also the parasitics associated with these paths. While always a completely symmetric layout would be the obvious choice, this might not be possible because of several layout constraints.

The amount of effort needed to control these parasitic effects is indeed considerable since extremely low-level geometric details of the layout of individual devices can have a major impact on the circuit performance. These parasitic elements cannot be fully predicted early in the design process, because the layout is not complete yet. Over-estimation of the parasitics results in wasted area and power, and under-estimation leads to specification non-fulfillment. It is then crucial that parasitic effects have to be taken into account during the design process[4] and that the selected layout synthesis method provides ways to minimize their impact. Furthermore, their effect must carefully treated not only during a retargeting process, but also during a migration process since the values of parasitic elements may vary from one fabrication process to another.

2.3 Coupling effects

Layout can also introduce unexpected signal coupling between the circuit nodes, which may inject unwanted electrical noise and even destroy the circuit stability due to unintended feedback [Rute02]. This capacitive coupling effect, known as *crosstalk*, may appear between two wires running in parallel over a long distance, or in two wires crossing at different levels. For two

4. At certain phases of the design process it is possible to roughly estimate the parasitic elements (via area and perimeter measurements) but this estimation is not sufficient.

metal surfaces which are sufficiently close and running in parallel, the coupling is directly proportional to the traveled distance, and inversely proportional to the wire spacing. Across adjacent levels, the capacitive coupling is proportional to the crossover.

A first solution is to avoid, whenever possible, both wire parallel running and wire crossover. When the circuit node is highly sensitive, a neutral wire, e.g., a ground line, can be placed in between, as illustrated in Fig. 3 [Raza01].

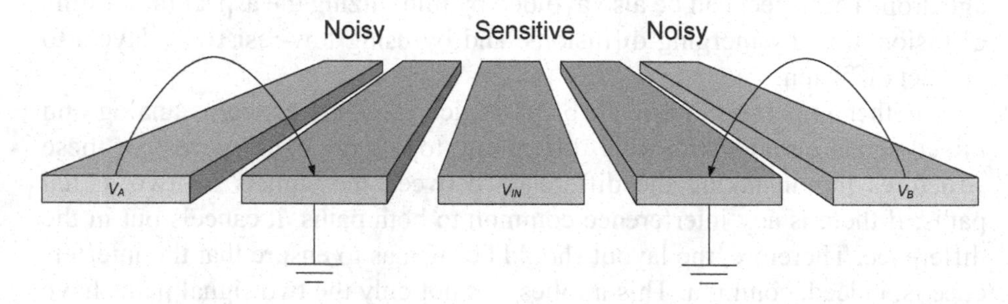

Figure 3. Shielding sensitive wires.

Capacitive and resistive coupling can also appear by means of substrate coupling [Stan94]. Since all the devices share the same substrate bulk, any noise signal into the substrate will propagate to every node of the circuit. This is particularly important in AMS SoCs, where analog circuits share a substrate with naturally peaking digital circuits. Common solutions rely on protective guard rings (providing that the substrate is not low resistive) and on the use of repeated substrate contacts to prevent it from de-biasing [Hast01].

2.4 Reliability

Reliability of an IC can be defined as the total time that an IC can provide perfect operation. This figure depends, at a high extent, on the quality of the IC layout. Reliability constraints, imposed and contemplated during the generation of the corresponding layout, are very important for a long-term function of the circuit. A serious source of reliability loss is electromigration, a physical effect caused by the imperfect nature of the materials used to fabricate the circuit. Electromigration is a phenomenon present in metal wires where extremely high current densities flow. For instance, electromigration in aluminum occurs when current densities approach 5×10^5 A/cm^2, which for a one-micron wide wire of 5000 Å thickness entails electromigration for a current of just 2.5 mA. Moving carriers collide with stationary metal atoms causing a gradual displacement of the metal. This causes a decrease in the metal‘s effective cross-sectional area and raises the current density seen by the remainder of the metal wire. As the conductor overheats, it eventually becomes an open circuit which may turn the circuit utterly unusable [Hast01]. Preventing electromigration from occurring can be attained by properly adjusting the wire width. Contact and via holes, making the current flow from geometries on different layers, should also be adjusted to minimize the resistance to such current flow. In order to do this, the number of holes of the wire must be increased [Wolf99], as illustrated in Fig. 4.

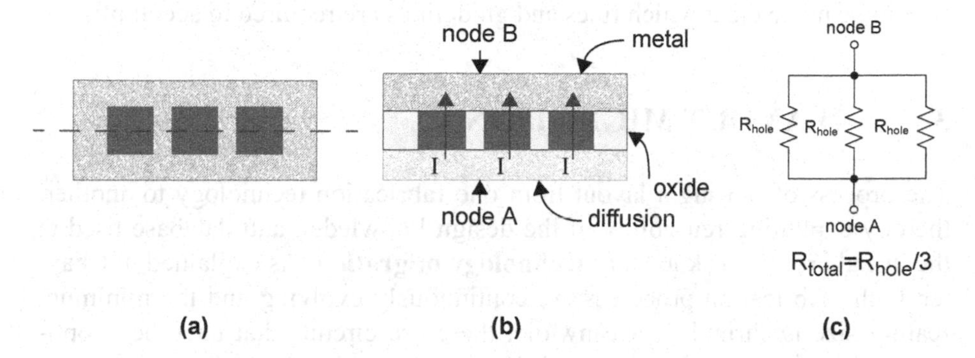

Figure 4. (a) Top view, (b) cross-section, and (c) schematic views of a diffusion-to-metal contact row.

When the performance specifications are modified, current densities may also change. For instance, suppose that, in the new design, a wire must carry a current of 10mA. If the wire width is 10μm and the maximum current density that it can support is 0.5mA/μm (carrying a 5mA maximum current) it results that the wire width must be adjusted to, at least, 20μm. Consequently, wire widths and the number of contacts placed may be now completely unsuitable for reliability purposes. It is thus necessary to enhance the layout synthesis process by taking a set of appropriate reliability measures.

2.5 Area occupation

Minimizing the area occupation is usually a design concern in AMS circuit design since it may lead to more integrated functionality and to eventually lower chip fabrication costs. Attaining a compact layout with minimal unused area can also improve the chip area usage. Therefore, when layout retargeting is required and changes in the circuit parameters result in changes in the circuit layout (small or large), both area and unused area should be kept as small as possible. Another important factor is the aspect ratio (i.e., width/height) of the circuit layout. In order to make the assembly of several circuit layouts easier, to attain aspect ratios closer to one is typically recommended. Nevertheless, in order to make the layout-reusable block less specific and more flexible, it should be feasible that the user can choose a different value for the aspect ratio or, sometimes much more critical, an upper bound for the layout width and height.

All the characteristics described above are critical to AMS layout design and, in this sense, a set of rules and guidelines should be followed to enhance the quality of the layout[5]. Actually, the relevant conclusion is that, whatever the method selected to create the layout-reusable AMS block, it has to efficiently cope with all these important issues and, as shown later in the chapter, should be made clear which rules and guidelines are required to accomplish it.

3 LAYOUT MIGRATION

The process of porting a layout from one fabrication technology to another, thereby exploiting reusability of the design knowledge and database used at the initial process, is known as **technology migration**. As explained in Chapter 1, the fabrication processes are continuously evolving and the minimum feature size is shrinking. Meanwhile, there are circuits that have been optimized for a given technology and that, to cope with this evolution and the associated improvement of the digital circuitry, need to be ported to new processes for still a long time in the future.

There are two types of problems related to technology migration. The first problem is the change of the circuit performance that may occur in the meantime, as the original design was targeted for a different technology (i.e., optimized for a set of transistors, resistors, and capacitor models). The second problem refers to the migration of the circuit layout database. The former problem has been the focus of the previous section (i.e., layout retargeting). This section deals with the latter problem: layout database migration.

Circuit layouts are created by arranging a set of geometric shapes, each shape made of a particular *mask layer* (e.g., polysilicon or different metal levels), to form the devices (e.g., transistors, resistors, capacitors) present in the circuit device-level description. Each fabrication process stipulates its own set of mask layers and its own set of layout design rules, according to which all the circuit's devices and interconnections must be laid out. Suppose a circuit layout made on one technology, T_1. The main problems arising when trying to port a layout from said technology to a different, goal technology, T_2, are [Cast02b]:

---

5. The reader is referred to Appendix A, where the most important rules and guidelines have been summarized.

1. **Variation of the geometric process parameters**. Foundries provide sets of design rules and guidelines which encapsulate the fabrication geometric constraints (e.g., likc the minimum feature size), and which the circuit layout must comply with. When the technology changes, these rules and guidelines may also change[6]. A violation of any of these rules may lead to complete invalidity of the circuit layout. Some of the most common design rules are illustrated in Fig. 5 and are explained below:

   - *Rules for device sizing and routing*: minimum width, length, and notch of each of the mask layers used to create the circuit devices as well as the routing wires.
   - *Rules for compaction*: minimum spacing, enclosure, extension, and overlap between a pair of mask layers.
   - *Grid rule*: as shown in Fig. 5, every single shape must be adjusted to a specified grid.

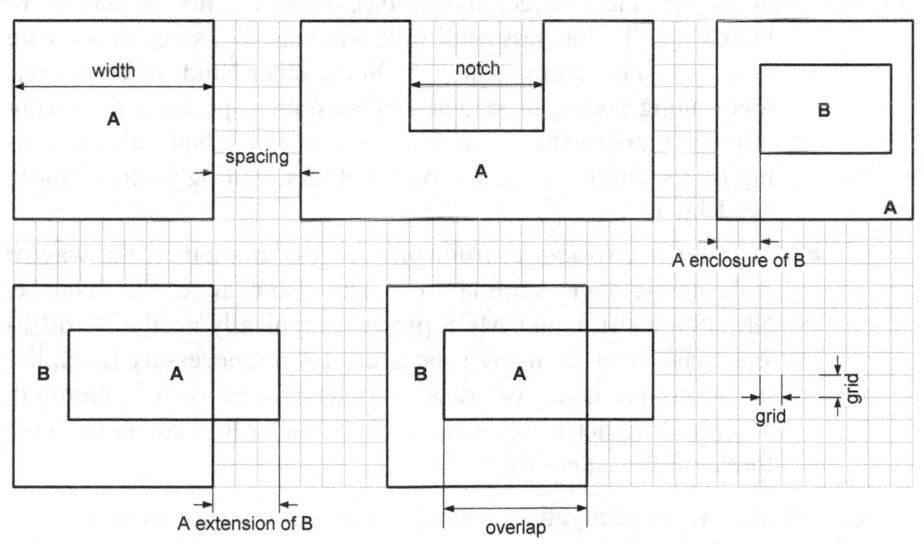

width: distance inside(A) to inside(A).
A spacing to B: distance outside(A) to outside(B) in different rectangles.
A notch: distance outside(A) to outside(A) in same polygon.
A enclosure of B: distance inside(A) to outside(B) (A contains B).
A extension of B: distance inside(A) to outside(B) (A may intersect B).
A overlap of B: distance inside(A) to inside(B).

Figure 5. Generic geometric design rules.

6. These rules and guidelines may even change as the technology itself upgrades.

2. **Variation of the electrical process parameters**. Electrical process parameters define the electrical characteristic of the process layer materials. Typical examples are the area and perimeter capacitance of poly-insulator-poly (PIP)[7] capacitors, the sheet resistance of the poly mask layer, the maximum current density of metal layers, etc. Calculations made in the technology T_1 may be totally impractical in the technology T_2. For instance, a $1\text{k}\Omega$ resistor was implemented in T_1 with a $100\mu\text{m} \times 5\mu\text{m}$ (length $\times$ width) poly mask layer, whose sheet resistance is $50\Omega/\square$. If this resistor is ported to T_2, where the sheet resistance of the poly mask layer is different, it is thus clear that the dimensions of the poly mask layer need to be updated.
3. **Variation of mask layers**. The set of available mask layers may vary from T_1 to T_2. Typical problems are:
   - *Number of routing layers*: it is not uncommon that T_1 and T_2 have a different number of metal routing layers (metal-one, metal-two, metal-three, and so on). There are not serious problems when T_1 has less routing layers than T_2, since every wire in T_1 has a counterpart in T_2. On the other hand, when T_2 has less routing layers, it becomes impossible to perform the layout migration, unless the original layout does not exhaust all the routing layers, and it uses as many, if not less, routing layers as those available in T_2.
   - *Device mask structure*: from one process to another, the way or style atomic devices are laid out may also change. For instance, NMOS transistors in CMOS processes typically need a P+ diffusion mask layer. In many processes, it is not necessary to explicitly draw this layer, whereas, in other processes, it is required. The problem then arises when moving the NMOS layout from the former to the latter process.

As with the layout retargeting issues, the layout-reusable analog block must be created so that layout migration can be seamlessly, rapidly performed.

---

7. *Poly* refers to common polysilicon layers present in almost all the fabrication processes. Standard processes use just one *poly* layer (both for active and passive devices) whereas mixed-signal processes feature two types of *poly* layers (one for transistor gates and another for resistors and capacitors). The insulator is thin oxide grown on the lower polysilicon layer.

4 ANALOG LAYOUT STRATEGIES

Once the two scenarios of analog layout reuse have been explained, it is necessary to examine which layout generation method is best suited to cope with the requirements of layout retargeting and migration.

A first classification of the layout generation methods for analog and mixed-signal ICs can be made with respect to the degree of layout customization that can be attained [Lamp99]. Layouts can be thus classified as either **full-custom** or **semi-custom**. Full-custom layouts are done in a hierarchical bottom-up fashion. There are no restrictions on the geometric characteristics of the blocks in the layout, and each of these blocks is designed manually. Consequently, resulting layouts are highly optimized for performance, power consumption, and area occupation. However, the resulting development effort is typically very high. On the other side of customization, semi-custom style provides a way to speed up the layout. When compared with full-custom, however, the standard cell approach results less flexible and both the gate array and sea-of-gates approaches yield layouts for which area occupation is not optimized. The improvement in turnaround time is obtained by imposing some restrictions or limitations on the physical design of the IC.

Within this semi-custom category, there are three different styles, namely the **standard cell** [Plet86], the **gate-array** (comprising the mask-programmable gate array –MPGA– and the field-programmable gate array –FPGA– styles), and the **sea-of-gates** styles [Mead80]. The standard cell approach consists of rectangular cells of fixed height and variable width, placed in rows, the space between the rows dedicated to cell interconnection. Although the individual standard cells may have been well optimized in area[8], this style imposes constraints on the global placement because the cells are generated with a predetermined height, their input/output terminals are placed at both sides of the cell, and their power supply connections are usually fixed at the top and the bottom of the cell. Perhaps more important is the fact that AMS design, with its huge variety of different required performances (i.e., topologies and device sizes) cannot be just covered with a finite library of predesigned standard cells. MPGA's are developed via a two-step process: first, the array structure is designed, second, the metal routing between the transistors and passive components (which customizes the array to obtain the intended design) is completed. In contrast, FPGA's customization is done via electrically programmable switches. The sea-of-gates style is similar to the MPGA one but without channels for routing, so it has to be done over the gates. The gate-array and sea-of-gates styles impose very important

---

8. Actually, each standard cell is done full-custom.

restrictions as well as several layout-induced performance problems on the analog circuits; in consequence, they are rarely used for analog designs [Lamp99].

The goal of the work presented here is to provide a suitable method for reusing analog and mixed-signal layout. The most suitable option, with respect to which layout style to use, is the full-custom approach, as it provides the optimum degree of customization and high-performance that most analog or mixed-signal designs require. Then, this review will immediately focus on the full-custom approaches. This review is by no means intended to be exhaustive (the interested reader is referred to excellent reviews on analog and mixed-signal layout generation in [Giel00] and [Rute02]). Its sole objective is to find out which of the layout methods is best suited to generation of layout-reusable analog blocks.

A typical flow of the full-custom layout process for analog and mixed-signal circuits is depicted in Fig. 6. The input to the layout process is a circuit description, typically a sized schematic. Technological information is used all throughout the layout process.

The first step consists in the generation of all the components of the circuit. At the cell level (e.g., an operational amplifier), these components are

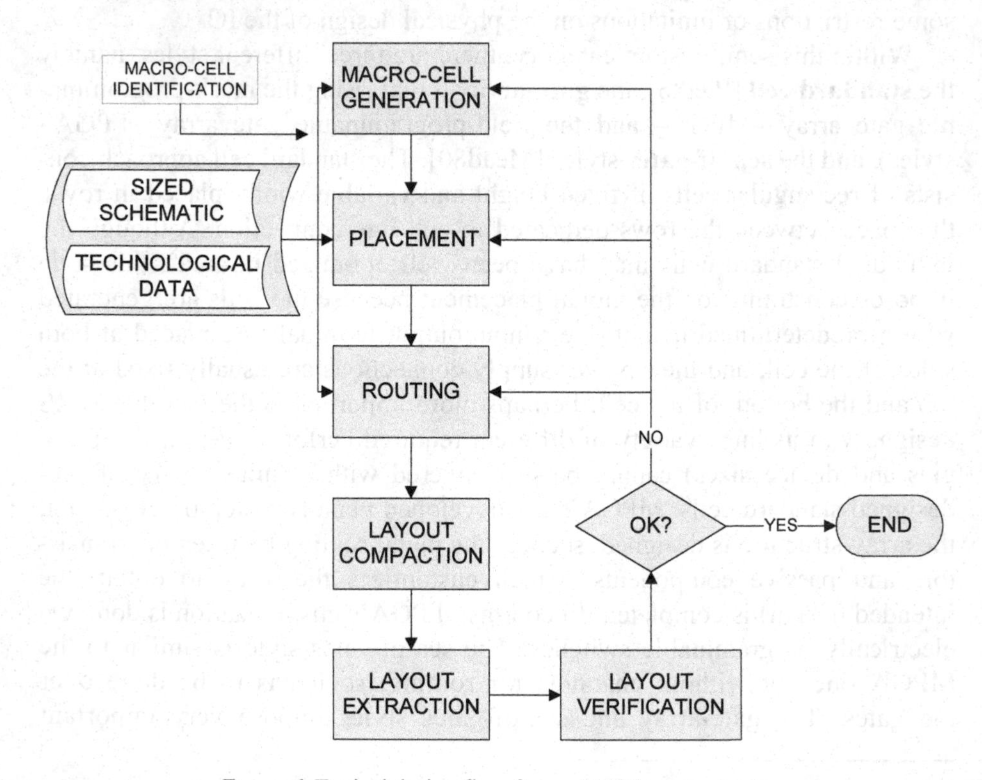

Figure 6. Typical design flow for an AMS layout process.

groups of one or more devices (e.g., mirror or cascode CMOS structures) known as *modules, structural entities*, or *macro-cells*[9]. Each macro-cell can be generated in several ways, all electrically equivalent, called *geometric variants* (e.g., a transistor differential pair may be laid out in a 1-dimensional or 2-dimensional common-centroid style) [Lamp99]. At higher levels, the layout components can be functional blocks as well (the layout of cells, modules, or sub-system circuit blocks) which have also been generated by using device groupings at the cell level. Macro-cell identification can optionally be used prior to macro-cell generation. The next step is the placement of every component, considering a wide set of analog and mixed-signal constraints to obtain a better result. Then, in the routing phase, the placed components are interconnected, in accordance with the circuit connectivity provided earlier in the flow. It is worth noting that the three previous steps, macro-cell generation, component placement, and component routing, do not take place always in that order. Sometimes it should be convenient to carry out placement and routing at the same time, since the problem of estimating how much space must be reserved around each component for wiring can be somehow alleviated [Prie97] [Prie01]. Other layout methods combine macro-cell generation, placement, and routing into a single and complex stage. After all components have been routed, a compaction of the whole layout may take place, but it also can be regarded as an integral part of the placement and routing phases. Whether compaction takes place or not, the resulting layout is then extracted to check that both circuit connectivity and circuit dimensioning have been faithfully mapped to the layout. Afterwards, the performance of the circuit layout (which unfortunately includes non-desirable effects) is verified and, in case there are unacceptable differences with the original circuit performance, one or more layout steps have to be modified, in a trial-and-error cycle, until an optimal layout is obtained.

Methods for generation of full-custom analog and mixed-signal layout focus either on automating one or more of the different steps involved in the process, or on providing the layout as a whole single process. Whatever the focus, these methods can be broadly classified into two different approaches: optimization-driven approaches and knowledge-driven approaches. Fig. 7 shows the taxonomy of these two approaches.

---

9. The terminology is derived from the digital design, where the layout blocks, called *macros*, are placed and routed to compose the circuit layout. Do not confuse modules here with the analog hierarchical level composed of analog blocks such as PLLs or DACs, which are commonly referred as modules.

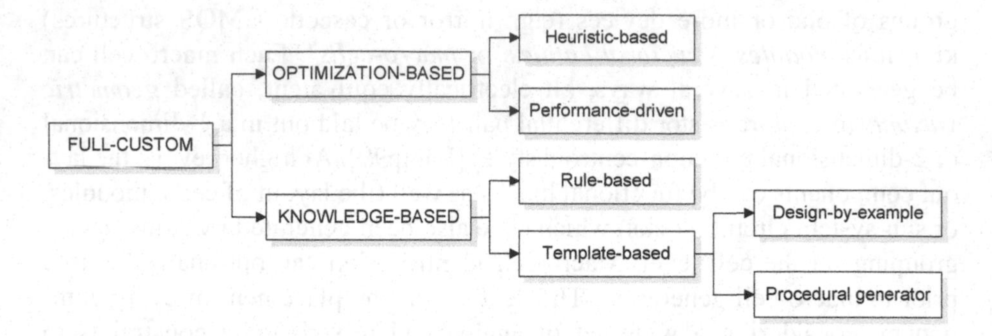

Figure 7. Taxonomy of CAD tools for layout generation of AMS circuits.

4.1 Optimization-driven approaches

Optimization-driven approaches aim at automatically generating the layout while striving to minimize the layout-induced errors by means of an optimization algorithm. Following a digital-like approach, placement and routing stages of the layout generation are carried out by such an optimization program according to a certain cost function. This cost function typically considers minimization of some design aspects such as area and net length, while penalizing violation of some analog design constraints, such as device mismatch, loading capacitances, and crosstalk. The quality of these optimization-driven tools is mainly determined by the efficiency of the optimization algorithm and the set-up of this cost function.

Depending upon the way of deriving the cost function and dealing with constraints on analog performance, two categories are usually considered [Lamp99]. The first group is composed of **heuristic-based** approaches. Layout-induced errors are taken into account by classifying nets according to their sensitivity and circuit function. Although optimizing the circuit performance, these approaches do not use a systematic method to generate net sensitivities and to handle performance constraints. Therefore, they may yield solutions that do not meet performance constraints after the layout is verified. Time-consuming layout-extraction-verification spins may be necessary. Besides, there is no way to identify which parasitics are mostly degrading the circuit behavior and, therefore, know which changes are necessary. Examples of these tools are ILAC [Rijm89], ANAGRAM [Garr88], KOAN/ANAGRAMII [Cohn91], and LADIES [Moga89]. Alternative approaches for placement of MOS transistors divide this step into device stacking and stack[10]

---

[10] A stack is a cluster of interconnected MOS transistors whose diffusion areas are merged to minimize parasitics, by using parallel side-by-side abutment of shared source or drain terminals

placement. This is done by using several heuristic algorithms rendering the circuit as diffusion graphs[11] of connected drains and sources [Mala95] [Basa96].

A notable improvement is accomplished by the other group of approaches, whose operation is based on **performance-driven** optimization of the circuit layout (also known as constraint-driven optimization). Unlike heuristic approaches, where no quantification of the performance degradation is done during the layout generation, performance-driven tools try to measure the layout-induced degradation on the circuit performance and keep it below desired margins. In this way, the impact of each layout parasitic is weighed out according to its effect on the circuit performance [Mala96]. The first contributions reported were to perform channel routing[12] [Chou90a] [Chou90b] [Chou90c]. In these works, the effect of layout parasitics are modeled by using sensitivities and, then, the performance constraints for the circuit are mapped to a set of constraints on the layout parasitics. Later approaches [Lamp95] [Lamp99] [Plas01] [Prie97] [Prie01] showed that this intermediate mapping can be skipped.

Routing tools such as ROAD [Mala90] [Mala93] and ANAGRAM III [Basa93], minimize the deviation from acceptable bounds on wire parasitics. Other tools reported are PUPPY [Char92] [Char94] for placement and SPARCS-A [Felt93] for compaction. The placement-then-routing approach may result in over-estimation of how much wiring space to leave for routing: too much yields empty space, too small turn the problem unsolvable. A solution, implemented via simultaneous placement and global routing for slicing-tree structures, has been reported in [Prie01].

The main advantage of optimization-driven tools is their *generality*: in principle, they can be applied to any analog or mixed-signal circuit [Giel00]. The drawbacks are the complexity of the optimization problem (even for the simplest problems, these are *NP-hard* problems [Shing86] [Leng88] [Prie01]), the difficulty of the cost-function set-up, and the large turnaround times[13].

---

11. A graph is a visual representation of the connectivity of a group of devices (MOS transistors in this case) by means of edges and vertices. See Appendix A for a detailed description of diffusion graphs and its properties.

12. Channel routing is performed when all cells have been placed such that the only area dedicated to cell interconnection are pre-assigned channels between cells created during the placement phase.

13. An exhaustive exploration of 1-dimensional placement of N cells to minimize net length involves N! different combinations. For N = 20 and one millisecond for each combination, it would take a 77-million years to exhaust the entire problem

4.2 Knowledge-driven approaches

These approaches try to store and exploit the knowledge required to create the AMS layout. This knowledge refers to the procedures that expert layout designers use to improve the quality of the layout, and spans a wide variety of techniques, from specific placement strategies to improve device matching (e.g., complex common-centroid arrays) and minimize the layout area, to routing techniques to minimize the loading effects, etc. Since this specific knowledge is to be stored and used whenever necessary, this approach is mainly intended to **reuse** previous experiences of expert layout designers.

Knowledge-driven approaches are specifically developed to generate the layout of fixed architectures/topologies. This means that the input information is not only a netlist of the sized circuit (see Fig. 6), but also a description of the layout as well as valuable layout knowledge to drive the layout generation. Knowledge-driven layout generation is not as complex as the optimization-driven one, as placement and routing are specified in advance.

There are two types of knowledge-driven approaches, namely rule-based and template-based approaches. **Rule-based** approaches store the layout knowledge in a customizable rule to be obeyed during layout placement and routing. A clear example of this approach is ALSYN [Bext93]. Its user-defined rules derive control information on the topology and parameters of the circuit, to guide the layout generation process. Rules are given, for instance, to identify current mirror macro-cells within a circuit, or to match capacitance values. The synthesis process of ALSYN starts with the rule application, which enables macro-cell identification and, afterwards, macro-cell generation. Then, the ALSYN's min-cut algorithm for placement and a maze-style router are applied. Generation of a three-opamp SC filter layout takes 50 seconds of CPU time[14]. Although every user can adapt the set of controlling rules to his/her own needs, the quality of the layout largely depends on the quality of this set of rules. Besides, the rules are difficult to formulate if they are intended to be general and context-independent. The specificity of the rules makes ALSYN suitable for only the limited set of circuits for which the rules have been formulated.

Template-based tools are also developed to best use layout designers' expertise. The underlying idea is to capture this expertise in a pattern or **template** that specifies all necessary device-to-device and device-to-wire spatial relationships [Lamp99] [Rute02]. Besides, it must capture analog-specific constraints like symmetry, device matching, and parasitic minimization [Dess01a]. To generate a circuit layout from this template, which is called *instantiation*, it is required to provide the value of a set of

[14] Performed on a Sun SPARC 2@40 MHz workstation with 16 Mb RAM.

electrical and geometric parameters (e.g., the transistor width and length, or the maximum current density allowed to flow on a certain layer type). The template can be generated either in a procedural or graphical way. The latter way consists in capturing the layout knowledge from a template previously laid out by an expert designer. A typical example is the **design-by-example** approach presented in [Conw92]. The example provided by the expert captures his/her knowledge (regarding device placement, routing wire trajectories, material types and widths, and position of macro-cell terminals). To generate a new layout, it is necessary to provide the required electrical parameters for each device, the sets of matched devices, and the geometric constraints (e.g., a desired aspect ratio). Starting from a fixed device placement, the tool derives all possible layouts (emerging from all possible device layouts, e.g., from different values of the number of unitary components of a MOS transistor). Then, an exhaustive optimization is executed to find the one that satisfies the specified geometric constraints. Finally, routing and compaction phases are carried out. This approach can produce good compact layouts in a moderate amount of CPU time (around 37 minutes for a 24-device operational amplifier[15]), but the layout templates have to be updated for each new fabrication process, which requires additional effort.

Capture of layout knowledge through templates can also be done by using **procedural generators** [Kuhn87]. Actually, this is the most mature layout technique used for analog circuits [Lamp99]. The mechanisms to describe these procedural generators can be specific layout languages such as BALLISTIC [Owen95] and ICEWATER [Harv92], or common spreadsheet interfaces [Hend93], but both approaches are intended to **code** the analog-specific layout knowledge into the software itself. The BALLISTIC language allows describing the layout template hierarchically by using fully-parameterized objects, and placement and routing are specified relatively. Several mechanisms can be used to adjust the position of the devices to account for size changes. This approach allows easy migration of the templates from one fabrication process to another without changing the code. Although the coding effort can be high, this effort needs no longer to be wasted when the device sizes and/or the fabrication process are changed. In addition, generation turnaround times are very low, since no optimization step is carried out. The process simply consists in the compilation of the coded template (when necessary) and the update of the device sizes as well as fabrication process parameters. Therefore, template-based approaches using specific layout languages are very useful when a circuit architecture/topology is to be

---

15. Performed on a Apollo DN4500@33 MHz workstation with 16 Mb RAM.

intensively reused. Procedural generation has been successfully applied to DACs [Neff95], switched-capacitor ADCs [Jusuf90], and filters [Yagh88].

Both for optimization-driven and knowledge-driven approaches, macro-cell generation is an important issue since the quality of the final layout depends also on the ability of the macro-cell generator to deal with analog-specific layout constraints, such as symmetry, device matching, or parasitic considerations. Most of the macro-cell generators follow the template-based approach explained above. Several macro-cell generators of MOS transistors (considering interdigitized, common-centroid structures) and passive components have been reported [Degr87] [Bruce96]. Recently, the approach in [Naik99] focuses more on the improvement of diffusion and routing parasitics, area efficiency, and matching considerations of the MOS transistor stacks. In [Sayed02], a macro-cell generator for arbitrary capacitor ratios in common-centroid unit-capacitor arrays is described that, based upon an oxide gradient model, takes into account capacitor mismatch. A generator for layout of capacitor and switch arrays for CMOS ADCs and DACs is presented in [Leme91]. Reliability constraints have also been applied to macro-cell generators in [Wolf99].

A different classification of the layout generation tools can be made according to the means of communicating analog layout information to the tool from the circuit synthesis process [Neff95] and vice versa. When the tool is loosely coupled, as it is the case of the optimization-driven approaches, the input to the tool is typically a sized circuit (netlist) with some additional rules indicating critical device pairs and nets to be matched. Then, the sensitivities of layout effects on the circuit performance are passed to the layout tools, and a layout is generated by limiting the adverse effects of layout-induced errors on performance using such sensitivities. When the layout tool is tightly coupled to the synthesis process and the layout generation approach is knowledge-driven, it may be possible to predict layout parasitics providing that the layout generation process is swift enough (or even without actually going to the layout step) and compensate for layout parasitics early in the circuit synthesis process [Onod90].

5 AUTOMATED LAYOUT GENERATION FOR DESIGN REUSE

The reusable circuit block is the key to a successful design reuse methodology that aims at shortening the overall design time while managing the inherent complexity of analog and mixed-signal circuits. To this end, capturing the design knowledge is essential. Thanks to these reusable blocks, the layout implementation, one of the most laborious tasks of the circuit design process, is considerably eased. Moreover, as explained in Chapter 2, a major concern of design reuse is, in order to speed up the design process, to bring in layout information and use it early in the circuit synthesis process therefore minimizing global design iterations.

Having this in mind, the most suitable layout generation approach to capture the design knowledge, thus making design reuse possible, is the knowledge-driven generation of procedural layout templates. Layout design knowledge can be efficiently stored within layout templates. The reasons supporting this conclusion are the following [Cast01] [Vanc01] [Cast02b] [Tang02] [Tang03]:

- *Layout templates are very efficient at handling design expertise*. Analog layout retargeting, as explained in Section 2, requires imposing several layout constraints based on accumulated design knowledge. These constraints cannot be easily considered by traditional placement and routing algorithms [Ocho96]. Layout templates are, on the contrary, structures where user-defined constraints are easily stored.
- *Layout templates permit searching for optimal block parameters while revealing the knowledge needed for evaluation of layout parasitics*. One main problem of traditional circuit design is to close the loop between circuit synthesis and layout generation (when layout-induced errors degrade the circuit performance, circuit design changes are accomplished via circuit re-synthesis [Giel00]). An important aspect of this problem is early evaluation of parasitics: in manual design, overestimation of parasitics results in wasted power and area, whereas underestimation of parasitics may lead to fatal performance degradation. Extraction of circuit parasitics within the circuit sizing process solves this problem. The extraction is performed by means of either layout generation or parasitic modeling. Heuristic-based or performance-driven approaches are currently too slow for layout generation within the circuit sizing process [Lamp99]. Using procedural layout template allows, on the other hand, fast generation of circuit layout since no time-consuming optimization algorithms are involved. Furthermore, as both placement and routing are specified in a procedural code, modeling the layout parasitics becomes also possible.
- *Layout templates ease placement*. The layout generation procedure is simplified because the positions of the blocks in the template are stored according to pre-defined relationships embodying constraints

from the layout expert that enhances the layout quality. Optimization-driven methods try to attain the same quality at the expense of time-consuming algorithmic techniques.

- *Layout templates can be straightforwardly ported*. Technology independence can be achieved by coding of the procedural template generator using symbolic process parameters and mask layers. Therefore, neither scaling methods [Mead80] nor complex compaction techniques [Apan98] need to be applied. The main advantage of procedural layout with respect to scaling and compaction methods is its higher precision and speed, respectively.

Despite these important benefits, procedural methods have two drawbacks:

- *Cost*: the effort to generate every new template may largely exceed the effort to create, manually, the corresponding full-custom layout.
- *Flexibility*: large changes of the circuit performance may lead to a dramatic degradation of the layout regularity, aspect ratio, or area usage.

As it will be shown in Sections 7 and 8, the cost of template generation can be largely palliated by using appropriate procedural functions for both placement and routing. Besides, intensive use of the same layout template compensates for the effort and time employed to create the layout template. As it will be explained in Section 7.3, flexibility of the layout templates can be increased by carefully planning both placement and routing.

In the work presented here, procedural layout methods have been used to implement the layout facet of analog reusable circuit blocks. The following sections develop the methodology and techniques used.

6 LAYOUT TEMPLATE: DEFINITION AND PROPERTIES

A general definition of a layout template is the following: *a layout template is a data structure that completely defines the physical implementation of a certain circuit architecture/topology without having detailed intelligence on actual device sizing*. Such a data structure can adopt several forms, ranging from a detailed graphic planning of the circuit layout to just some basic guidelines, such as "place all the devices side by side". Nevertheless, the most important factor common to all types of layout templates is that **experience and knowledge from expert layout designers can be stored in an orderly**

systematic way. Therefore, *designer*'s *expertise* on analog layout can be reused when needed.

As said above, layout templates are just descriptions of a circuit topology. This architecture/topology describes only which are the circuit components and how they are connected. The layout template does not contain information about a specific circuit sizing or the fabrication process: quite the opposite, the layout template must be as generic as possible. Design reuse is, as already explained, the action of using the available template when a change of the circuit sizing is required or a different fabrication process is used, or both. The consequence is that the layout template must be a fully *parameterized* entity. The properties of a parameterized layout template (PLT) are the following:

1. **Parameterized components**. The layout template can adapt different performance specifications –or, in other words, different device sizing– because each of the circuit design parameters (e.g., transistor width and length) is a parameter of the template itself. For the sake of illustration, let us consider a simple CMOS operational amplifier as the one depicted in Fig. 8(a). Its corresponding *floorplan* (a partitioning of the floor rectangle into smaller ones which determines the topology of the layout) in Fig. 8(b) shows how the transistors are arranged in three different blocks B_1, B_2 and B_3. Each block is a parameterized component and the transistor fundamental parameters (namely width w, length l, and multiplicity m) are the parameters controlling the shape of such blocks.
2. **Relative placement**. The location of every single block in the layout template must be a function of the location and dimension of the rest of its neighboring blocks. An optimal way of describing the structure of a layout template placement is by means of *corner-stitching* data structures and *constraint graphs* [Oust84] [Marp90] [Wimer88]. Fig.9 illustrates this type of description. The entire plane of the block layout is represented explicitly with rectangles called *tiles*. These tiles represent physical layers (e.g., metal or polysilicon), primitive devices (transistors, resistors, capacitors, inductors) and connectors (contacts and vias), an arrangement of primitive devices, and a cell or other hierarchically higher circuit layout (see Section 2.1 of Chapter 1 on page 14 for a definition of the different hierarchical levels). The set of tiles define vertical and horizontal line segments or cuts. Each tile is linked to the rest of tiles by a set of pointers, called *corner stitches,* at two of their four corners, and related *geometric constraints*. As illustrated in Fig. 9, these stitches are at the bottom-left

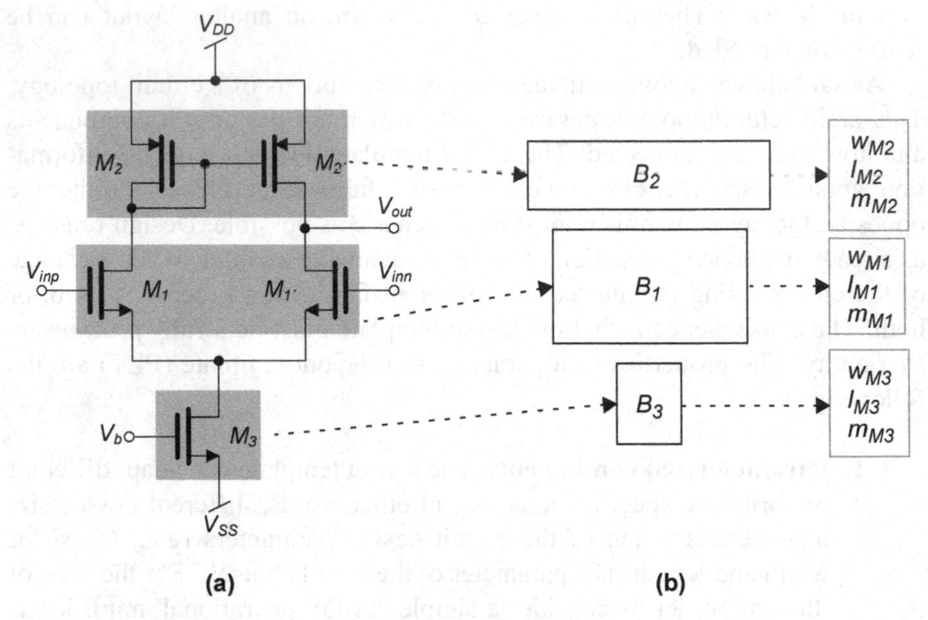

Figure 8. (a) Simple opamp schematic and (b) its layout floorplan.

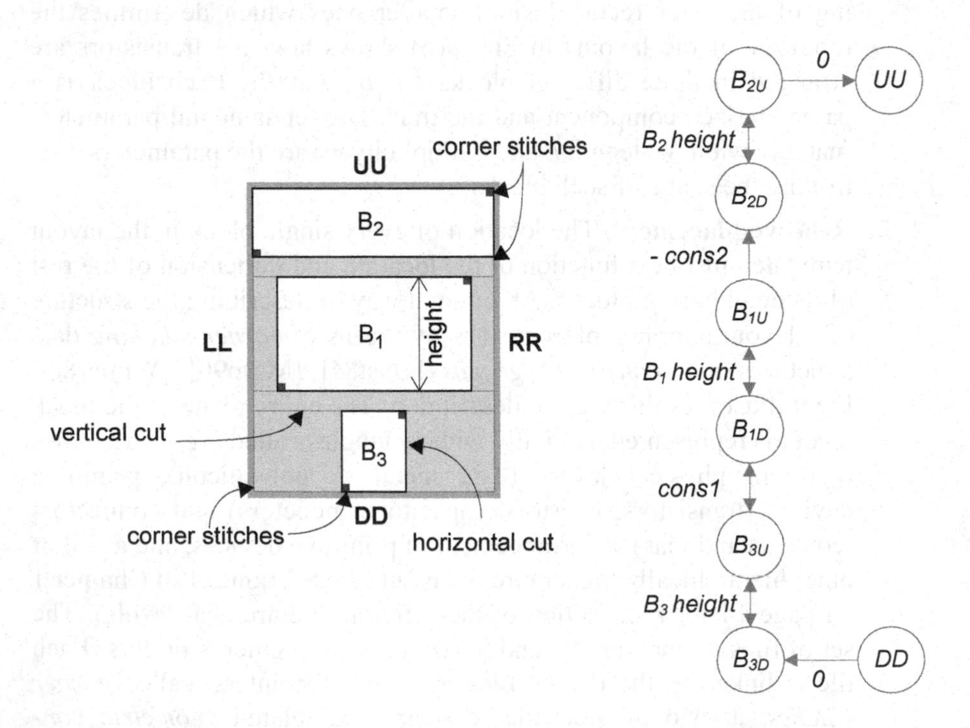

Figure 9. A three-block layout template and its vertical constraint graph.

corner and at the top-right corner. Each stitch represents two coordinates, horizontal and vertical, so each tile is defined by four coordinates, left (*l*) and down (*d*), for the bottom-left stitch, and right (*r*) and up (*u*), for the top-right stitch. In this way, the arquitecture's/topology's floorplan can be represented by two planar graphs $G_H(\mathbf{V}, \mathbf{E})$ and $G_V(\mathbf{V}, \mathbf{E})$, called horizontal and vertical graphs respectively, as follows. A vertex V in $G_H(\mathbf{V}, \mathbf{E})$ (or $G_V(\mathbf{V}, \mathbf{E})$) represents a vertical (horizontal) cut of the floorplan. The vertices are ordered according to the distance of the corresponding cuts from LL (DD), the leftmost (bottom-most) side of the floorplan rectangle, until the rightmost side RR (the top-most side UU) are reached. Two vertices v_i and v_j in G_H (G_V) are connected by an arc e_{ij} directed from the former to the latter if there is a sub-rectangle in the floorplan whose left (bottom) and right (top) edges lie on the corresponding vertical line segments, respectively. Through this representation, geometric constraints between the tiles can be easily established by assigning each arc e_{ij} a weight w_{ij}. These constraints arise as consequence of (1) process design rules, (2) connectivity (to ensure that two tiles remain electrically connected after layout retargeting/migration), and (3) analog specific rules (see Section 2). Fig.9 illustrates the vertical constraint graph of the operational amplifier template floorplan. Weights cons1 and cons2 represent two hypothetical restrictions between position of tiles B_3, B_1, and B_1, B_2, respectively. Two-sided arrows mean "equal", while one-side arrows mean "higher than" for positive constraints (e.g., cons1) or "lower than", for negative constraints (e.g., −cons1). For the example in Fig.9, two equations can be derived:

$$B_{1d} - B_{3u} = \text{cons1} \quad (4)$$

$$B_{2d} - B_{1u} < \text{cons2} \quad (5)$$

3. **Relative routing**. As with relative placement, the physical implementation of the connections between all the circuit tiles must be stored in a relative way. Continuing with the previous example, Fig. 10 depicts two different *instances*[16] of the operational amplifier layout template. Note that the routing used to connect B_1, B_2, and B_3 must be defined as a function of the block placement, of block dimensions, of block pin positions, and, to avoid wire crossing over, of the location of other routing wires.

---

16. An instance of the parameterized layout template is an individual example of the template, each instance having different values of the template parameters.

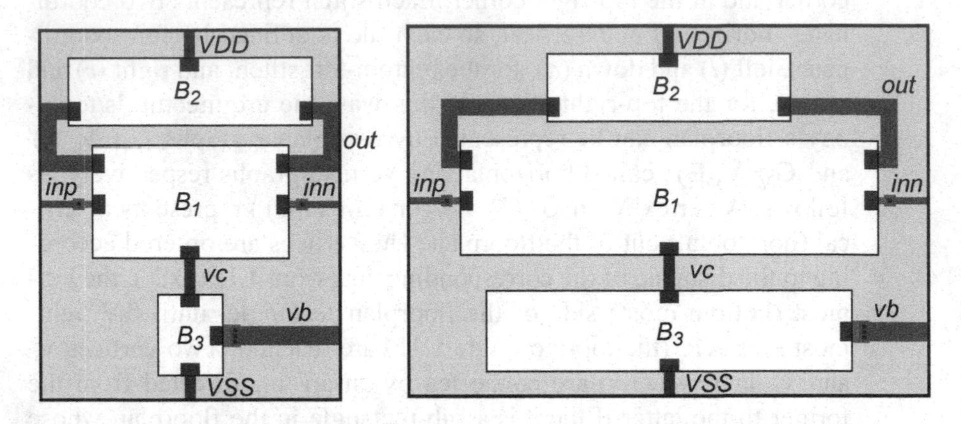

Figure 10. Two examples of the opamp layout template.

4. **Technology independence**. To ease the reuse of designers' experience in analog layout, and because of the continuous shrinking pace of integration technologies, layout templates have to be easily ported from one fabrication process to another. To achieve this, any reference to a particular fabrication process in the layout template has to be completely avoided, and turned completely generic. That is, all mask layers and relative placement and routing have to be stored in a process-independent way. When the circuit layout is implemented in a particular technology, it must be able to adapt to both the technological design rules and the set of layout layers. In this way, technology-specific information is stored away from the actual template generator and the template can be ported to newer processes by feeding the new process information to the template generator.
5. **Hierarchy**. The layout template is the physical implementation of a circuit at any hierarchical level, be this the system, sub-system, module, or cell level. Therefore, the layout template may contain lower hierarchical levels within. For the sake of illustration, Fig. 11 shows the hierarchy of a PLL, a module-level circuit. Every building block layout in the PLL, namely the filter, the VCO, the phase detector, and the gain block, is implemented by using the template approach. These templates are in like manner made of other building blocks. The gain block, implemented with an operational amplifier, is composed of a current mirror, a differential input pair, and a single transistor. Suitable procedures are therefore required for transmitting down the parameters of the *parent* block to its immediate hierarchically lower building blocks, correspondingly called *child* or *leaf components*. Note that said *child components* of a cell-level parent block are not

layout templates themselves as defined here, but parameterized *device-level blocks*[17] (e.g., the differential input pair) composed of parameterized *device primitives* (e.g., the single transistor). The actual difference between layout template blocks and parameterized device-level layout blocks considered here is that, although these blocks may also feature a lower hierarchical level (e.g., the primitive MOS in the current mirror), they are not treated independently at the synthesis stage. For instance, the opamp is sized as a whole, and no further division into smaller, separately sized building blocks is needed. In any case, the underlying method to create both layout templates and parameterized device-level layout blocks and primitives is the same, with the exception made of some important concerns coming from their different levels of design complexity. These concerns will be explained in due course.

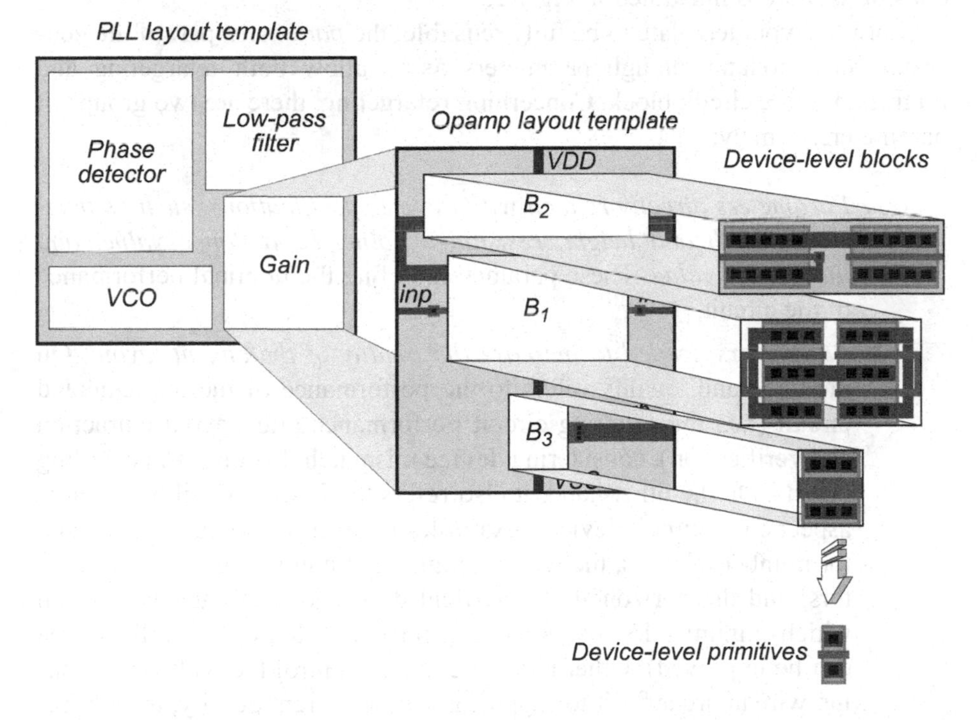

Figure 11. Illustration of the layout template hierarchy concept.

[17] Previously referred as macro-cells.

7 CREATING THE LAYOUT TEMPLATE

The parameterized layout template, whose properties have been described in the previous section, must be coded into a software structure that allows reuse, retargeting and migration, of the circuit layout. From a practical point of view, this data structure can be implemented by using the resources of well-known programming languages, such as LISP, C, or C++. Reuse of the template is therefore accomplished through multiple execution calls to the generator represented by such a structured software code. Actually, the code of the parameterized template is written as if the layout was being constructed, and the effort is made just once to obtain many different instances of the layout of the reusable block.

The generic structure of the code for a generic parameterized layout template, fully independent of the actual programming language used and considered here is illustrated in Fig. 12.

For the layout template to be fully reusable, the *parameter* part of the generator must contain enough parameters as to allow both retargeting and migration of the circuit block. Concerning retargeting, there are two groups of parameters, namely:

1. *Parameters directly related with device specifications, such as transistor width and height, resistance value, capacitance value, and inductance value*. These parameters define the nominal performance of the circuit.
2. *Parameters needed to increase the quality of the circuit layout.* On the one hand, quality refers to the performance of the implemented circuit (i.e., the resulting circuit performance after layout extraction and verification), considering device mismatch, loading, and coupling effects. On the other hand, it also refers to circuit reliability and area aspects. For circuit devices, examples of such parameters are transistor number of folds, the width, length, and number of strips of resistors, and the horizontal and vertical dimensions of capacitors (with which minimization of parasitics, current matching as well as area can be improved). Other parameters may control the width of a routing wire in order for it to support a certain current density, the separation and shielding of critical routing wires, the number of contacts, etc. In short, although these parameters do not define the nominal performance of the reusable circuit block when a retargeting process is carried out, they can improve the overall quality of the layout.

```
PARAMETERS
{
Retargeting parameters
Migration parameters
Other parameters & var.
}
LEAF COMPONENTS
{
Leaf-Component#1 {
     Parameters link
     Terminals}
Leaf-Component#2 {
     ...}
...
Leaf-Component#N {
     ...}
}
PLACEMENT
{
...
}
ROUTING
{
...
}
CONNECTIVITY
{
...
}
```

Figure 12. Generic structure of the code of a generic parameterized layout generator.

Regarding technology migration, the parameters required are the following:

1. *Parameters defining each possible design rule involved, an example of which is shown in Fig.5.* It is worth noting that true technological independence of the layout template is reached only if every design rule, from largely different fabrication processes, is considered.

Technological parameterization becomes critical when processes from different families (foundries) are considered.

2. *Parameters defining process layers used*. Each circuit device is laid out by assembling a set of pre-specified layers, whose number and kind change from process to process. The PLT generator works with generic layers to be particularized for the target technology. *Void* layers are also used for those layers that do not have a counterpart in the generic fabrication process. Examples of this situation are those processes that do not use N+/P+ implant layers (which are usually added during the fabrication phase). Likewise, every single layer (together with its corresponding design rules) should be considered for seamless technology migration.
3. *Parameters defining the electrical characteristics of the mask layers, such as the sheet resistance, the area and perimeter capacitance, or the maximum current density*. Note the relationship between these parameters and those concerning retargeting.

An important difference between retargeting parameters and migration parameters is that, whereas the former are specific for a layout template, the latter are common to all the layout templates (i.e., they share the same design rules, layers and electrical parameters).

In addition to retargeting and migration parameters, the layout may have other parameterized characteristics that provide it with extra flexibility. By way of example, consider two different parameterized MOS transistor primitives, one for NMOS transistor and another for PMOS transistor. This extra flexibility is also useful when considering different implementation styles for the primitives, such as PIP, MIM (metal-insulator-metal), and MOS-type capacitors. All different, possible layers and layer arrangements should be altogether considered to develop this kind of primitive such that the various types and styles can be combined into a single parameterized entity having a *variant* parameter upon whose selection the desired primitive is correctly instanced.

After the *parameter* section, the *leaf components* section contains an enumeration and description of all the leaf components in the layout template. Depending on the complexity of the circuit block for which the template is being created, these leaf components can be layout polygons (for primitives), primitives (for device-level blocks), device-level blocks (for templates), or even other templates. Every parameter of each leaf component must be here either linked to one of the parameters in the *parameter* section, or set to a fixed value. In the same way, migration parameters must be linked. The connection points or terminals by which each leaf component connects to other leaf components must be also specified here. Specifically, it is important to define the position of each terminal with respect to the leaf component itself in order to route the leaf components more easily.

Once all the leaf components have been completely specified, they are placed by following the floorplan detailed in the *placement* section. This floorplan can be derived from both the horizontal and vertical constraint graphs, as illustrated in Fig. 13. For each leaf component, two corner stitches are defined, either as independent values (when the position of the leaf component does not depend on the rest of the leaf components) or as a function of the size and position of other components. In this latter case, the constraint graph provides a simple way to obtain such a function. In this placement section, the orientation of the leaf component is also specified within the four different possible values (0°, 90°, 180°, and 270°).

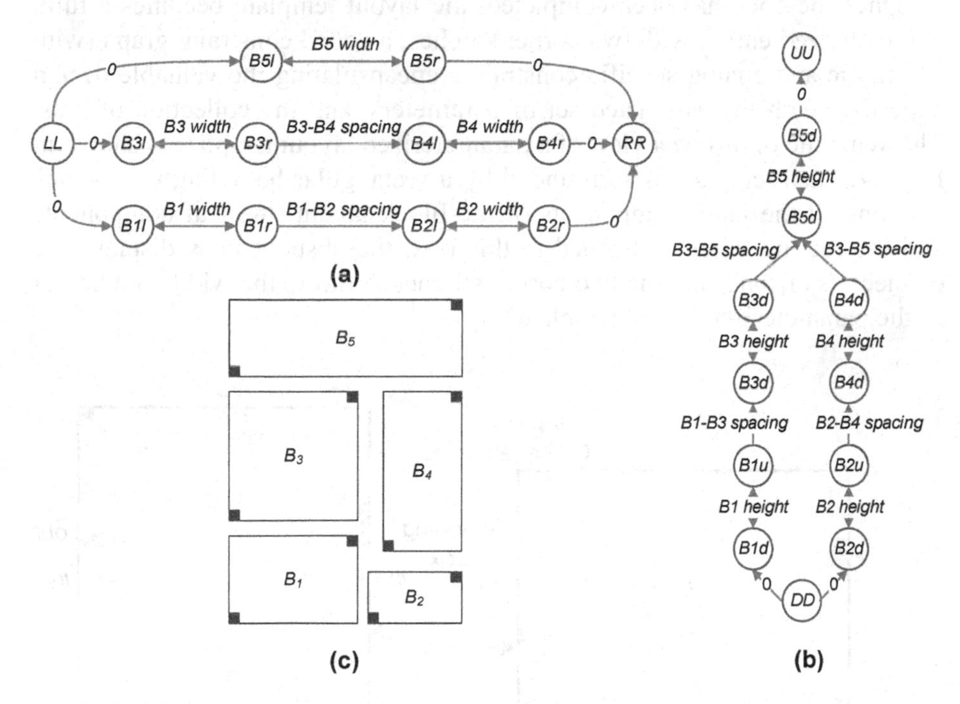

Figure 13. (a) Horizontal constraint graph; (b) vertical constraint graph; (c) a sample instance of the layout template.

The following part of the code covers the interconnection of all the leaf components. The routing phase is thus done by using the parameterized position of the leaf components' connection points, previously defined in the *leaf component* section. Again, the constraint graph helps to parameterize the routing wires.

The last part of the template generator is devised to specify which and where are the ports (called *pins*) through which the layout template can be connected to other templates to compose a hierarchically higher layout template.

As it will be explained in detail below in this same section, the code described above is complemented with a set of geometric and database procedures with which the parameterization, relative placement and routing, hierarchical nature, and technological independence properties, can efficiently be incorporated in our layout templates. Moreover, the structure described above and the procedures can be implemented in many different ways and in many different design environments. Below, we also present a methodology to generate and use layout templates that is fully embedded in a well-known, commercial design environment.

Once the code has been completed, the layout template becomes a fully parameterized entity, with two corner stitches, a coded constraint graph (with geometric and analog-specific constraints encapsulating the valuable design expertise) with the associated set of parameters, and the collection of pins. The symbolic or *abstract* view of a parameterized layout template is shown in Fig. 14. Each template is surrounded by a rectangular bounding box, which contains all the inner components of the block layout (the leaf components and the routing wires). Attached to this box, the abstract view displays the connectors or pins, and the two corner stitches defining the width and height of the parameterized layout template.

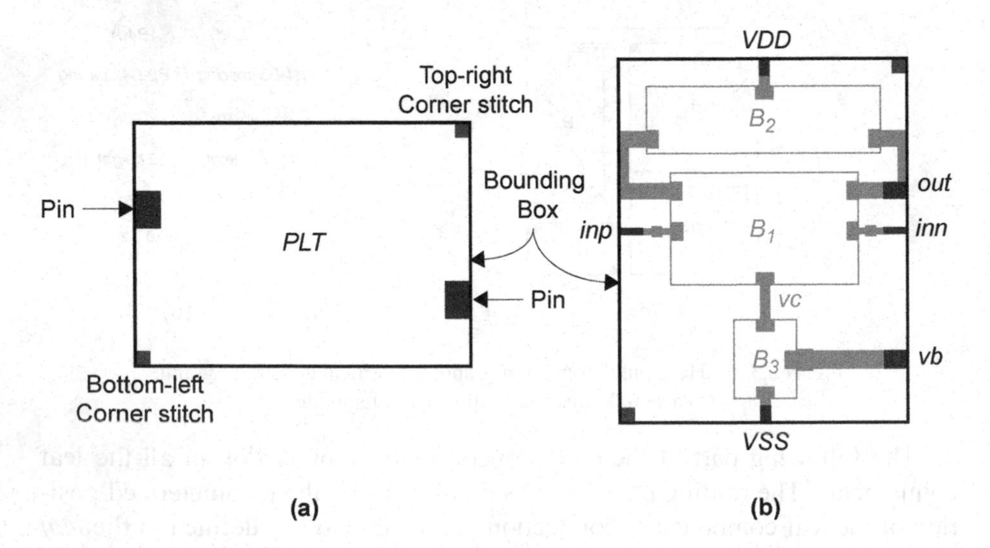

Figure 14. (a) Abstract view of a PLT; (b) abstract view of the opamp PLT.

A key component of layout templates is, as inferred from the previous code description, the constraint graph. A carefully devised constraint graph translates into a carefully devised floorplan, which is critical to the flexibility of the parameterized placement since the relative position of each leaf component remains constant during the retargeting/migration process. A careful floorplan is also crucial to ease the parameterization of the routing wires. When several leaf components change, the wires connecting them must adapt to every geometric change according to the new leaf component parameters or different design rules. If the wires are parameterized, just in the same way as the relative positions of the leaf components, the designer has full control over the layout characteristics, but the routing process may become tremendously complex if a proper placement is not made.

Figure 15 illustrates this point with the same example of Fig. 13. Suppose that a retargeting or migration process involves that blocks B_1-B_4 change to adopt the sizes illustrated in Fig. 15(b). If the constraint graphs of Fig. 13 were used to generate the parameterized placement and routing, the resulting layout, in Fig. 15(b), will suffer from several serious faults, all pointed out in Fig. 15.

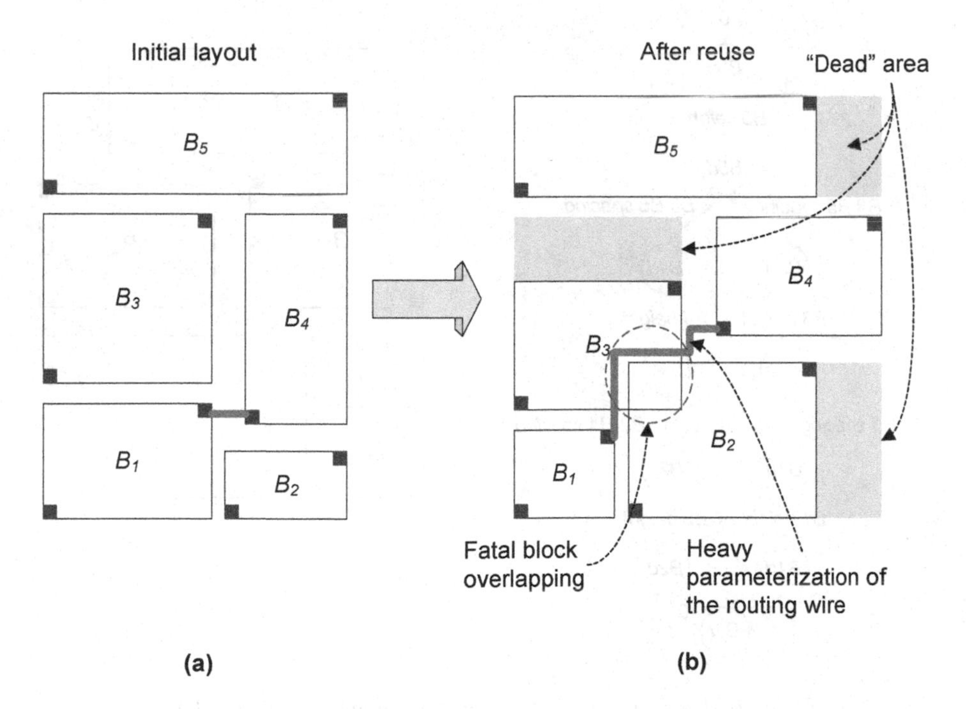

Figure 15. Impact of a poorly-devised floorplan.

The most important fault is perhaps block overlapping. This fault may cause violation of many design rules and, consequently, completely invalidate the design. A fault of this nature is caused by shortcomings of the constraint graph. In this very example, the vertical constraint in Fig. 13(b) does not take into account that both blocks B_3 and B_4 must be placed, accordingly to the horizontal constraint graph, atop both blocks B_1 and B_2. Therefore, the graph must be modified as shown in Fig. 16, where four constraints between corner stitches *B1u*, *B2u*, *B3d*, and *B4d* prevent these blocks from overlapping. The improved layout is also shown in Fig. 16.

Another fault of poorly devised floorplan is the resulting heavy parameterization of routing. The routing wire shown in Fig. 15[18] becomes very difficult to parameterize (it depends on the size and position of blocks B_1, B_2, B_3, and B_4), no matter the vertical constraint graph in Fig. 13 or Fig. 16 is used.

A solution to ease this wire parameterization is illustrated in Fig. 17. Now, the horizontal graph includes constraints to center all the blocks with respect

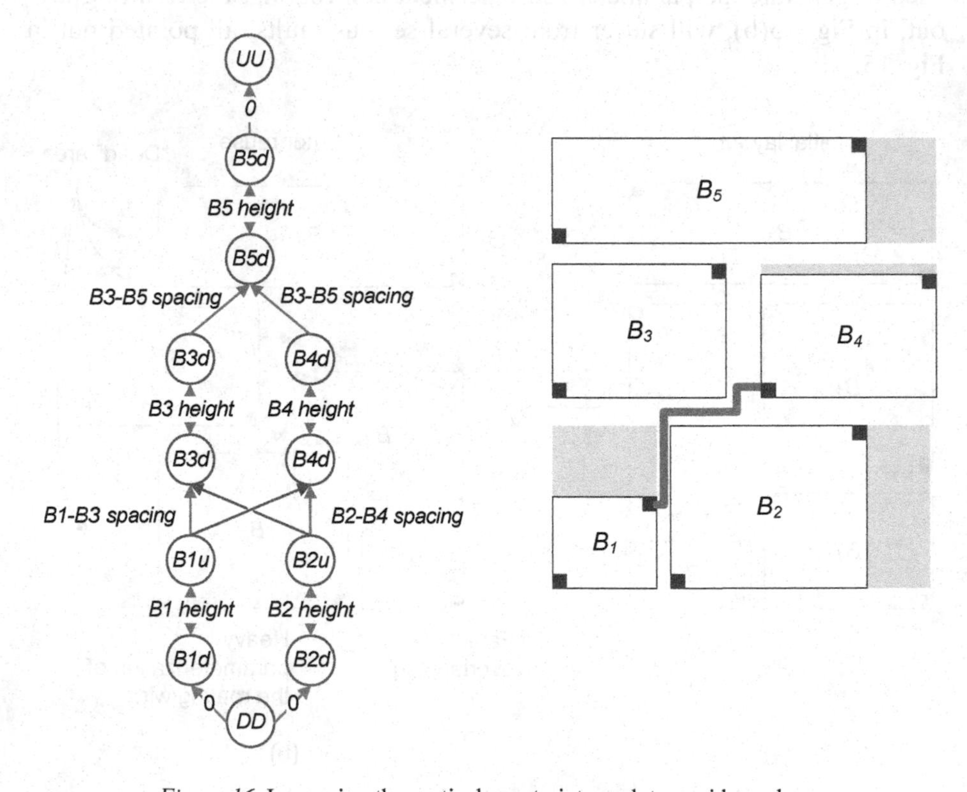

Figure 16. Improving the vertical constraint graph to avoid overlaps.

[18] For simplicity, the rest of the routing wires are not drawn.

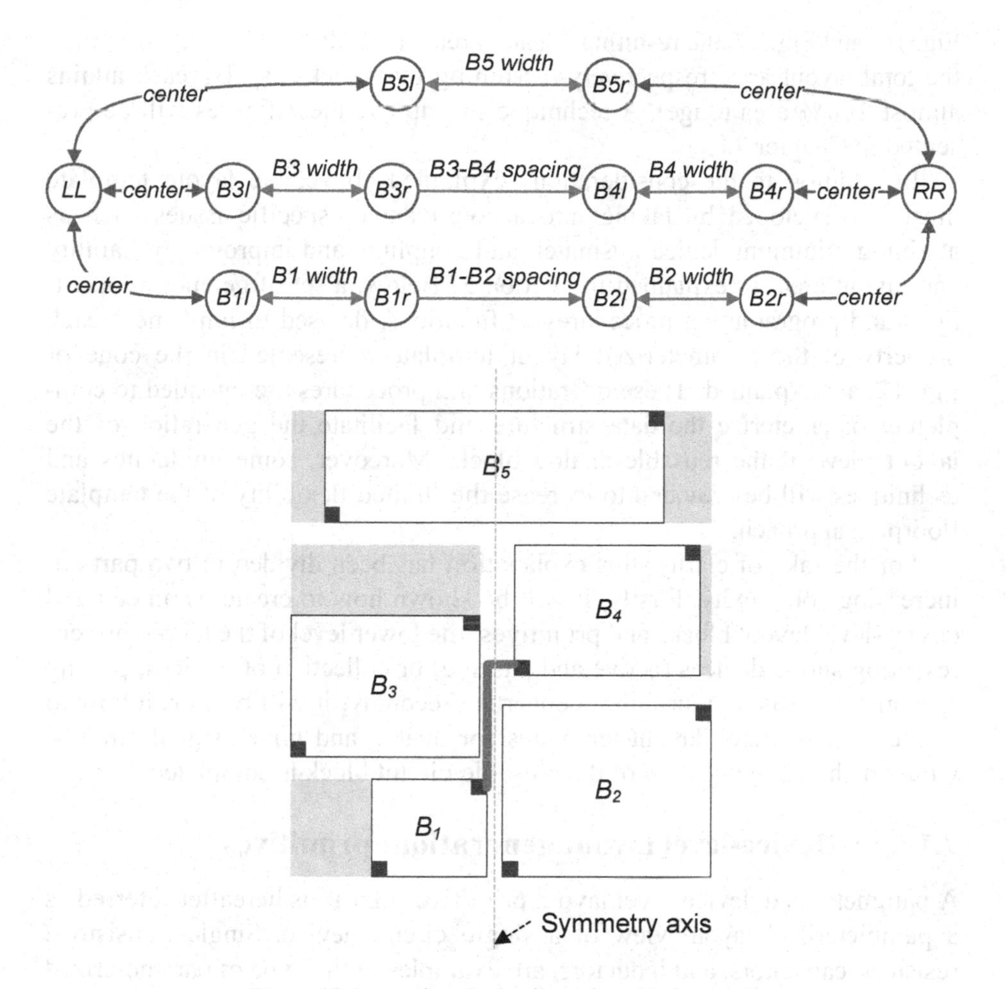

Figure 17. Improving the horizontal constraint graph.

to a vertical symmetry axis. The *center* constraint places the blocks such that the distance from the template left-most side *LL* to the blocks B_1, B_3, and B_5 equals the distance from the blocks B_2, B_4, and B_5 to the right-most side *RR* of the template. This constraint turns out to be very helpful for fully differential circuits, where symmetries of block positions and, therefore, routing wires are to be preserved throughout the circuit reusing process. As shown also in Fig. 17, the resulting wire depends only on size and position of blocks B_1 and B_2.

The last fault is, as pointed out in Fig. 15, the resulting unused or "dead" area (gray zones in Fig. 15, Fig. 16, and Fig. 17). Although it may not affect circuit performance, this unused area results in poorly compacted layouts, eventually yielding wasted chip area. For the examples shown in Fig. 15,

Fig. 16, and Fig. 17, the resulting "dead" area was 18.08%, 15.47%, 16.34% of the total layout area, respectively (assuming that blocks B_1-B_5 each attains almost 100 % area usage). A technique to improve these figures will be presented in Chapter 7.

In addition to the considerations explained above, the layout template must be developed by taking into account analog-specific issues such as attaining minimum device mismatch and coupling, and improving reliability and layout area, as explained in Section 2. Below, a set of geometric operations and programming procedures or functions, devised to implement each property of the parameterized layout template represented in the code of Fig. 12, are explained. These operations and procedures are intended to completely parameterize the data structure and facilitate the generation of the layout view of the reusable analog block. Moreover, some guidelines and techniques will be provided to increase the limited flexibility of the template floorplan approach.

For the sake of clarity, this explanation has been divided in two parts of increasing complexity. Firstly, it will be shown how to create parameterized device-level layout blocks and primitives, the lower level of the hierarchy representing single devices (active and passive) or collection of devices, putting special emphasis on reusability concerns. Secondly, it will be shown how to create parameterized layout templates for analog and mixed-signal circuits, with which the layout view of the reusable circuit block is completed.

7.1 Device-level layout generation: primitives

A parameterized device-level layout primitive (PDLP) is hereafter referred as a parameterized layout view of a single circuit device. Single transistors, resistors, capacitors, and inductors, are examples of this type of parameterized layout. A parameterized device-level layout block (PDLB) is, on the other hand, a collection of such primitive devices with specific connectivity. Differential pairs, cascode structures, current mirrors (all made of transistors), or resistors and capacitors groups, are examples of this type of parameterized layout. Both types of parameterized layouts are eventually placed and routed to make up a hierarchically higher and more complex layout template. An important difference between PDLBs and layout templates is that the circuit being implemented with the PDLB is not typically treated separately in the synthesis phase as it is done with the circuit represented by the layout template. For instance, the differential input pair of a comparator circuit is not sized separately from the rest of the circuit. Another difference with respect to layout templates is that the components of parameterized device-level primitives and blocks may overlap. That is, whereas layout template components are not allowed to overlap in order for the constraint graphs to define correctly a slice floorplan, the constraint graphs for PDLPs and PDLBs relate position and size of mask layers and primitive layouts. Strictly speaking, there is no floorplan for PDLPs and PDLBs, but either a mask layer arrangement (for PDLPs) or an arrangement of both mask layer and primitives (for PDLBs).

Both types of layout blocks, primitives and PDLBs, are commonly referred in the related literature as *macro-cells*. The first two steps of PDLP generation are: (1) to decide how many polygons are involved and which layer each of these polygons is made of and, (2), to devise an arrangement of the layout polygons (i.e., to devise a vertical and horizontal constraint graph). These two steps are critical to the flexibility of the parameterized layout of the primitive. This flexibility aspect concerns, first, the number of fabrication processes the layout can be ported to, and, second, the number of *variants* the primitive can implement. By way of example, Fig. 18 illustrates different variants of a MOS transistor: unitary, multi-fingered, serpentine, and annular.

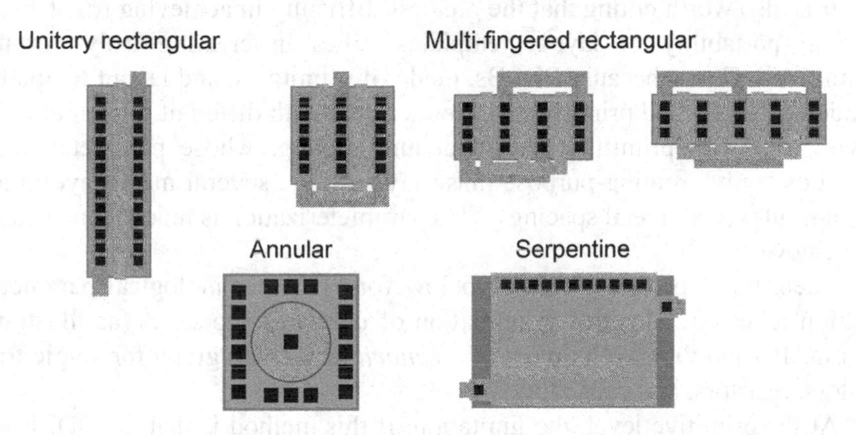

Figure 18. Several variants of the MOS transistor primitive.

In the following, we first deal with which reuse-related issues (mainly focusing our attention to technology migration) have to be taken into account at creating fully parameterized device-level primitives. Afterwards, how to implement these reuse-related concerns is explained.

7.1.1 Reuse: migration issues

For a primitive to be truly *portable* to a large number of different processes, it is necessary to examine in detail the implementation styles of the primitive and its variants in as many different processes as possible. Then, the PDLP generator is written according to not only the needed mask layers, but also to the complete set of design rules that may be involved. Fig. 19 illustrates a comparison of different implementation styles of polysilicon resistors in four different CMOS processes (from four different foundries), two of them featuring a 0.35-µm minimum gate length, and the other one belonging to a 0.25-µm minimum gate length process. As it can be noticed, there are 11 different layers involved, from which only two of them, namely the polysilicon and contact layers, are common to the four fabrication processes. Additionally, in some processes, there are several implementation styles for the polysilicon resistor (intended to obtain different values of resistance for the same geometric parameters of the resistor strip –width and length– and different nonlinear behavior of the resistance). Furthermore, the complexity of parameterizing such a structure increases as every mask layer involved has its own set of design rules, which relate it to the rest of the layers of the poly resistor and, in some cases, with other mask layers[19]. For instance, there are not less than 20 different design rules for the poly resistor structures shown in Fig. 19. It is thus manifest the complexity of achieving full portability of primitive devices.

It is also worth noting that the greatest difficulty in achieving true technological portability of layout templates relies in creating truly portable primitives. This is because PDLBs, made of primitives, and layout templates, made of PDLBs and primitives, no longer deal with different implementation styles, but with primitive placement and routing, whose parameterization requires "only" routing-purpose mask layers (e.g., several metal layers) and design rules (e.g., metal spacing). This parameterization is indeed much easier to achieve.

Then, the approach presented below for PDLP technological parameterization relies on a detailed examination of unrelated processes (as illustrated in Fig. 19) and the development of a *generic* constraint graph for single transistors, resistors, and capacitors.

At the primitive level, the limitation of this method is that the PDLP will only be successfully ported to any of the examined processes and, with a non-zero probability of arising migration problems, to related processes. Anyhow, most of these problems can be solved, as explained below, by adding new design rules and mask layers to the PDLP generator.

7.1.2 Reuse: retargeting issues

In order to improve the layout of any primitive device, there is a set of few basic, important rules to be taken into account. These rules can be collected from many sources (e.g., [Hast01] [Tsiv96] [Raza01], to name just a few) and have been made available at the end of this book in Appendix A. For PDLBs and templates, these rules involve device matching, loading and coupling

[19] In the case of the poly resistor, there are fabrication processes where any type of metal routing is not allowed over the resistor.

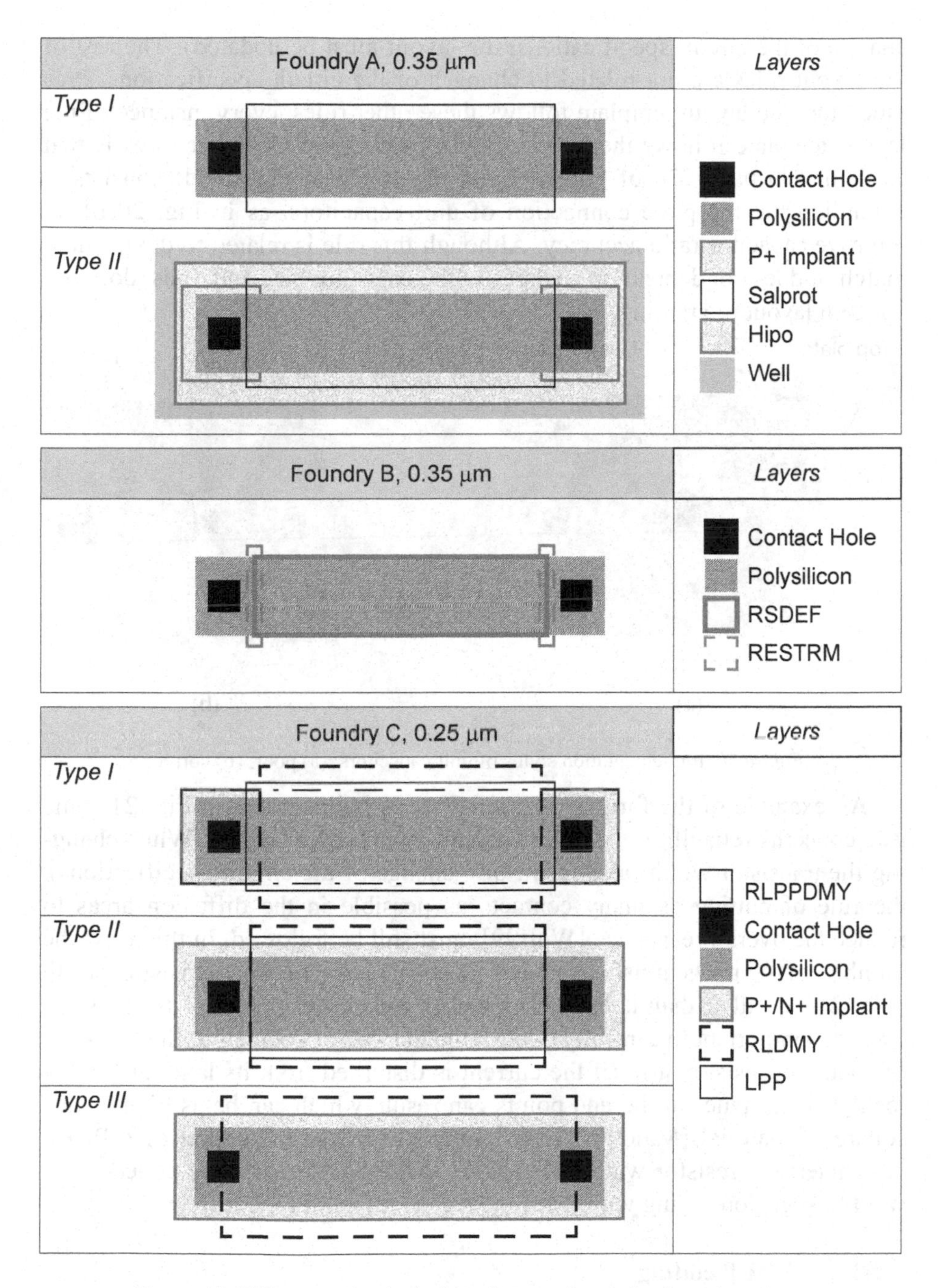

Figure 19. Different technological implementation styles for poly resistors.

effects, reliability, and area. We are here interested in those rules that need to be carefully "coded" when considering layout retargeting (i.e., when, due to a change of the circuit specifications, the layout must be updated). The rest of the layout rules are not related to changes of the circuit specifications. Provided that the layout template follows these other rules, every instance of the layout template follows them as well. An example of the latter rules is that explained in page 226 of [Tsiv96] and illustrated in Fig. 20. It consists in extending the top-plate connection of unit capacitors, as in Fig. 20(b), to improve capacitor ratio accuracy. Although this rule is related to device mismatch, it does not depend on changes in the capacitor size and, thus, does not concern layout retargeting.

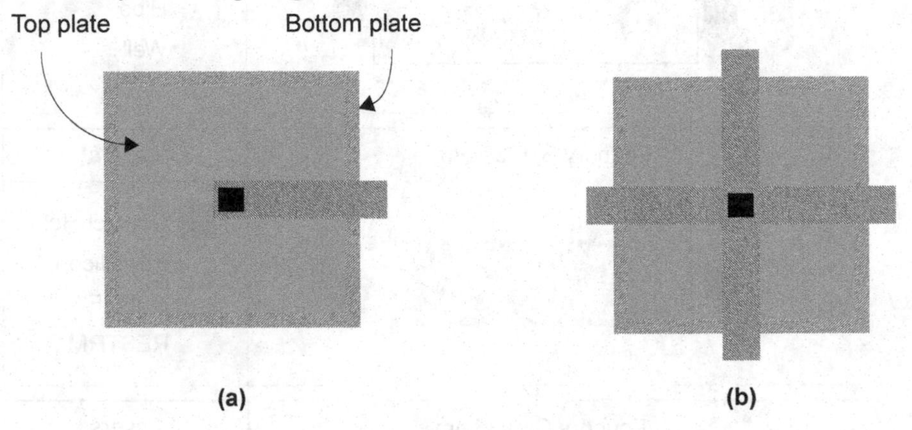

Figure 20. Implementation styles of unit capacitors: (a) poor; (b) correct.

An example of the former rules for PDLPs is illustrated in Fig. 21. This rule concerns reliability of MOS transistor and resistor layouts. When changing the transistor width (as a result of a change of the circuit specifications), the rule of putting as many contacts as possible in the diffusion areas to reduce the overall resistance [Wolf99] must still be followed. In this way, the number of contacts must be coded as a function of the transistor width [Fig. 21(c)] rather than keeping one single contact [Fig. 21(b)]. In the same way, it is important to carefully control the number of contacts at the two ends of each unit resistor strip (if the current is disturbed from its laminar flow, a localized resistance at the end points can result, which can be as high as one square of material [Malo94]). This number is again a function of a PDLP parameter: the resistor width. Therefore, it must be necessarily coded if we want to keep complying with the rule [Fig. 21(d) and Fig. 21(e)].

7.1.3 PDLP coding

The generator of the parameterized device-level primitive must be created considering all the reusability concerns explained above. The code shown in Fig. 12 (page 151) serves as guide to explain how to create such a generator.

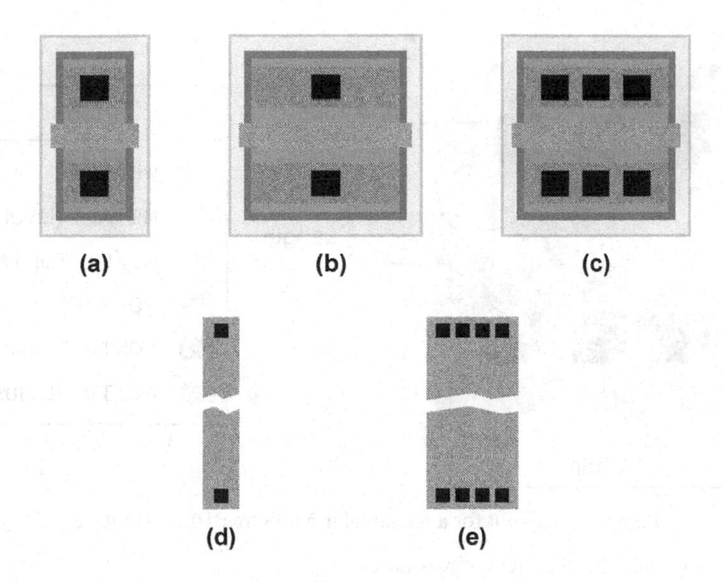

Figure 21. Reliability rules for MOS transistors and resistors.

To thoroughly illustrate this, let us consider the example of a rectangular MOS transistor, whose generic mask layer arrangement is shown in Fig. 22. Each layout rectangle is defined by two corner stitches.

The retargeting parameters for this basic example are the following:

- The channel **width** of the transistor.
- The channel **length** of the transistor.

Regarding technology migration, it is necessary to take into account all the mask layers listed in the legend in Fig. 22. Likewise, the related generic design rules to consider are those shown in Table 1.

An additional parameter to consider is the type of transistor channel. The required adjustments in the mask layer arrangement to accommodate this parameter are the following: if the specified type is PMOS, the N+ Implant and Well layers are used; for the NMOS type, only the P+ Implant is necessary[20]. With the retargeting, migration, and variant parameters completely specified, all the leaf components, which for this example are all rectangular mask layers, are detailed. Afterwards, the leaf components are to be placed and the terminals or connection points routed. Before this, however, it is helpful to devise the constraint graphs very carefully. These graphs are shown in Fig. 23 and Fig. 24.

[20] This is valid for N-well processes. Equivalent adjustments can be made for P-well or twin-well processes.

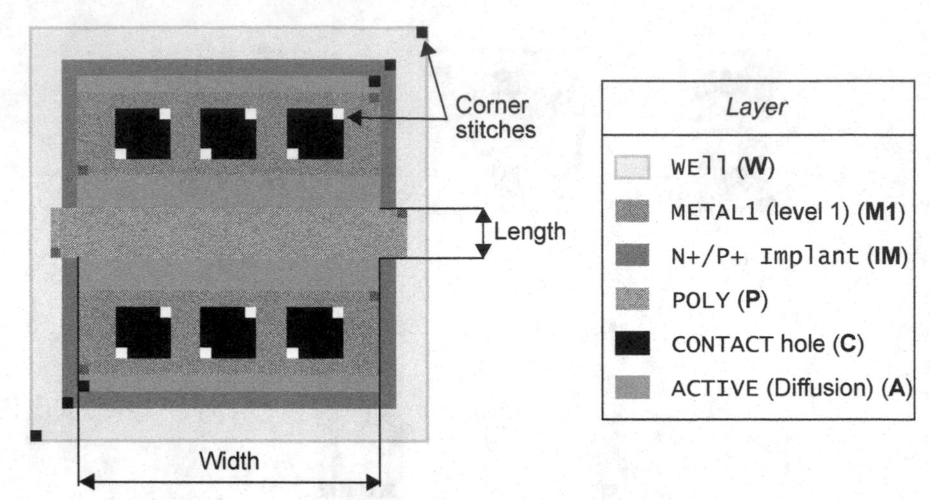

Figure 22. Layout for a rectangular MOS transistor primitive.

Table 1. Design rules for the MOS transistor.

Rule	*Description*
DR1	Minimum WELL enclosure of ACTIVE of a PMOS transistor
DR2	Minimum N+/P+ IMPLANT enclosure of ACTIVE
DR3	Minimum CONTACT spacing to POLY
DR4	Minimum POLY (gate) extension of ACTIVE
DR5	Minimum POLY width
DR6	Minimum METAL1 enclosure of CONTACT
DR7	Minimum ACTIVE enclosure of CONTACT
DR8	Fixed CONTACT size
DR9	Minimum CONTACT spacing

The description of the leaf components, for this case of primitive devices solely made of mask layers, is composed of the following items[21]

- Coordinates of the corner stitches (labelled as u, d, l, and r)
- Layer type, indicating a type of manufacturing material (labelled as LAYER).

---

21. For PDLBs and templates, the leaf component description is composed of the parameter link and the terminal definition. The parameters of a mask layer are the coordinates of its corner stitches and the layer type. Therefore, the leaf component description in PDLP generators corresponds to the listed items.

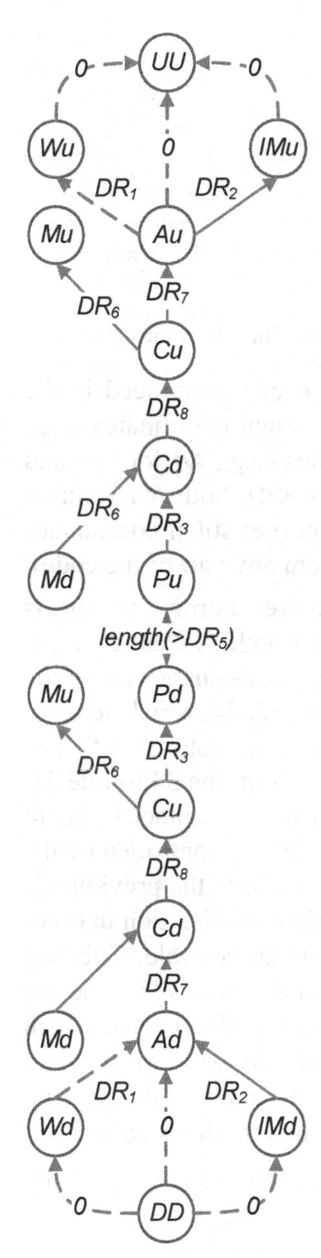

Figure 23. MOS transistor vertical constraint graph.

Therefore, the leaf component section of the MOS transistor PDLP code is as shown with the pseudocode of Fig. 25. Each layer has five main variables: four of them for the two corner stitches defining the position (on the 2D layout surface of the device layout) and size of the layer, and one last variable defining the manufacturing material or mask layer. In this code sample, the bold-type is used to differentiate the parameters of the PDLP. For retargeting, these parameters are **`length`**, **`width`**, **`variant`**. For migration, there are the mask layer names **`WELL`**, **`IMPLP`**, **`IMPLN`**, **`ACTIVE`**, **`GATE`**, **`CONTACT`** **`METAL1`**, and the design rules **`DR1`** to **`DR9`**. Whenever the PDLP is reused, it is necessary to specify the name, (for mask layers), numerical value (for the **`length`** and **`width`** parameters, and the design rules), and choice (for the **`variant`** parameter). As said above, migration parameters can be updated without user interaction and a technique to do so is explained below in Section 8.1.

It is important to note that, via the constraint graphs and a set of geometric and database procedures –that can be used both in the leaf component and the placement sections– it is possible to successfully develop any parameterized layout (i.e., any parameterized device-level primitive, block, or layout template).
The procedures are the following:

- **Stretching**: this procedure changes the size of a polygon. This change can be horizontal, vertical, or both. For instance, width of `WELL_LAYER` is controlled by the parameter `width`, whose change causes a horizontal (following the mask arrangement of Fig. 22) stretching or shrinking of the polygon. Stretching is not only used to change the size of an object[22], but also to move groups of objects while stretching others, which turns out to

[22] An object can be either a mask layer, a device-level primitive, a device-level block or a layout template.

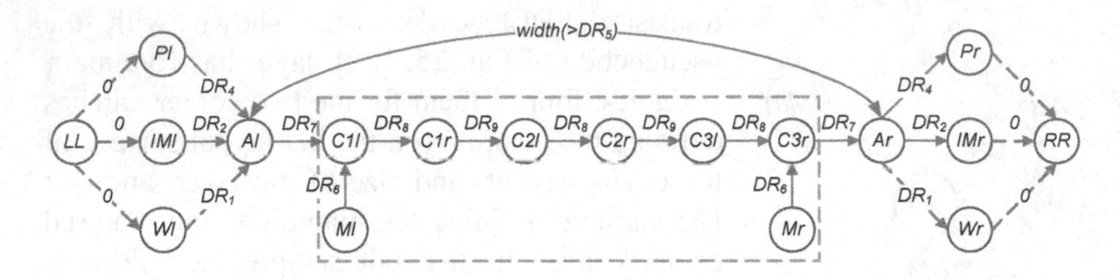

Figure 24. MOS transistor horizontal constraint graph (instance with 3 diffusion contacts).

be very useful in the placement phase. As it can be noticed in the pseudo-code of Fig. 25, some of the corner stitch coordinates (e.g., `Cd`) depend on other layers' stitch coordinates (e.g., `Ad` for `Cd`) and the retargeting parameters (e.g., `length` for `Cd`), both determining position rather than size. Therefore, all the corner stitch coordinates should be global variables (i.e., accessible from any part of the code).

- **Repetition**: a repetition procedure is used to create arrays of objects in the horizontal direction, vertical direction or both. A clear example is shown in Fig. 25 for the repetition of contacts, as suggested by the dash line of the horizontal constraint graph in Fig. 24, which actually represents a 3-contact instantiation of the MOS transistor PDLP. The repetition is implemented with two for-loops, from line 34 to line 51. The first loop performs a vertical repetition of the contact layer, to allocate contacts at both diffusion areas. This loop is controlled by the counter `ROW` (from 0 to 1). A second for-loop (within the previous) is used to repeat the contact layer along the horizontal direction in order to fill the diffusion area of as many contacts as possible. This has been stated for reliability purposes, as explained above. The counter is `COLUMN`, ranging from 0 to `NCOL`, the number of contacts, which has been previously defined (in the `PARAMETER` section) and is a function of the width parameter. This number can be easily deduced from Fig. 24, by noting the constraints translate into the equation

$$\mathtt{width} \leq 2 \cdot \mathtt{DR7} + \mathtt{DR8} \cdot \mathtt{NCOL} + \mathtt{DR9} \cdot (\mathtt{NCOL} - 1) \quad (6)$$

Therefore `NCOL` is:

$$\mathtt{NCOL} = \mathtt{floor}\left\lceil \frac{\mathtt{width} + \mathtt{DR9} - 2\mathtt{DR7}}{\mathtt{DR8} + \mathtt{DR9}} \right\rfloor^{23} \quad (7)$$

---

23. `floor(x)` rounds the element `x` to the nearest integer towards minus infinity. Additionally, it is possible to center the set of contacts with respect to the transistor gate, by adding `(width - (NCOL·(DR8+DR9)+2·DR7-DR9))/2` to `Cl`.

COND. INCL.

```
1--   if variant=="PMOS" then
2--      WELL_LAYER {▷ well mask layer leaf component
3--         do Wd    ←   max(DR1,DR2)-DR1
4--         do Wl    ←   max(DR1,DR2,DR4)-DR1
5--         do Wu    ←   Wd+DR1+DR7+DR8+DR3+length+DR3+DR8+DR7+DR1
6--         do Wr    ←   Wl+DR1+width+DR1
7--         do LAYER   ←   WELL
8--      }
9--      do IMPL_LAYER   ←   IMPLP
10-   else
11-      do IMPL_Layer   ←   IMPLN
12-   ▷ end variant if (*)
```

```
13-   N_P_IMPLANT_LAYER {▷ well mask layer leaf component
14-      do IMPd    ←   max(DR1,DR2)-DR2
15-      do IMPl    ←   max(DR1,DR2,DR4)-DR2
16-      do IMPu    ←   IMPd+DR2+DR7+DR8+DR3+length+DR3+DR8+DR7+DR2
17-      do IMPr  ←   IMPl+DR2+width+DR2
18-      do LAYER   ←   IMPL_layer
19-   }
20-   ACTIVE_LAYER {▷ well mask layer leaf component
21-      do Ad   ←   IMPd+DR2
22-      do Al   ←   IMPl+DR2
23-      do Au   ←   Ad+DR7+DR8+DR3+length+DR3+DR8+DR7
24-      do Ar   ←   Al+width
25-      do LAYER   ←   ACTIVE
26-   }
27-   GATE_LAYER {▷ well mask layer leaf component
28-      do Pd   ←   Ad+DR7+DR8+DR3
29-      do Pl   ←   max(DR1,DR2,DR4)-DR4
30-      do Pu   ←   Pd+length
31-      do Pr   ←   Pl+DR4+width+DR4
32-      do LAYER   ←   GATE
33-      do TERMINAL   ←   G}
```

REPETITION

```
34-   for ROW   ←   0 to 1
35-      for COLUMN   ←   0 to NCOL-1
36-         CONTACT_LAYER {▷ well mask layer leaf component
37-            do Cd   ←   Ad+DR7+(DR8+DR3+length+DR3)·ROW
38-            do Cl   ←   Al+DR7+(DR8+DR9)·COLUMN
39-            do Cu   ←   Cd+DR8
40-            do Cr   ←   Al+DR8
41-            do LAYER   ←   CONTACT
42-         }
43-         COLUMN   ←   COLUMN+1 ▷ end COLUMN for
44-      METAL1_LAYER {▷ well mask layer leaf component
45-         do Md   ←   Cd-DR6
46-         do Ml   ←   Al+DR7-DR6
47-         do Mu   ←   Cu+DR6
48-         do Mr   ←   Cr+DR6
49-         do LAYER   ←   METAL1
50-         do TERMINAL   ←   D,S}
51-   ROW   ←   ROW+1 ▷ end ROW for
```

(*) Comments are preceded by a ▷ symbol.

Figure 25. Leaf components declaration of the MOS transistor code.

In the same way, the `METAL1_LAYER` must be vertically repeated to complete the contact covering at both sides of the gate. Note that this layer is stretched according to the number of contacts (i.e., the width of the transistor).

- **Parameterized Layer**: this procedure is vital to implement the technology independence property of layout templates. It consists in aliasing the manufacturing material each polygon is to be made of. Since each fabrication process may have a different identifier for the available mask layers, it is possible to use the PDLP by simply assigning each mask layer the corresponding layer alias in the technology where the PDLP is being instanced. For instance, the `N_P_IMPLANT_LAYER` in the code sample of Fig. 25 is parameterized in order to act as P+ or N+ implant layer to implement PMOS and NMOS transistors respectively.
- **Conditional Inclusion**: this geometric procedure is used to include or exclude an object (a mask layer in this case) from the instance of a PDLP. In the code sample of the MOS transistor, the procedure is represented by the `if` statement. Only when the selected MOS type is P the `WELL_LAYER` must be present and, accordingly, the `IMPL_LAYER` becomes the corresponding one. Therefore, it is possible to further extend the number of technologies where the parameterized template can be ported by applying conditional inclusion procedures on layers making up several variants of a device-level block (e.g., resistors in Fig. 19).

A multi-fingered transistor[24] can be straightforwardly coded by just repeating the vertical constraint graph in Fig. 23 to form the transistor *stack*, as shown in Fig. 26, and add the metal and poly mask layers to connect drains, sources, and gates.

Terminals or connection points are defined over the entire mask layer (see Fig. 25). In this way, the `GATE_LAYER` leaf component becomes the gate terminal –line 33– whereas the pair of metal strips are the drain and source terminal –line 50.

Neither placement nor routing should be required for primitive parameterization, since, first, the leaf components are straightforwardly placed in the leaf components section and second, there are no components to be interconnected. Finally, pins can optionally be placed at the previously defined

---

24. Multi-fingered transistors can be also regarded as parameterized device-level blocks, since they are actually made of transistor primitives. It is, however, easier to code a multi-fingered transistor from mask layers than from transistor primitives.

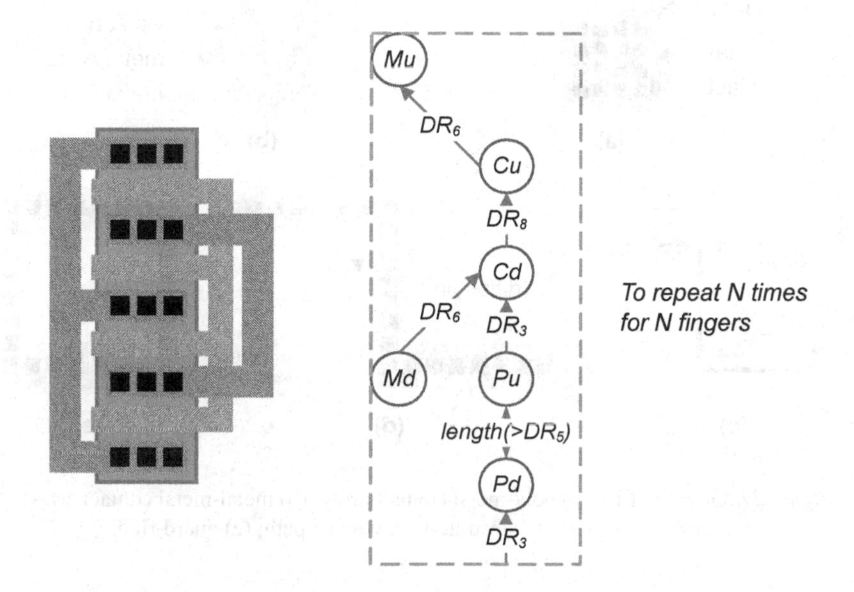

Figure 26. Vertical constraint graph of a transistor stack used for multi-fingered transistors.

terminals. The specific procedure to define a pin depends upon the final design environment where our parameterized layouts are generated.

The previous PDLP example is a component of the layout templates at the lowest level of the layout hierarchy (the leaf components are all plain layout rectangles). Apart from the unit MOS transistor, resistor, and capacitor, there is a set of other PDLPs, which are very useful components in order to obtain completely reusable blocks at higher levels of the hierarchy. This set is illustrated in Fig. 27. It is composed of:

- A contact array (poly, metal, and contact mask layers and related design rules as migration parameters; the number of contacts in the vertical and horizontal direction as retargeting[25] parameter).
- A via array (via and interconnected metal mask layers, and related design rules as migration parameters; the number of vias in the vertical and horizontal directions as retargeting parameter).
- Contact paths, whose conceptual layout realization is shown in Fig. 27(c) (the retargeting parameters are the coordinates of each corner point) for substrate and well contact (involved mask layers and design rules as migration parameters)

[25] Strictly speaking, these PDLPs are not retargeted since they are not present in the circuit schematic. We preserve, however, this name for the sake of clarity.

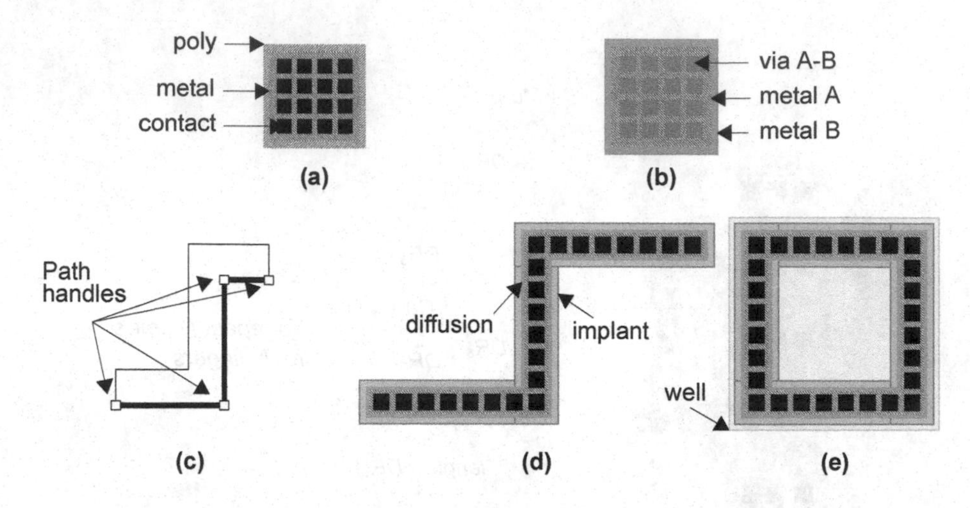

Figure 27. Other PDLPs: (a) poly-metal contact array; (b) metal-metal contact array; (c) conceptual path; (d) substrate-well contact path; (e) guard-ring.

- Guard-rings (the retargeting parameters are the guard-ring height and width; the migration parameters are that of the substrate/well path).

An example library of PDLPs is shown in Fig. 28. It envisages different types of passive devices [Gray01], the unit MOS transistor, and the contact PDLPs explained above.

7.2 Device-level layout generation: blocks

The parameterized device-level blocks are layout blocks used to compose functional circuits, such as opamps, comparators, bandgap references, etc. Fig. 29 illustrates two examples where PDLBs are used. The clusters of devices highlighted (two capacitors in Fig. 29(a) and two current mirrors, a differential pair, an a two-transistor stack in Fig. 29(b)) are parameterized layouts which, individually shaped and placed, turns the assembly of the circuit layout into an easier task. Besides, these blocks can be designed to improve device matching, to minimize loading and coupling effects, to enhance design reliability, and to optimize the area occupation.

As said above, these blocks are usually referred in the related literature as *modules*, *macro-cells*, or *structural entities*. As with layout primitives, in this section we will firstly focus on those features of the analog layout that should be preserved when dealing with either a performance retargeting or a technology migration. Then, the coding of these few but fundamental features will be explained.

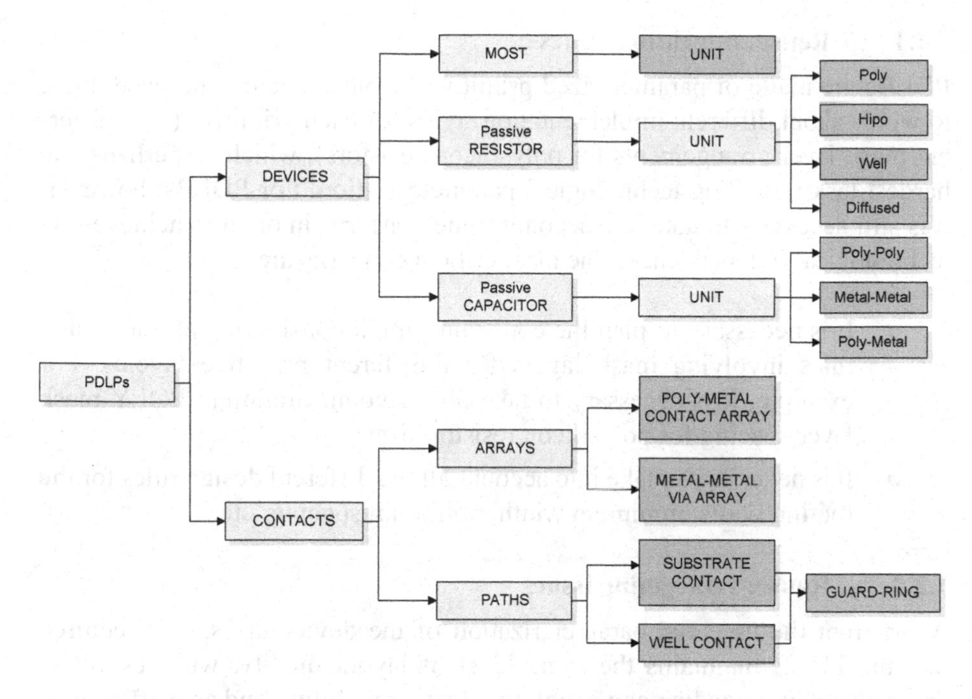

Figure 28. Example library of parameterized device-level layout primitives.

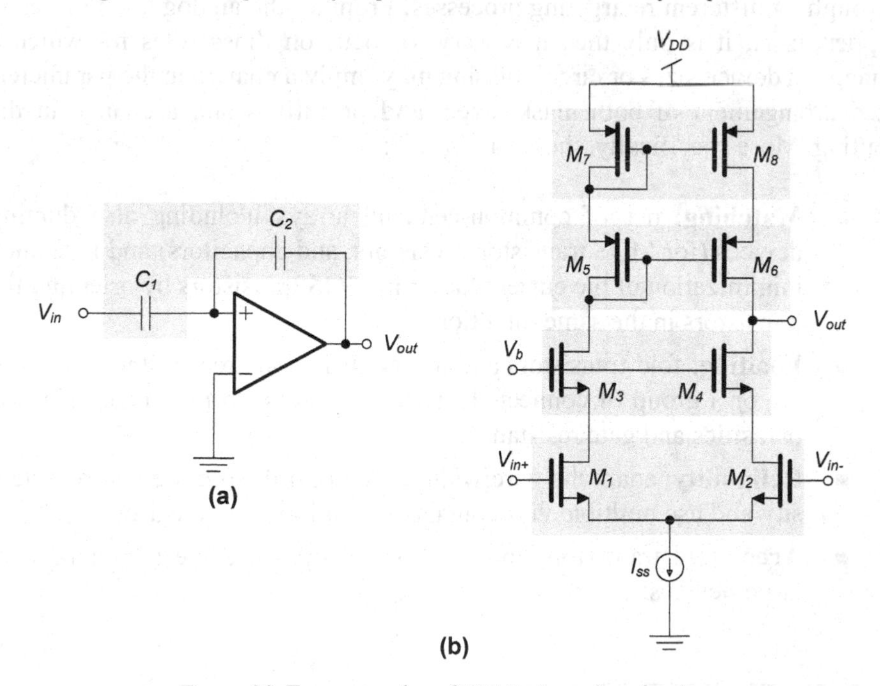

Figure 29. Two examples of PDLBs in analog circuits.

7.2.1 Reuse: migration issues

PDLBs are made of parameterized primitive layouts. There is no need, then, to worry about different implementation styles for each primitive (e.g., different mask layer arrangements for polysilicon resistors), which is, perhaps, the hardest task regarding technological parameterization. For PDLBs, however, it is still necessary to take into account some concerns in order to achieve true technological independence. The most critical concerns are:

- It is necessary to plan the constraint graph considering all the design rules involving mask layers from different primitive layouts. For example, it is necessary to take into account minimum POLY mask layer spacing for polysilicon resistor arrays.
- It is necessary to take into account all the different design rules for the routing wires: minimum width, minimum spacing, etc.

7.2.2 Reuse: retargeting issues

Apart from the essential parameterization of the device sizes, it is required that the PDLB maintains the same level of layout quality, with respect to device matching, loading and coupling effects, reliability, and area efficiency, throughout different retargeting processes. From all the analog layout rules in Appendix A, it is only then necessary to focus on those rules for which a change in device sizes or circuit biasing may imply a change in the parameterized arrangement of both mask layers and primitives and a change in the routing. More specifically, these rules are[26]:

- **Matching**: use of common-centroid arrays including also dummy devices (for MOS transistors, resistors, and capacitors) and mismatch minimization of the current factor in MOS transistors by orienting the transistors in the same direction.
- **Loading**: fold transistors and merge diffusion areas (either for a single or a group of connected MOS transistors) to minimize diffusion parasitics and gate resistance.
- **Reliability**: adapt the wire width to support the required current density and use multiple vias/contacts to minimize ohmic drops.
- **Area**: fold transistors and resistors to optimize the aspect ratio of large devices.

---

[26] These rules are explained in detail in Appendix A.

Adapting common-centroid arrays to new device sizes may imply changing the distribution of the arrayed components completely. Consider, for example, that the capacitance ratio of two capacitors must change to obtain different circuit performances. This new ratio may require partitioning the capacitors in a different number of unit capacitors and, therefore, generate a completely new common-centroid array. Therefore, whereas technological parameterization was the critical issue in PDLP generation even limiting the technology migration success of layout templates, now retargeting parameterization of PDLBs may limit the retargeting scope of layout templates. If PDLBs are not flexible enough to adapt new device sizes while using common-centroid arrays, folding, and merging techniques, and wire width adjustment, the layout template will not be, consequently, flexible enough to cover a wide range of circuit performances. In this case, circuit synthesis should be "constrained" to obtain only performances translating into sizes the layout template can implement.

7.2.3 PDLB coding

This section describes how the structure of Fig. 12 on page 151 is used to code a parameterized device-level block layout. In the *parameter* section, both the retargeting parameters and the migration parameters must be specified. As said above, retargeting parameters should include, not only device sizes (e.g., transistor width and length, or capacitance values), but also those parameters which help improving the flexibility of the block layout. Examples of these parameters are the number of transistor fingers and the current flowing through the routing wires. Migration parameters include every migration parameter of every leaf component in the PDLB as well as those of the PDLB itself and described above for leaf component placement and routing. Other variables that may be included here refer to the variant of the PDLB, such as the type of transistor stacking to be realized.

Since any PDLB contains both mask layers, mainly used for routing, and parameterized primitives, the leaf component definitions to insert in the PDLB code are for:

- Declaration of mask layers.
- Declaration of parameterized device-level primitives.

Mask layer declaration has already been explained and illustrated in Fig. 25. Fig. 30 shows two examples of PDLP declaration[27]. The first example

[27] Bold type is used for the PDLB parameters.

```
1--    LEAF COMPONENT#N {
2--        name MOST
3--        do MOST_width  ←  width
4--        do MOST_length  ←  length
5--        do MOST_WELL  ←  WELL
6--        ...
7--        do MOST_POLY  ←  POLY
8--    }
```

```
1--    LEAF COMPONENT#N {
2--        name POLYRES
3--        do POLYRES_width  ←  width
4--        do POLYRES_length  ←  R·(width-weff)/(Rsq·m)
5--        do POLYRES_RESLAYER  ←  POLY2
6--        ...
7--        do POLYRES_METAL  ←  METAL1
8--    }
```

Figure 30. Two examples of leaf component declaration in a PDLB.

corresponds to a MOS transistor primitive and the second one corresponds to a polysilicon resistor primitive.

The MOS transistor declaration includes the name of the primitive layout and the relation between the MOS transistor parameter and the parent block (our PDLB) parameter. These parameters include both retargeting parameters (lines three and four) and migration parameters (the rest). The POLYRES primitive is used to implement a horizontal array of vertical resistor strips, and it illustrates two important aspects. First, in line 4, the parameter relationship is now a more complex function of the parent block parameters[28]:

$$\mathrm{length_{strip}} = \frac{R \cdot (\mathrm{width} - w_{eff})}{R_{\square} \cdot m}^{29} \tag{8}$$

with R being the total resistor width, width is the poly strip width, m is the number of strips, and $R_{\square}$, and w_{eff} are two technological (migration) parameters. The second important issue illustrated by the POLYRES declaration is that layer parameterization allows defining the correct correspondence between technologies. In this way, if the resistive layer used (**POLY2**) is different from the gate poly layer (which is usually **POLY1**), the layer assignment in line 5 allows the poly resistor to get the correct layer name.

---

28. For m resistor strips that are connected in series, the total resistance is $R=(m \cdot \mathrm{length_{strip}} \cdot R_{\square})/(\mathrm{width}-w_{eff})$.

29. Although not explicitly shown, $\mathrm{length_{strip}}$ must be adapted to the process layout grid. Besides, to comply with resistor matching, Rules A.1.2.3 and A.1.2.9 in Appendix A should be taken into account.

The symbol "← " in Fig. 30 corresponds, in fact, to a database procedure described below:

- **Parameter inheritance**: this is a database procedure that lets a leaf component inherit or use one or more parameter values from the parent block in which it is placed. This allows hierarchically generating nested parameterized layouts as well as maintaining complete control over all parameters.

Placement and routing of the PDLBs leaf components are done according to the corresponding vertical and horizontal constraint graphs. The same geometric and database procedures, namely stretching, repetition, and conditional inclusion, used for leaf component description of primitive layouts explained above, can be used here. Both for PDLP and mask layer leaf components, it is only needed to specify one of the corner stitches and the rotation value, since the remaining stitch is either set by parameter inheritance (i.e., the leaf component width and height is a function of the parent block parameters) or kept constant.

At this point, it is necessary to clarify the following concern regarding placement and routing of certain device-level blocks, on the one hand, and performance retargeting issues, described above, on the other. Parameterized generators for device arrays, such as one- or two-dimensional common-centroid structures, or simpler one-dimensional interdigitized device stacks, can be regarded as two-component programs. A first component is an algorithmic sub-program that takes the circuit device sizes and the circuit netlist, and finds optimal values of the geometric parameters (e.g., number of fingers and the finger distribution) optimizing, at the same time, one or more analog layout concerns (e.g., device matching or parasitic and area minimization). For instance, common-centroid capacitor arrays need to attain, as best as possible, four geometric properties, namely coincidence, symmetry, dispersion, and compactness (see Appendix A for further details). In addition to the previous four properties, transistor stacks have to deal with proper unit transistor orientation to minimize the mismatch of the current factor (see Appendix A for further details). Fig. 31 illustrates this latter example. The algorithm takes transistor sizes and netlist, and returns the optimal number of folds for each transistor and the precise finger arrangement to implement an one-dimensional common-centroid array. Examples of dedicated array algorithms have been presented in [Gatti89] [Mala95] [Bruce96] [Peas96] [Basa96] [Naik99], for transistors arrays, and [Yúfe91] [Sayed02], for common-centroid capacitor arrays.

The second component is the layout generator itself. This sub-program takes the outputted values of the first sub-program (e.g., number of fingers

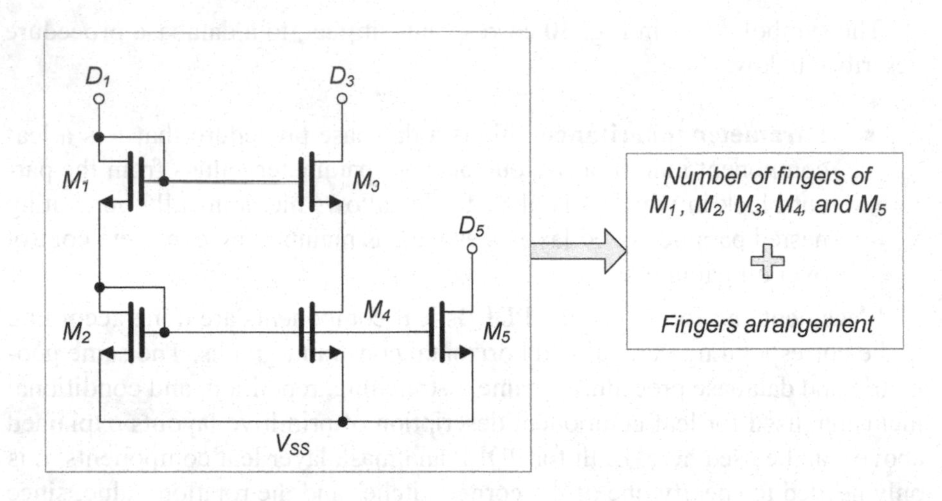

Figure 31. Transistor stacking algorithm: input and output.

and finger arrangement of the example above), and generates the corresponding layout of the device array, by placing and routing primitives and mask layers according to pre-defined constraint graphs.

From this point of view, the parameterized device-level block layout described here, takes the role of the second sub-program component. In this way, the PDLB retargeting parameters are designed to be the output parameters of the stacking algorithms. Besides, it is better to separate the array algorithm execution from the layout instantiation phase since the circuit sizing process may need reliable information on layout parasitics. To accelerate this sizing process, it might be required to avoid actual layout generation, but, instead, to obtain the output information from the algorithm execution. Therefore, both components, algorithm and instantiation of the PDLB layout, should be better separated. Sometimes, however, it is possible to devise a placement such that there is no need to carry out a previous algorithmic optimization, but rather rely on the parameterization of placement and routing to comply with retargeting issues, such as device matching and loading effects.

In any case, either if the PDLB needs a previous algorithmic optimization or if the PDLB is parameterized directly from the device characteristics[30], three of the four geometric and database procedures explained above, namely stretching, conditional inclusion, and repetition, can be used for placing the PDLB leaf components.

---

30. It is also possible to create a PDLB for device arrays without any supporting algorithm. In this case, the flexibility of the parameterized layout might not be as optimal. For instance, it is always feasible to create a fixed-ratio common-centroid capacitor array. Thus, capacitance ratios different from the implemented one are not feasible.

For the parameterized routing of the leaf components of a PDLB, two different methods can be used. The first consists in using rectangular[31] routing mask layers, such as different levels of metal layer or polysilicon layer, together with contact and vias PDLPs (see Fig. 28) to compose each segment of each routing wire in the layout. In such a case, a S-shaped routing wire should be composed of three segments plus starting and ending contacts/vias, as illustrated in Fig. 32(a). Besides, each mask layer should be defined by two corner stitches. The drawback of this method is the increased parameterization complexity since every wire segments as well as every contact and via should be completely specified. The second method is a typical programming resource to reduce the complexity of any code, and it consists in capturing all the elements of a routing wire into a single function. As illustrated in Fig. 32(b), a routing function lays out a shape that, as with PDLP paths, is defined by a number of handles determining the "flow" of the routing wire. Besides, it also includes contact or vias primitives at its starting and ending points.

Either by using a pre-coded routing function or several segments and via/contact primitives, in order to facilitate performance retargeting (reliability above all) and technology migration, the routing must be performed taking the following aspects into account:

- The mask layer definition (name) must be parameterized for correct technology migration. Here it is important to notice, as said above, that, since the number of routing layers change from one technology to the next, low-level metal layers (e.g., **METAL1** and **METAL2**) should be better used. Although using several metal levels should compact the layout, it also restricts the migration scope to those processes with equal or higher number of metal levels[32].
- Wire width must be parameterized for two reasons: first, to comply with minimum width design rules; second, to prevent electromigration from happening in order to attain a reliable circuit. Thus, the width of the wire must be adapted to accommodate the current density flowing through the wire. This reliable width, $W_{\text{layer}}^{\text{rel}}$, is computed as:

$$W_{\text{layer}}^{\text{rel}} = \frac{I}{I_{max,layer}} \quad (9)$$

---

[31] In case of Manhattan layouts (where all shape edges are parallel to either the X or the Y axis), only rectangular shapes are permitted. Otherwise, 45-degree shape edges are admissible.

[32] It is however worth noting that migration usually takes place towards more modern processes with higher number of metal levels.

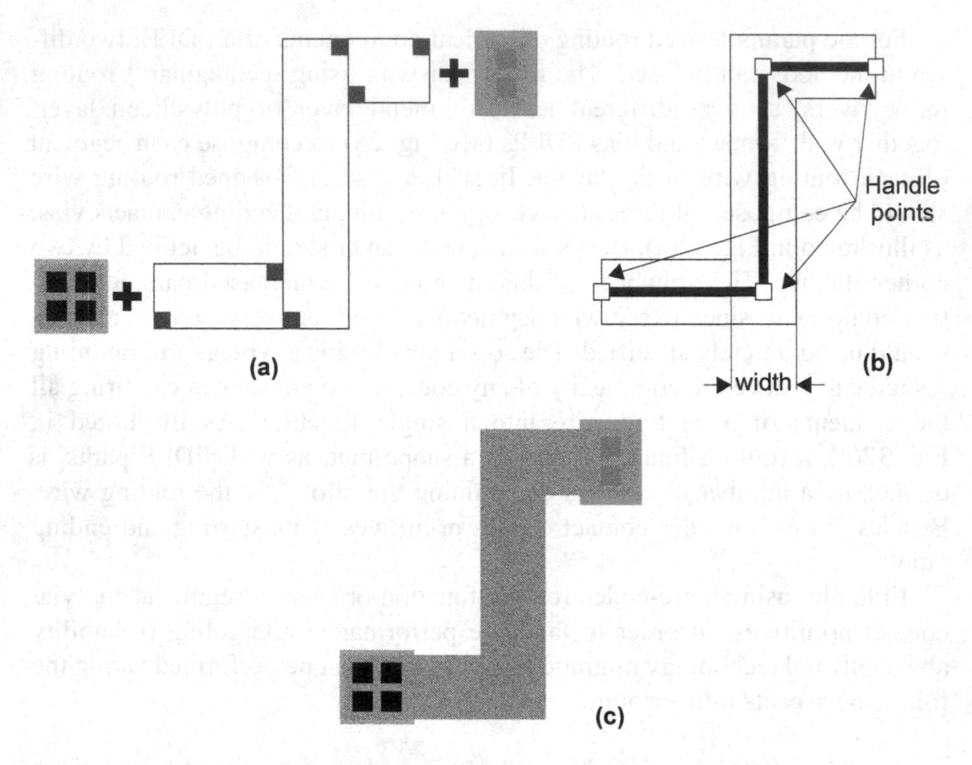

Figure 32. Two methods for leaf component routing: (a) segment by segment; (b) with routing functions; (c) example.

where I is the current to flow through the wire and $I_{max,layer}$ is the maximum current density of the routing mask layer, expressed in Ampere per meter. To comply with these two factors, the wire minimum width, W^{min}, must be computed as the maximum between the wire process minimum width, W^{min}_{layer}, and the reliable width in Eq. (9):

$$W^{min} = \max\left(W^{min}_{layer}, \frac{I}{I_{max,layer}}\right)^{33} \qquad (10)$$

- If the wire is to have a via/contact PDLP at one or both end-points, the number of vias/contacts making possible the flow of current between different geometries on different layers, must be, for the same reasons as the wire width, adapted to the current flow. Not only the current flow is unimpeded, but also the ohmic drop decreases if the number of holes (vias/contacts) is maximized. The minimum number of holes, N_{holes}, is given by the following expression:

33. Note that all layout dimensions must be multiple of the process grid.

$$N_{holes} = \mathtt{ceil}\left(\frac{I}{I_{max,hole\ type}}\right)^{34} \qquad (11)$$

with I being the flowing current and $I_{max,hole\ type}$ the maximum current in a via/contact connecting the different layers, usually expressed as Ampere per hole.

- Each handle point must be defined by the X and Y coordinates. These values are functions of the leaf component placement and pin positions.
- A good rule of thumb to ease parameterization of any template-based layout is that the wire encloses the via/contact PLDP completely. As illustrated in Fig. 33, parameterizing the L-shaped `METAL2` wire with respect to the bottom `METAL2` (to observe minimum `METAL2` spacing design rule) is harder because of the "bulge" in `METAL2` (which is caused by the design rule for minimum enclosure of the hole by the `METAL2` mask layer). It should be then necessary to account for the horizontal position of the L-shaped wire with respect the via, as illustrated in Fig. 33(a), (b), and (c). Enclosing the via completely, however, eases the parameterization since, as shown in Fig. 33(d), there is no need to consider different horizontal arrangements. The drawback is the increase of the parasitic capacitance to ground of the wire as a result of a larger wire area. If such an increase is not acceptable (probably because the wire is connected to a critical node), the best solution is to keep individual wire widths but consider, for spacing parameterization, that the wire width is the maximum of all segment and contact/via widths. This solution is illustrated in Fig. 33(e). Fig. 33(f) shows another critical scenario that can cause design rule violation since vias fail to comply with the minimum `METAL2` notch. The solution involves the automatic detection of such a situation, then "filling in" the notches with the corresponding mask layer, as shown in Fig. 33(g). The increase in parasitic capacitance is negligible as minimum notch values are very small.

In short, the routing function should consist in a multi-segment wire with parameterized mask layer name and optional starting and ending via/contact PDLPs. The routing flow is guided by $N_S + 1$ handle points, with N_S being the number of segments. Besides, the wire width and size of the via/contact PDLPs (i.e., the number of horizontal and vertical holes, of the primitive) should be either input parameters of the routing function, or automatically

---

[34] The function `ceil(x)` returns the nearest integer number towards plus infinite.

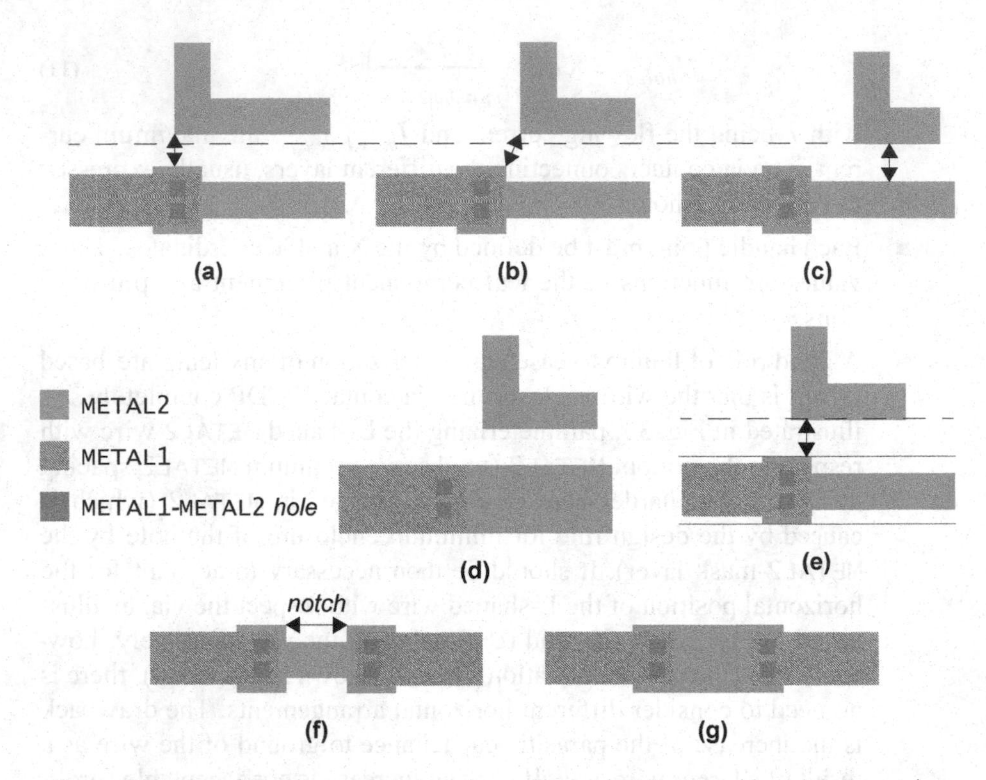

Figure 33. Rules of thumb: parameterize wire spacing as in (e) and avoid mask layer notches.

updated for new values of the flowing current, I, which becomes, thus, the input parameter. Finally, critical notches should be detected and fixed.

Figure 34 shows an example of a routing function call. The symbols "< >" means that the content is optional, and bold-type face is used for to highlight PDLB parameters. As it can be seen, each routing wire is defined by n handles (with two coordinates per handle). The `INIT_HOLE_type` and `END_HOLE_type` indicates the type of primitive to use (e.g., `POLY1-METAL1` contact array, `METAL2-METAL3` via array, and so on), and `size` refers to the number of horizontal and vertical holes. A relevant issue here is that all handle coordinates as well as maximum width of the wire (as explained in Fig. 33(a)), should be accessible for the rest of the PDLB in order to realize a relative parameterization of placement and routing. What is more, the starting and ending handles may be leaf component terminals, defined in the leaf components section of the code.

Figure 35 shows an example of routing wire. The wire, implemented in the `METAL1` layer, must sustain a 5mA current. Table 2 lists the maximum allowed current density for several routing layers and hole layers for a 0.35-µm process. From this table, the width of the wire example is 5µm, higher than the `METAL1` minimum width. If there is a `POLY1-METAL1` contact array

```
1--     routing_ID {
2--        <current  ←  I>
4--        layer_name  ←  LAYER_name
5--        <layer_width  ←   width>
5--        handle_1  ←  x_ID_1,y_ID_1<,INIT_HOLE_type><,size>
6--        handle_2  ←  x_ID_2,y_ID_2
7--        handle_3  ←  x_ID_3,y_ID_3
8--        ...
9--        handle_n  ←  x_ID_n,y_ID_n<,END_HOLE_type><,size>
10-     }
```

Figure 34. A generic routing function call.

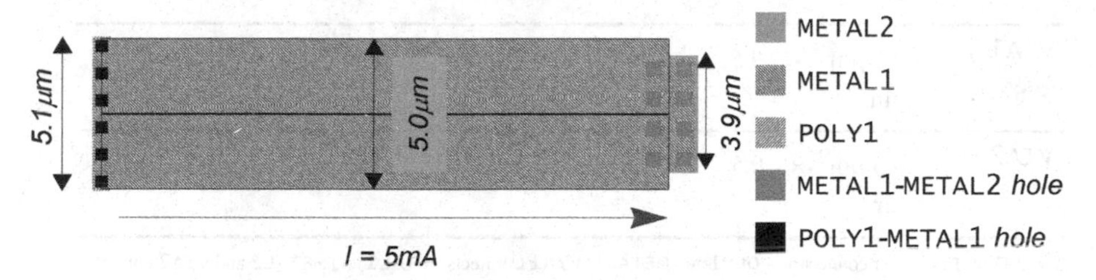

Figure 35. Example of routing function instance.

at the starting point, and a METAL1-METAL2 via array at the ending point of the wire, these must have six and eight holes, respectively. The exact shape of the arrays must be selected such that their widths approach the wire width. For the POLY1-METAL1 contact array, six holes can be arranged as 2×3, 3×2, 1×6, and 6×1, with the first number of each configuration establishing the array width. By using the array size equations in Table 2, the width of each configuration can be computed. It turns out that the best configuration of the POLY1-METAL1 contact array is 6×1, whose dimensions are $5.1\mu m \times 0.6\mu m$. For the METAL1-METAL2 via array, the best configuration is 4×2, whose size is $3.9\mu m \times 1.9\mu m$ [35].

To plan the routing phase, i.e., how many routing wires are necessary and their actual layout, it is useful to use classical stick diagrams [Saint02]. The geometric procedures for shape stretch and move, repetition, and conditional inclusion, explained in Section 7.1, are also used at this routing phase, to provide the PDLB with more and more flexibility. As shown above in Fig. 34, the database procedures for parameter inheritance and layer name parameterization are also used.

---

[35] It is also possible to increase the number of vertical holes to fill in the wire entirely. Thus, a 5×2 array, whose width is 4.9μm, could be used as well.

Table 2. Example of maximum allowed DC-current densities.

Mask layer	Minimum width	Array size	Maximum allowed current density
POLY1	0.3 μm		0.5 mA/μm
METAL1	0.4 μm		1.0 mA/μm
METAL2	0.5 μm		1.0 mA/μm
METAL3	0.5 μm		1.5 mA/μm
CONTACT[a]	0.4 μm × 0.4 μm	N_V·0.9-0.3 [b]	0.9 mA/hole
VIA1	0.5 μm × 0.5 μm	N_V-0.1	0.6 mA/hole
VIA2	0.5 μm × 0.5 μm	N_V-0.3	0.6 mA/hole

a. CONTACT is a layer connecting POLY1 and METAL1; VIA1 connects METAL1 and METAL2 and VIA2 connects this latter layer with METAL3.

b. We consider an array of $N_v \times N_h$ holes, with N_v and N_h being the number of vertical and horizontal holes, such that $N_{holes} = N_v \cdot N_h$ is given by Eq(11). The numerical factors account for the hole minimum size, the hole spacing, and the enclosure of the hole array by the two mask layers involved.

An example library of PDLB for designing analog circuit layouts, is shown in Fig. 36. The blocks on the right (filled with darker color) correspond to actual implementations of the device arrays. As said above, array generators can be built either for fixed device ratios (e.g., fixed capacitive ratios), which reduces the complexity of the generator, or for completely generic device ratios, which, in addition to increasing the complexity of the generator, requires a complementary algorithm to solve the device arrangement problem, as it was illustrated in Fig. 31.

The number of parameters of each PDLB is variable. The parameters may range from basic device sizes (e.g., type, width, and size of the transistors in a differential pair) to extra parameters such as custom position of the block pins, optional use of a guard-ring, optional use of dummies, biasing current, etc.

The next section completes the current analysis of layout template generation for analog and mixed-signal circuits.

7.3 Layout template generation

Layout templates are the procedural layout views of circuits defined at the cell, module, or even sub-system and system levels, of the AMS hierarchy. At

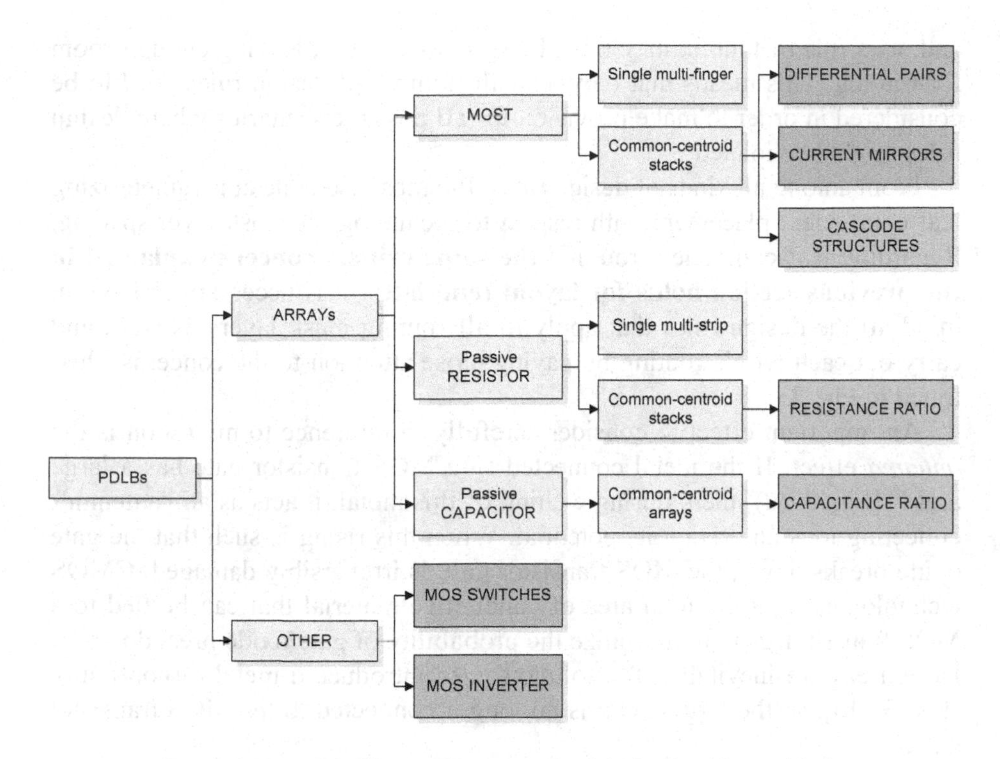

Figure 36. Example library of parameterized device-level layout blocks.

the cell level, the circuit layout template is made out of device-level blocks and primitives, and plain mask layers; at the module level, the layout template contains cell templates, device-level blocks, device-level primitives, and plain mask layers. The sub-system and system level circuit layouts are arranged in a similar fashion.

Next, the most critical issues to achieve fully reusable layout templates, regarding either performance retargeting and technology migration, are firstly discussed. Thereafter, the section covers the coding aspects of the layout template generator in order to reach that goal. Many of the aspects described in the previous two sections concerning both PDLPs and PDLBs, still apply for layout templates.

7.3.1 Reuse: migration issues

As we go up the analog layout hierarchy, technological parameterization of templates becomes a less complex task. In the two levels below, those of device-level blocks and primitives, technology migration may be rather involved since there would be several mask layers in a very small portion of area, therefore being necessary to consider quite a few design rules into account. For layout templates, however, most of the components are PDLBs, and, as a rule of thumb, they must be well spaced out, leaving enough room for routing. This means that only a small number of design rules need to be considered in order to make provision for all possible scenarios where design rules might be violated.

From among all kinds of design rules, the most used one at parameterizing leaf component placement with respect to technology is mask layer spacing. Regarding leaf component routing, **the same critical concern explained in the previous section holds for layout templates**: it is necessary to have in mind all the design rules that apply to all routing mask layers as well, and carry out each wire's routing by paying close attention to the concerns illustrated in Fig. 33.

An important effect to consider carefully in reference to migration is the *antenna* effect. If the metal connected to a MOS transistor gate has a large area[36] [Fig. 37(a)] then, during etching of this metal, it acts as an "antenna", collecting ions and rising its potential. When this rising is such that the gate oxide breaks down, the MOS transistor gate is irreversibly damaged. CMOS technologies limit the total area of conductive material that can be tied to a MOS transistor gate, to minimize the probability of gate oxide breakdown. If large areas are inevitable, the solution is to introduce a metal discontinuity [Fig. 37(b)], so the large metal is no longer connected to the MOS transistor gate.

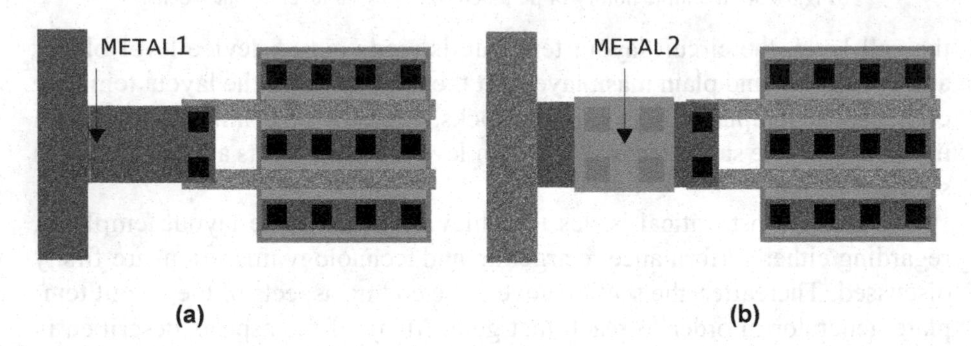

Figure 37. (a) Antenna effect susceptive layout; (b) metal discontinuity to avoid antenna effect.

Since the total permitted connected area changes from process to process, and the layout template designer knows there is a MOS transistor gate that might be connected to a large area of conductive material, then, the conductive discontinuity should be added to the technology parameterization. It is worth noting that the verification of the actual connected area of conductive

36. Recall that, unlike PDLPs or PDLBs, layout templates may occupy a large portion of area, therefore having their leaf components distributed and possibly interconnected with large routing wires.

material should be carried out after every change of both the fabrication process and the set of device sizes.

7.3.2 Reuse: retargeting issues

As layout templates are composed of device-level blocks and primitives, many of the analog retargeting concerns have already been considered and worked out as explained in earlier sections. At dealing with layout templates is, however, when valuable expertise both on circuit and layout design becomes essential. The guidelines to follow now are more of heuristic nature than before and, therefore, the construction of template generators must be closely guided by the coded rules capturing the designer's expertise. Therefore, in addition to the rules for analog layout design (available in Appendix A) that apply commonly to whatever the circuit performance is or the device sizes are (e.g., use dummy devices, or place devices away from power devices), there are other layout guidelines whose fulfillment implies a certain amount of expert knowledge of the circuit. These guidelines are the following[37]:

- **Loading**: a critical concern regarding parasitic elements is related to matched devices, with special emphasis on balanced, fully differential circuits[38]. Not only matched devices have to be as identical as possible, but also the layout-induced parasitics at each device node must equal each other. The technique that expert analog layout designers follow is to place and route blocks in a symmetrical fashion. That is, the devices of the matched networks are placed on either side of a symmetry axis and the routing of the symmetric nodes is made such that the routing wires on either side are mirror images. The circuit and the layout template in Fig.38 illustrate these points. The fully differential opamp has a horizontal symmetry "splitting" the circuit in two halves. Placement and routing of this opamp has to be performed, thus, following such symmetry axis, as illustrated in Fig. 38(b). An important early advice is that PDLBs have input and output pins

---

[37]. Matching, coupling and reliability-related issues, as described in Section 2, have all been considered when dealing with PDLPs and PDLBs.

[38]. Fully differential circuits are widely used because they have some advantages over their single-ended counterparts, as larger output swing and lower sensitivity to noise. Fully differential circuits are *balanced* if they are composed of two perfectly matched networks on either side of a symmetry axis, the outputs being identical with respect to ground. Even-ordered nonlinearities are not present in the differential output of a balanced circuit, but if the two networks fail to match each other, the advantages over single-ended circuits diminish [Gray01].

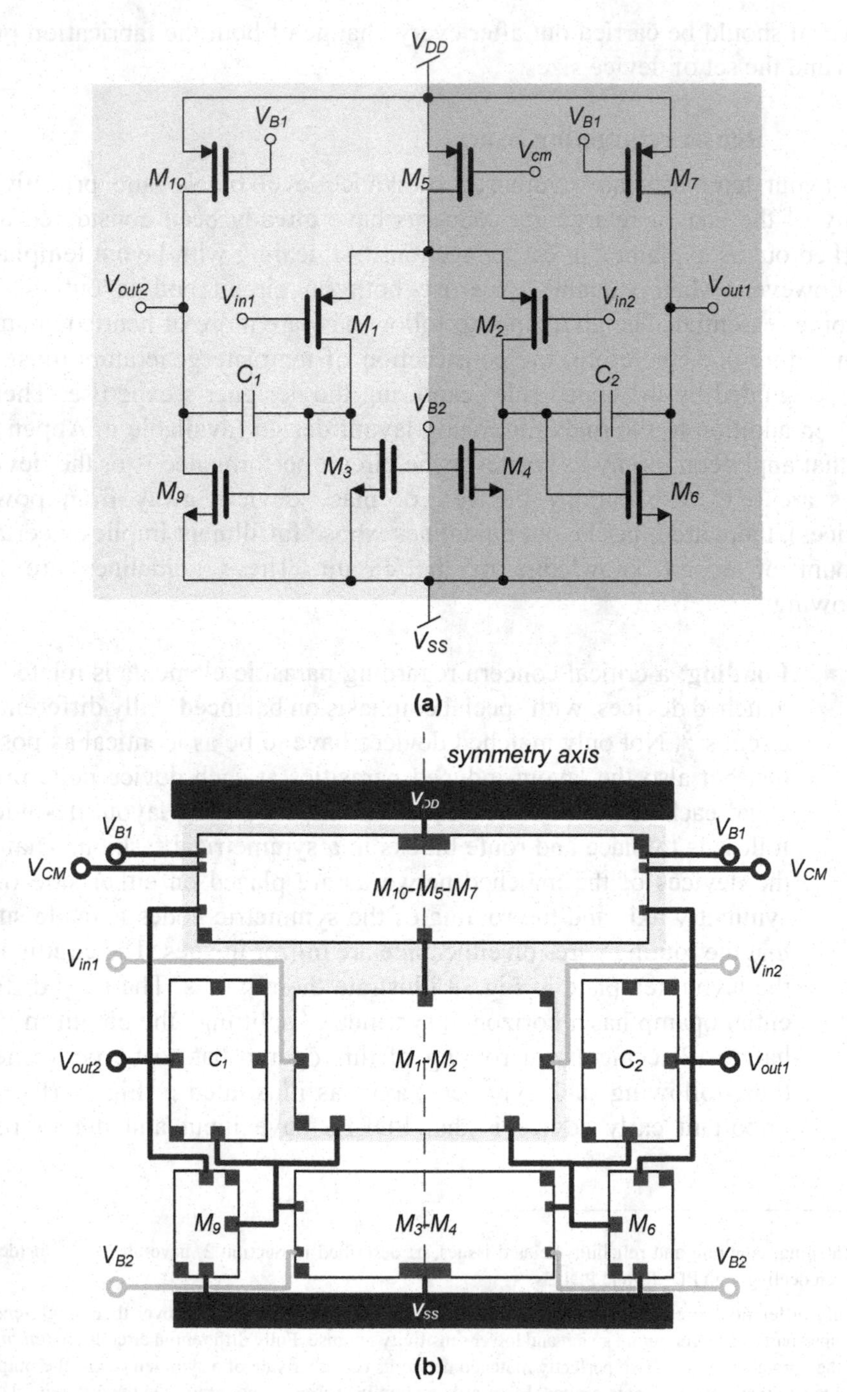

Figure 38. (a) A fully differential opamp circuit, and, (b), the abstract view of its layout template.

placed symmetrically, as the MOS transistor differential pair implementing M_1 and M_2 in Fig. 38. In this way, fully symmetric routing is easier to achieve.

- **Area**: another retargeting concern is maximization of the area density every time the layout template is instanced. Poor area densities imply a useless area when other layout templates are put together to make the whole analog or mixed-signal sub-systems. Note that such concern only does make sense if circuit specifications have been addressed. Since the positions of the template leaf components are relative to each other, certain values of the template parameters may yield poor area density figures, as it was illustrated in Fig. 15 on page 155. Parameterized placement has to be performed, therefore, in such a way that, for all parameter values, the resulting unused area is minimal. Design knowledge results, here, highly valuable. This knowledge can be used to specify, for instance, which devices in which leaf components are more sensitive to performance changes, so they are placed where its impact on the instanced layout is minimal. Fig.38 serves also as illustrative example. Consider that the opamp's performance scope is such that the width of transistors M_5, M_{10}, and M_7 ranges from a small to a large value. Consider also that, because of these width changes, the PDLB implementing the transistors also shrinks and enlarges horizontally in like manner. A solution is to "absorb" the impact that this PDLB has on layout by matching changes of this block to changes of other similar blocks or joint changes of a number of blocks. This solution is illustrated in Fig. 38(b): size changes in M_1-M_2 and C_1-C_2 as well as size changes in M_3-M_4, M_6, and M_9 may, altogether, equal the changes in M_5, M_{10}, and M_7, and, thus, maximize the area density.

7.3.3 Layout template coding

Writing the code of a circuit layout template generator, either at the cell, module, sub-system or system levels, implies gathering up all available design knowledge and transforming it into parameterized structures able to efficiently deal with performance retargeting or technology change processes. Part of such design knowledge is given with common-use layout rules for analog circuits, as described in Appendix A. Another type of design knowledge is circuit-specific and it entails designer's experience, typically accumulated over years. Methods to combine both sources of knowledge without excessively increasing the complexity of template generator coding, is described here. These methods make use of the same geometric and database procedures used with device-level primitives (PDLPs) and blocks (PDLBs).

For retargeting-related issues, namely loading and area concerns, a solution is to use *slicing floorplans* [Otten82] [Wong86]. This type of floorplan has been used for analog circuit layout [Koh90] [Conw92]. A slicing floorplan is obtained when the layout tiles are arranged such that the layout area is recursively divided into horizontal and/or vertical slices, as shown in Fig. 39(a)[39]. A non-slicing floorplan is shown in Fig. 39(b).

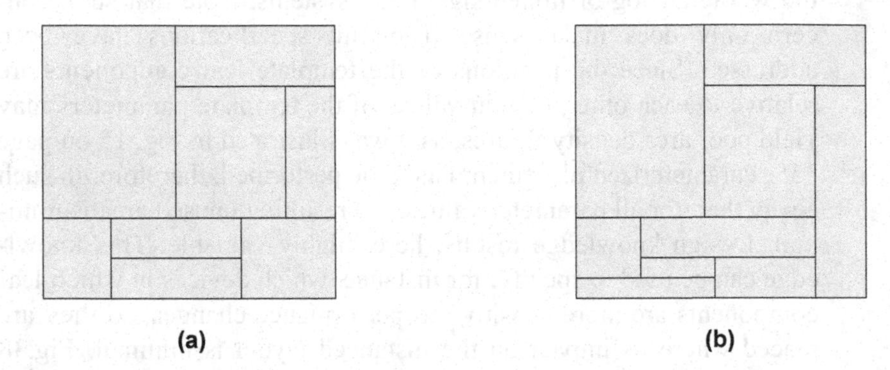

Figure 39. (a) A slicing floorplan; (b) a non-slicing floorplan.

Although non-slicing floorplans are a more general representation that can describe all kinds of tile packing, slicing floorplans have important advantages over non-slicing [Prie01], namely:

- **a.** Placement can be more easily specified by the relative positions of the layout tiles, since the hierarchy of slicing structures is better defined.
- **b.** It yields more compact layout instances.
- **c.** It also allows evaluating more easily other characteristics of the circuit layout such as routing[40].

Note that slicing floorplans are being considered only for parameterized layout of circuits at the cell, module, and sub-system levels. Since the tiles in a slicing floorplan are not allowed to overlap, this style could not be used at lower levels of the layout hierarchy.

Slicing floorplans ease parameterizing leaf component placement since positions can be specified relatively. Depending on the relative positions of the connected leaf components, leaf component routing, however, may still be rather complex. To further ease placement, and, more critically, leaf component

---

39. Routing channels are not shown.

40. This advantage is very useful at including parasitic extraction in the circuit sizing process, as it will be described in Chapter 7.

routing parameterization, a particular type of slicing floorplan is proposed here [Cast00]. This type of floorplan representation consists in arranging groups of connected cells into vertically stacked slices. The length of the routing wires is thus minimized. Furthermore, parameterizing them becomes easier since the wires are laid out by using a single routing channel and, most of the times, they are straight-line wires (i.e., one-segment wires). In case of connections required between leaf components of different slices, bringing the wire outside the slices and using vertical routing segments helps simplifying routing parameterization. For further reducing the parameterization complexity, leaf components connected only to leaf components in the same slice should be placed at the inside positions of the slice, providing that template symmetries are not spoiled. Fig. 38 and Fig. 40 below illustrate this type of floorplan. Input and output circuit connections as well as biasing connections (e.g., supply and bias voltages and bias currents) can also be accommodated by means of the side routing.

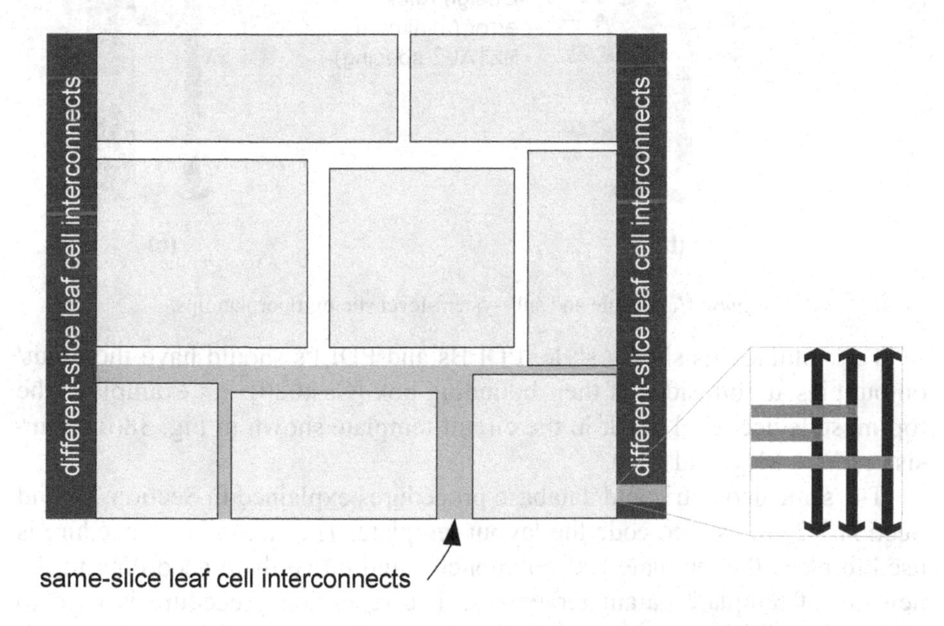

Figure 40. Slicing floorplan for easier leaf component placement & routing parameterization.

For module- or sub-system-level layout templates, it may be useful to place several cell-level templates of the kind above described side by side or one atop another. Interconnection of the cell-level template is carried out by using the vertical routing wires, as illustrated in Fig. 41(a). To avoid design rule errors during template reuse as in Fig. 41(b), and as long as increasing parasitics are not critical, a good solution is to widen the routing wire to accommodate enough room for mask layer spacing, as shown in Fig. 41(c).

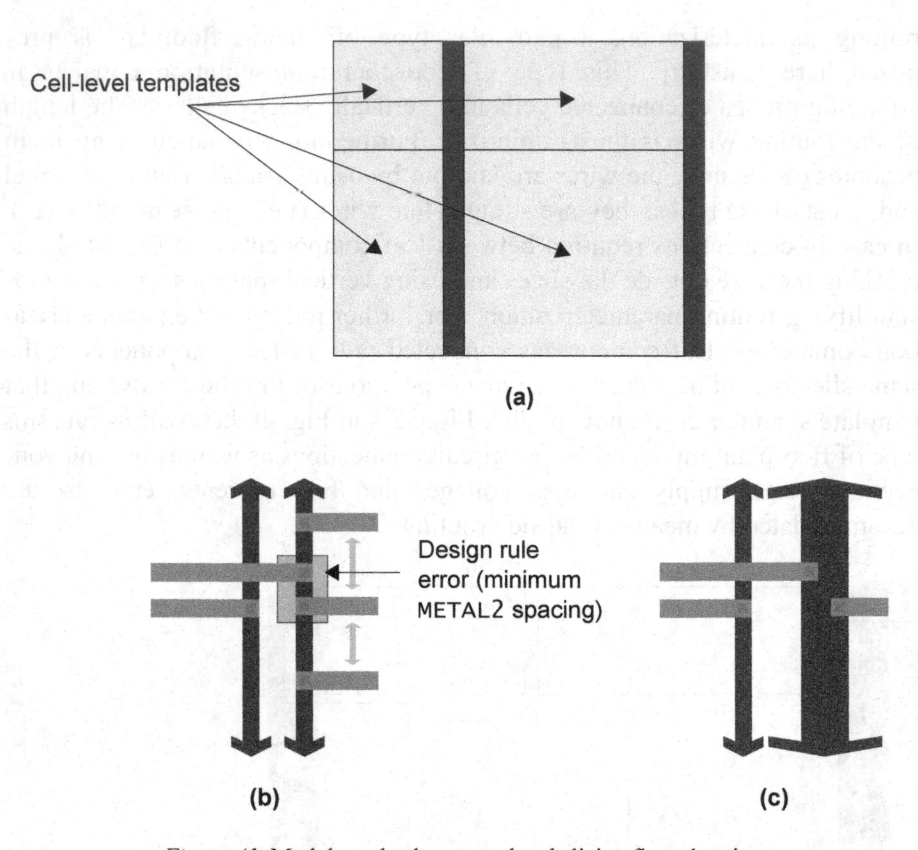

Figure 41. Module and sub-system-level slicing floorplan tips.

To facilitate this slicing style, PDLBs and PDLPs should have the input/output pins at both sides of their bounding box [Cast02b]. An example is the top-most device-level block in the circuit template shown in Fig. 38(b) (transistors M_5, M_{10}, and M_7).

The same geometric and database procedures explained in Section 7.1 and Section 7.2 are used to code the layout template. The stretching procedure is used to place the template leaf components and move them according to the new set of template parameter values. The repetition procedure is used to place many different instances of the same leaf component. With the conditional inclusion procedure, a leaf component can be selectively placed provided a condition holds. For instance, checking the area of the conductive material tied to a MOS transistor gate against the corresponding design rule can be used to conditionally include a discontinuity and, therefore, avoid the antenna effect described above. Mask layer names can be parameterized by means of the parameterized layer procedure. Finally yet importantly, the parameter inheritance procedure helps to define the template hierarchy.

To illustrate the layout template coding process, let us consider the example shown earlier in Fig. 38(a). First, the floorplan of the circuit is worked out with the help of designer's expertise and following the slicing floorplan style described above. The result has been shown in Fig. 38(b) and is reproduced here in Fig. 42.

Afterwards, vertical and horizontal constraint graphs can be designed to guide the placement and routing parameterization. Fig. 43 shows the graphs for this circuit opamp example. Constraint *center* in the horizontal graph is used to center each slice and, thus, preserve the horizontal symmetry of the circuit. For this, the distance from the left-most side (*LL*) of the template to the left side of the left-most leaf component at each slice, must equal the distance from the right-most side of the template (*RR*) to the right side of the right-most leaf component. W_i and H_i stand for width and height of the i-th leaf component, respectively[41]. Each of the remaining constraints c_1, c_2, c_3, and c_4 are defined to ensure compliance with the process design rules and to provide enough room for routing. For instance, the vertical separation between slices 1 and 2 requires that:

1. The separation between the mask layers in leaf components B_1, B_2 and B_3, B_4 has to comply with the corresponding minimum spacing rules. A good approach is to consider only the external mask layers of

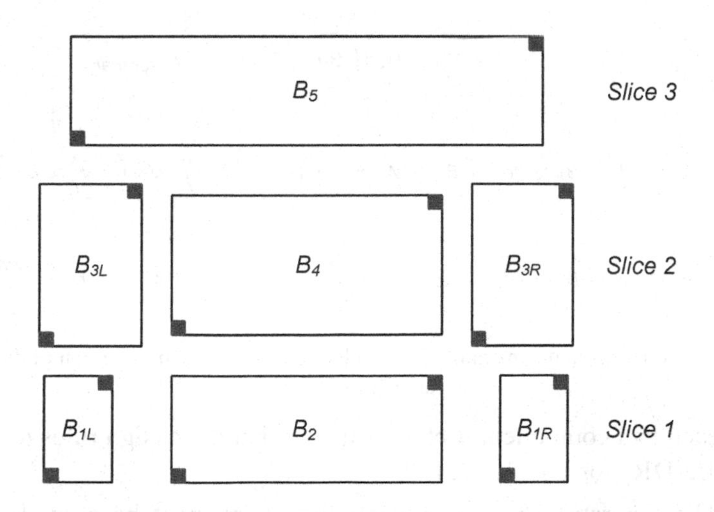

Figure 42. Layout floorplan of the opamp example in Fig. 38.

---

41. Recall that these are not true constraints for the *i*-th leaf component size, but are just used to compute the placement of other leaf components at the right or at the top of the *i*-th leaf component.

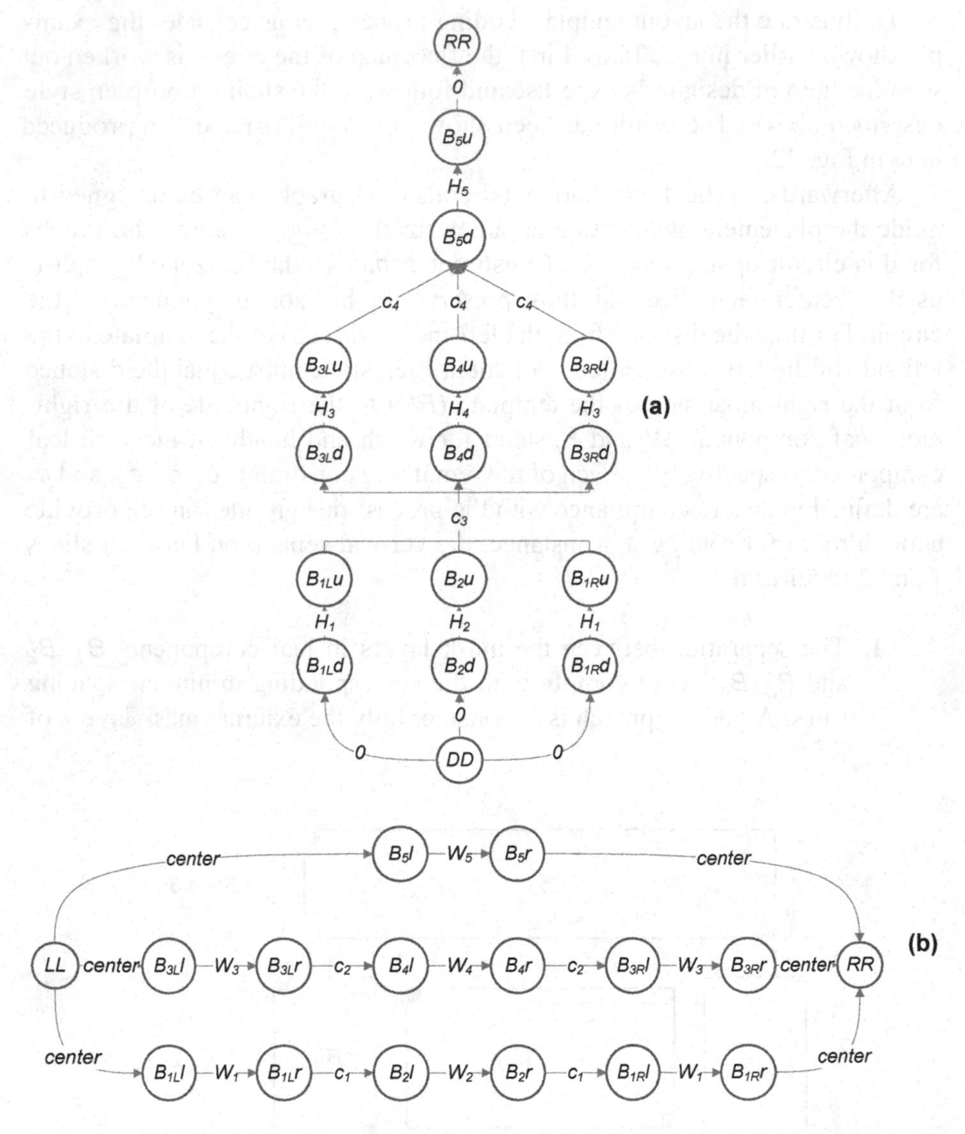

Figure 43. (a) Horizontal constraint graph and, (b), vertical constraint graph of the opamp example in Fig. 38.

each leaf component. Let us consider that the design rules to consider are DR_i for $i = 1, \ldots, n$.

2. The separation has to be such that there must be enough room to accommodate two horizontal metal wires, as illustrated in Fig. 44. Suppose that wire widths are w_1 and w_2, wire separation is w_s, and the minimum spacing distances from the metal mask layer to the

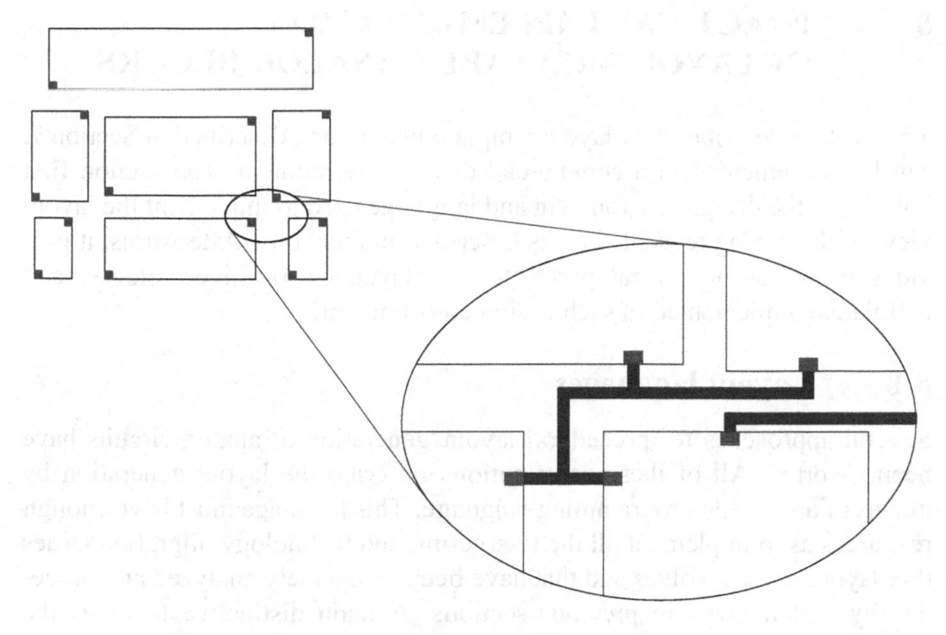

Figure 44. Coding the slice placement for the opamp example in Fig. 38: routing considerations.

external mask layers of the leaf components B_1, B_2, B_3, and B_4 are DR_{Bi}, for i = 1, ..., 4 (see Fig. 43(a)). Therefore, c_3 can be formulated as:

$$c_3 = \max(DR_1, \ldots, DR_n, H) \quad (12)$$

where H is:

$$H = w_1 + w_2 + w_S + \max(DR_{B1}, DR_{B2}) + \max(DR_3, DR_4) \quad (13)$$

Once the constraint graphs have been completely defined, it is straightforward to write the template generator. The various parts of the generator code are developed in the same way as for PDLPs and PDLBs. Routing functions can also be used to connect the leaf terminals.

So far, the template description has been completely independent of any circuit design environment where analog and mixed-signal circuits are designed partially or entirely. The making of layout templates as described in Sections 6 and 7, is likewise independent of the layout language, layout procedures, and layout tools such framework provides. The next section describes an implementation of the above described layout template parameterization and generation methods.

8 PRACTICAL IMPLEMENTATION OF LAYOUT-REUSABLE ANALOG BLOCKS

This section describes how layout template generators, described in Section 7, can be implemented in a commercial design environment. The section first introduces the design environment and language used to implement the layout view of the analog reusable blocks described in this book. Afterwards, it provides examples of several parameterized layouts (primitives, blocks, and templates) implemented in such design environment.

8.1 Layout languages

Several approaches for procedural layout generation of analog circuits have been reported. All of these contributions describe the layout generation by means of a versatile programming language. This language must have enough resources as to implement all the retargeting and technology migration issues that layout reuse involves and that have been extensively analyzed and procedurally systematized in previous sections. A main distinctive factor is the specificity of the programming language used. On the one hand, there are procedural layout approaches based on common-purpose languages such as C or C++. Examples are the high-level[42] languages CAIRO [Dess01a], GENLIB [Grei94], LAYLA [Lamp99], CYCLONE [Rant02], and MSL [Samp03]. On the other hand, specific-purpose languages are limited to a particular circuit design environment[43] such as BALLISTIC [Owen95], written in the Mentor Graphics® LX language for the GDT® environment [Ment90], or SKILL [Barn90] [Skill04], the programming language of the *DFII* environment from Cadence®.

Since one main goal of any design methodology is to achieve widespread acceptance from the design community, it is clear that specific-purpose approaches have more chances of success than common-purpose ones since design environments are commonly-used, well-known resources all over the semiconductor industry and high-level languages cannot be employed on such commercial design environments. Another advantage of working in a commercial design environment is interactivity, thanks to which the layout can be at any time customized by the designer to improve its quality.

---

42. A high-level language is a programming language which provides some level of abstraction above the low-level language (e.g., machine code, assembly language, or C/C++ language) in which it is based.

43. A design environment is a system for supporting the design, from system description to layout realization, of analog, digital, or mixed-signal circuits. As defined in Chapter 2, this system is composed of a number of CAD design flow tools managing the input/output data of CAD point tools.

For these reasons, we have used SKILL from the *DFII* environment to generate layout templates as explained in previous sections and hereby demonstrate the validity of the presented design reuse methodology at the layout level. In this environment, layout templates are based upon the concept of parameterized cell (pcell) [Virt00b] [Virt00a].

Coding a layout template into SKILL follows a similar structure as the one shown in Fig. 12 on page 151[44]. Layout templates can be created either by using a dedicated user interface where the geometric and database procedures are graphically applied to a collection of mask layers and other pcells, or by directly writing out the SKILL code of the pcell. The graphic method, however, may result rather involved for complex layout parameterization especially if technology migration is also a goal (a MOS transistor primitive parameterization requires more than 20 graphic operations). Writing SKILL code to bring about the same parameterization provides higher flexibility for creation of complex designs and an easier way to maintain and upgrade the layout template code.

In addition to the built-in SKILL functions to create parameterized layouts, several additional high-level SKILL functions were developed to further accelerate the writing of layout templates generator code. To give but a few examples, the functions are used to obtain the corner stitch absolute coordinates of the leaf components in the template (**GetLeafCellCS** function), intended to accelerate the coding of the constraint graphs of the layout template[45], to get the oordinates of a leaf component pin whose name is known (**GetPinCS** function), to procedurally perform routing just as described in Section 7.2.3 (**RoutingFunction** function), or to create a leaf component pin with a specific mask layer (**MakePin** function). A very important feature is that these functions allow to swap primitives or blocks (e.g., diffusion resistors to poly resistors) at no additional coding effort since the size and pin positions of the new leaf component are captured on-the-fly. This proves very effective at particularly difficult migration processes where changes in the devices are required.

The layout template SKILL file uses generic design rules as well as generic mask layer names, as it was illustrated in the code example of Fig. 25 on page 167. As shown in Fig. 45, a numerical value of each design rule and mask layer identifiers (e.g., layer names) must be supplied every time the

---

44. A slight difference, however, exists. Although parameter declaration and leaf component routing are likewise separate parts of the pcell SKILL code, leaf component description and placement are carried out at the same time. Nevertheless, this does not entail further loss of generality.

45. Otherwise, the generator designer should have to compute the corner stitch coordinates as a function of both the leaf component parameters and the migration parameters, which would imply to use and update the same function at all levels of the hierarchy where the leaf component is used.

template is instanced. Provided that the design rule *pairs* and the mask layer *pairs*, i.e., the generic design rule or mask layer name and the corresponding actual value at the targeted process, are both available, these can be automatically read off and each migration parameter correctly adapted every time the layout template is ported.

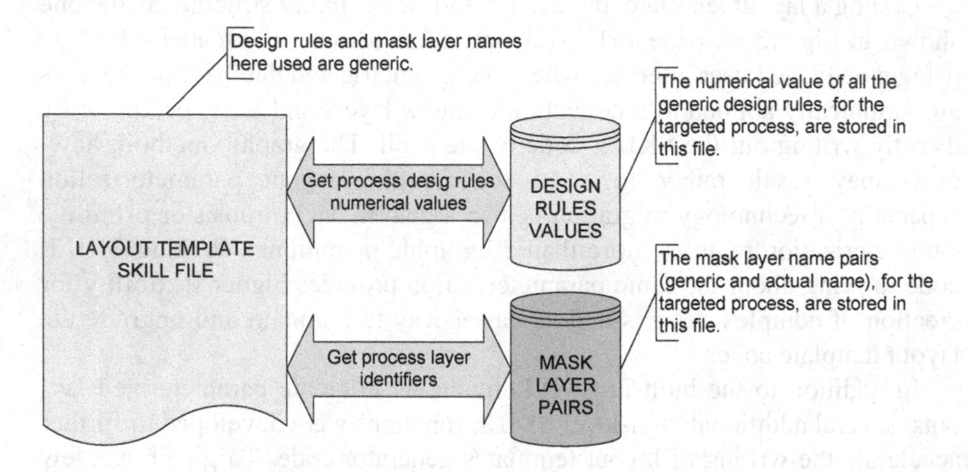

Figure 45. The migration parameter inputting process.

8.2 Implementation examples

The following are examples demonstrating the capabilities of the explained layout template approach to deal with the requirements of the design reuse methodology. The examples below show parameterized layout views implemented, as described in Sections 7.1, 7.2, and 7.2, in the *DFII* commercial design environment from Cadence®.

Figure 46 illustrates several examples of parameterized device-level primitives. Fig. 46(a) and (b) show a MOS transistor in a 0.35-μm technology, while Fig. 46(c) shows the same PDLP in a 0.5-μm technology. A folded transistor is depicted in Fig. 46(d). Two examples of guard-ring PDLPs are shown in Fig. 46(e) (the right-most guard-ring is used to enclose devices in a well on CMOS technologies). Several PIP capacitor primitives are shown in Fig. 46(f). A poly resistor (in a 0.35-μm technology) is shown in Fig. 46(g), while a *hipo* resistor (in a 0.5-μm technology) is depicted in Fig. 46(h).

Several and different instances of parameterized device-level blocks are shown in Fig. 47 to Fig. 51. Fig. 47 illustrates the MOS transistor differential pair for different technologies (i.e., different migration parameters) and different set of retargeting parameters (for the sake of simplicity, only the width, length, and multiplicity parameters are shown).

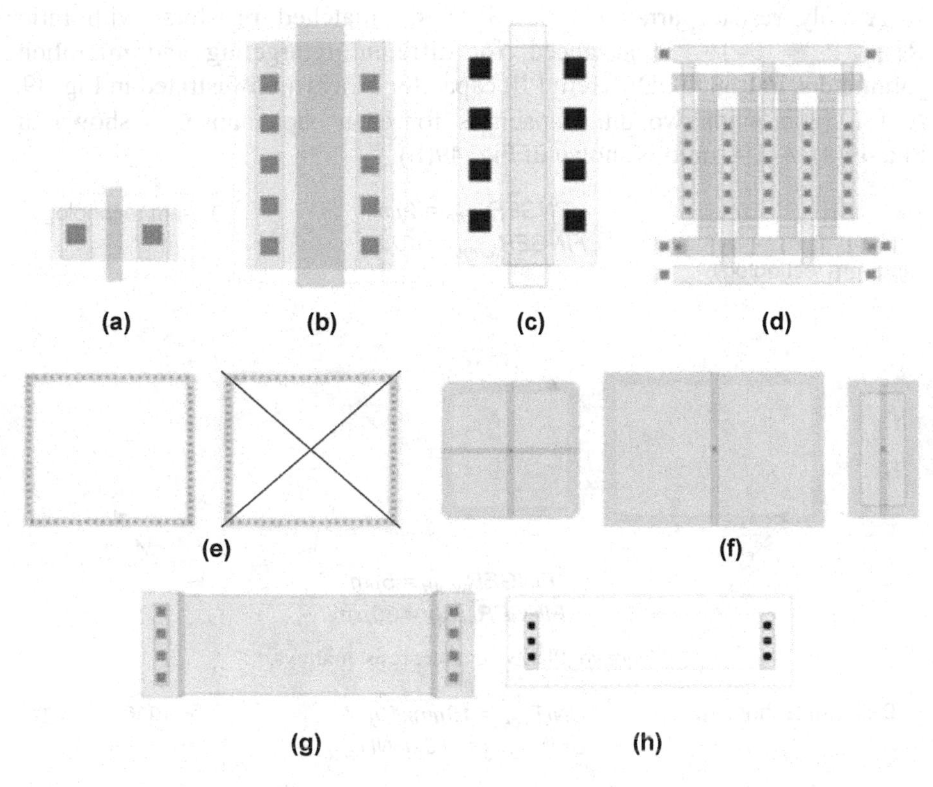

Figure 46. PDLP examples.

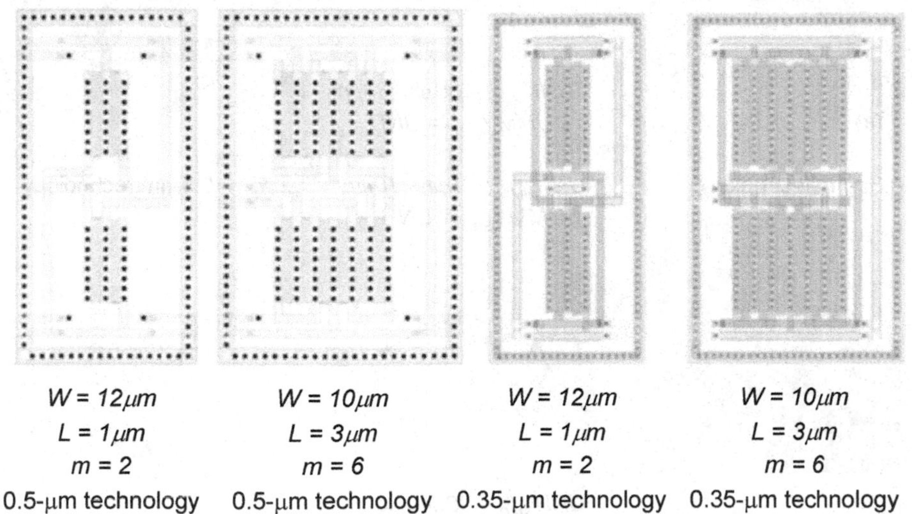

Figure 47. PDLB example: CMOS differential pairs.

A poly resistor array composed of six matched resistors, with ratio $R_{1-5}/R_6 = 1/16$, is instanced, for different retargeting and migration parameters, in Fig. 48. Different PIP capacitor arrays are illustrated in Fig. 49. A 1:1 ratio, with two unit capacitors for each capacitance, is shown in Fig. 49(a). A 1:4 ratio is shown in Fig. 49(b).

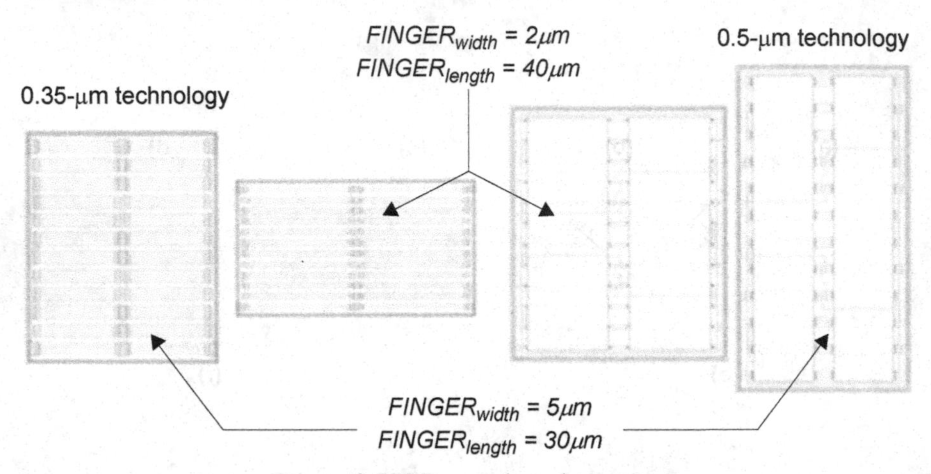

Figure 48. PDLB example: resistor arrays.

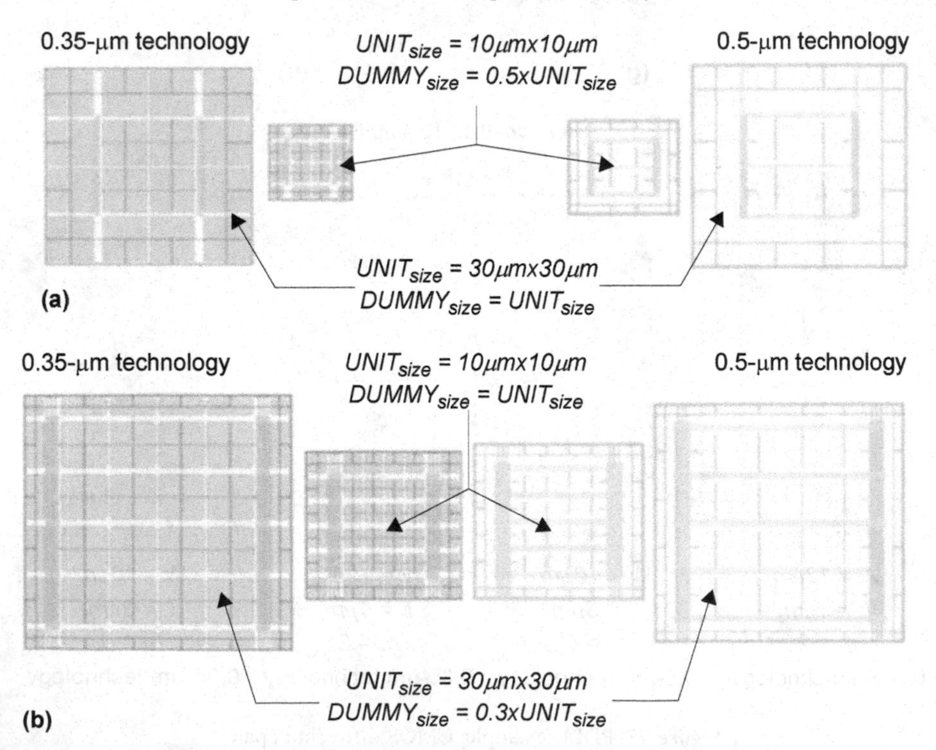

Figure 49. PDLB example: PIP capacitor arrays.

A PDLB implementing a MOS transistor cascode array, whose schematic is depicted in Fig. 50(a) (both for the NMOS and PMOS variants), is shown in Fig. 50(b). There are two variants for this PDLB. The first one, for all four transistors sharing the same number of fingers, consists in a one-dimensional common-centroid array, with two transistor stacks separated by a backgate contact. The second variant, for a differing number of fingers, consists in three different transistor stacks: the center stack implements M_1 and $M_{1'}$, whereas the stacks at both sides implement, respectively, M_2 and $M_{2'}$. Note the symmetric position of the PDLB pins to facilitate template placement and routing as explained in Section 7.3.3 and illustrated in Fig. 40. Instances with different retargeting and migration parameters are shown in Fig. 50(c).

A similar PDLB, this time implementing the MOS transistor current mirrors in Fig. 51(a), is shown in Fig. 51(b). The block has only the 1-D common-centroid variant. Several PDLB instances are shown in Fig. 51(c).

Last but not least, the following two figures illustrate layout templates. Fig. 52 shows several instances of the layout template of the CMOS comparator in Fig. 52(a). The instances correspond to different technologies as well as to different retargeting parameter values.

The layout template in Fig. 53(b) implements the fully differential opamp core[46] in Fig. 53(a). The shaded devices in the opamp schematic are the leaf components of the opamp layout template, as indicated in the figure. The layout floorplan follows the guideline explained in Section 7.3.3. This opamp layout template has been instanced for different technologies and different values of the retargeting parameters [Fig. 53(c) and (d)].

The reader is referred to Chapter 6, where other layout template examples at cell, sub-system, and system levels are shown.

Last but not least, it is necessary to comment on the drawbacks of the layout-reusable block as solution to the layout synthesis problem in the design reuse methodology. These drawbacks are the layout template high generation cost and its incomplete flexibility.

The generation cost of a layout template refers to the time it takes to go from floorplan design to layout coding and debugging. This time depends on two factors: the size of the circuit and the parameterization scope. The latter factor accounts for the number of generic mask layers and design rules involved, and the number of leaf component and template variants considered, which is in direct relation to the number of retargeting parameters and renders placement and routing parameterization more and more complex. In the course of the analysis of parameterized layouts, we have realized that achieving technological parameterization is hardest at the lower levels of the

---

46. The common-mode feedback circuitry is not shown.

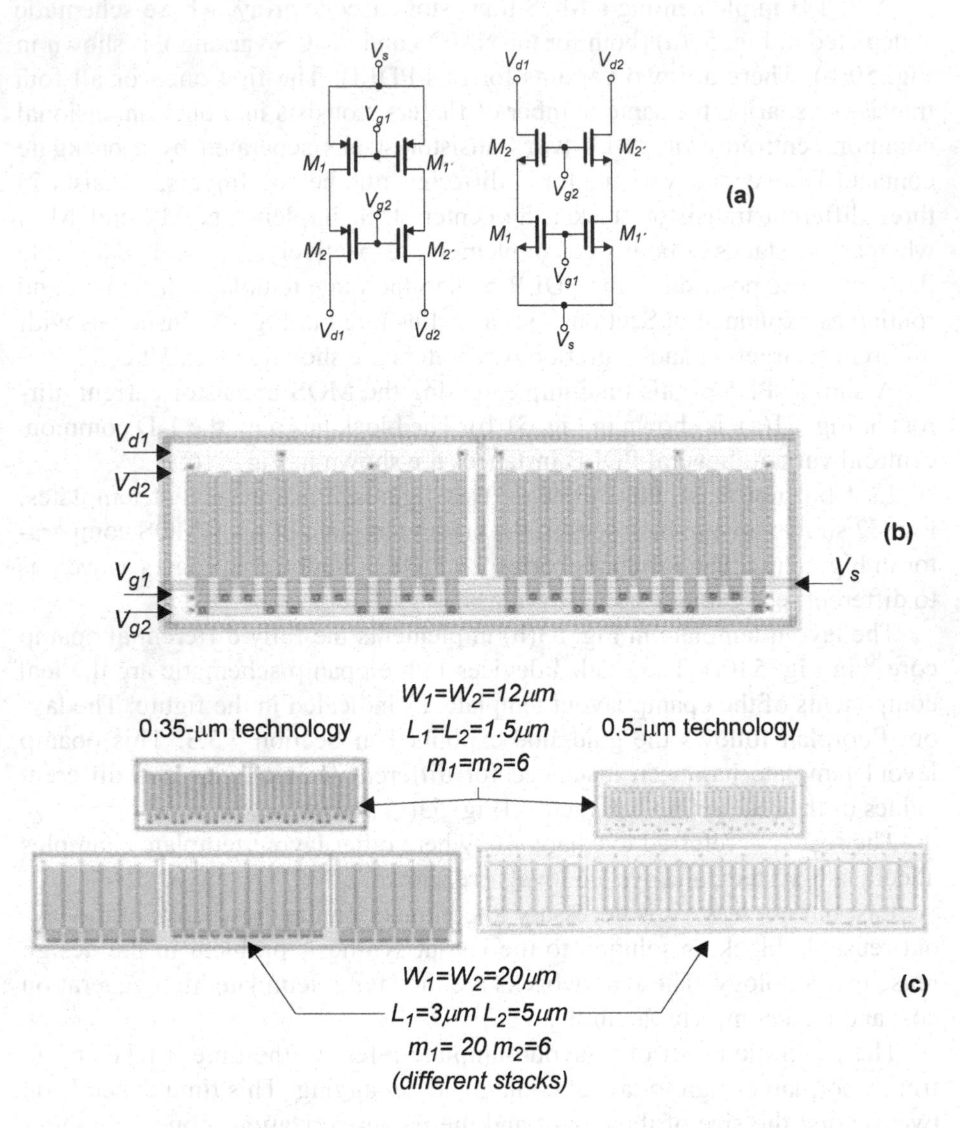

Figure 50. PDLB example: CMOS cascode arrays.

layout hierarchy, i.e., for parameterized device-level primitives and blocks. Parameterizing a layout template to deal with migration is much easier since most of the technological considerations have already been taken into account for the template leaf components. Routing mask layers and design rules, and parameterization of leaf component spacing are the only concerns at the template level. Dealing with retargeting parameterization gets, on the other hand, very much complex at the template level, since leaf component placement and

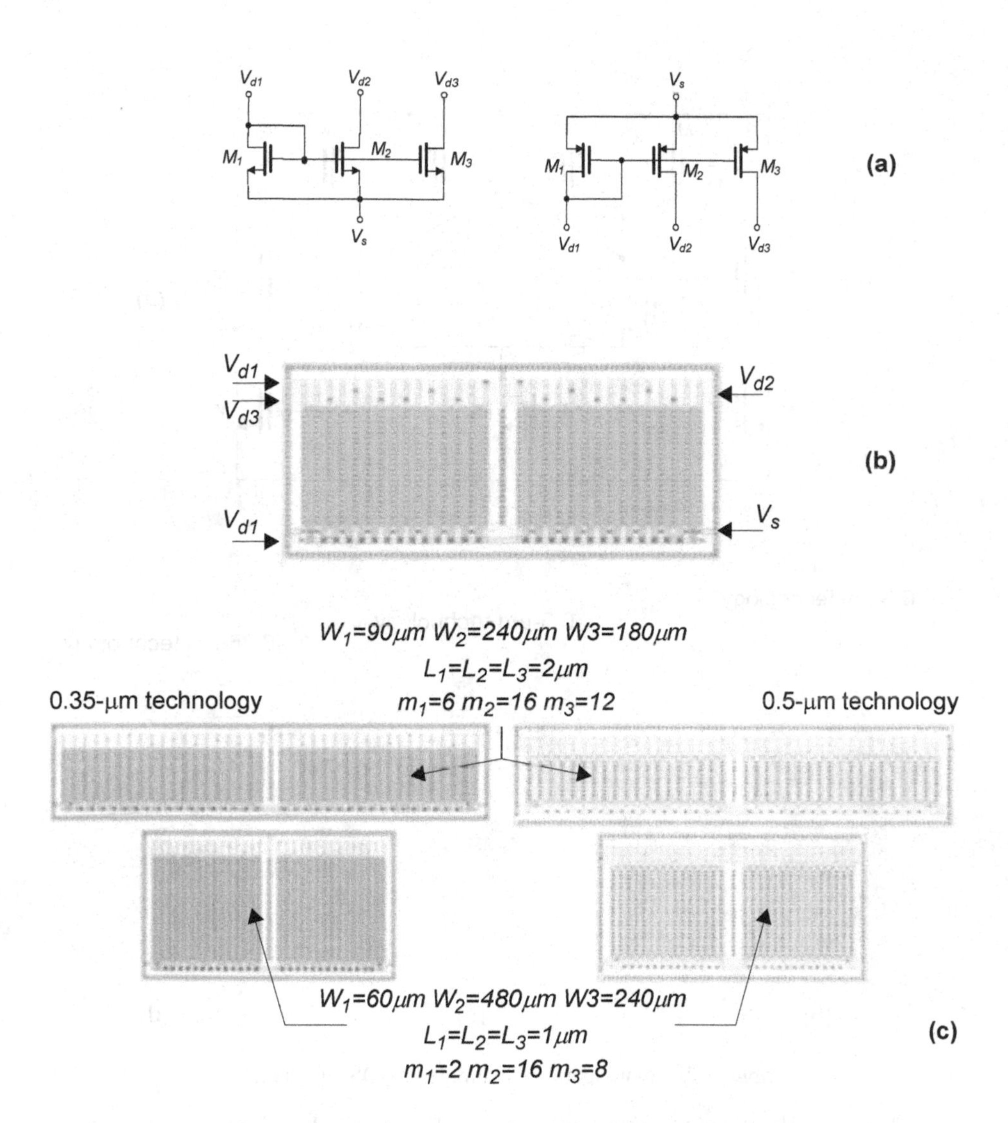

Figure 51. PDLB example: CMOS current mirror arrays.

routing may often require transforming available designer's expertise into code, which is not always straightforward to carry out.

Therefore, whereas technological parameterization is critical at the primitive and block layout levels, collecting and coding design knowledge is the most critical factor at the template level. A reduction of the template generation time can be, however, successfully achieved by means of a large, well-provided library with parameterized device-level blocks and primitives. Moreover, the high-level SKILL functions described in the preceding section

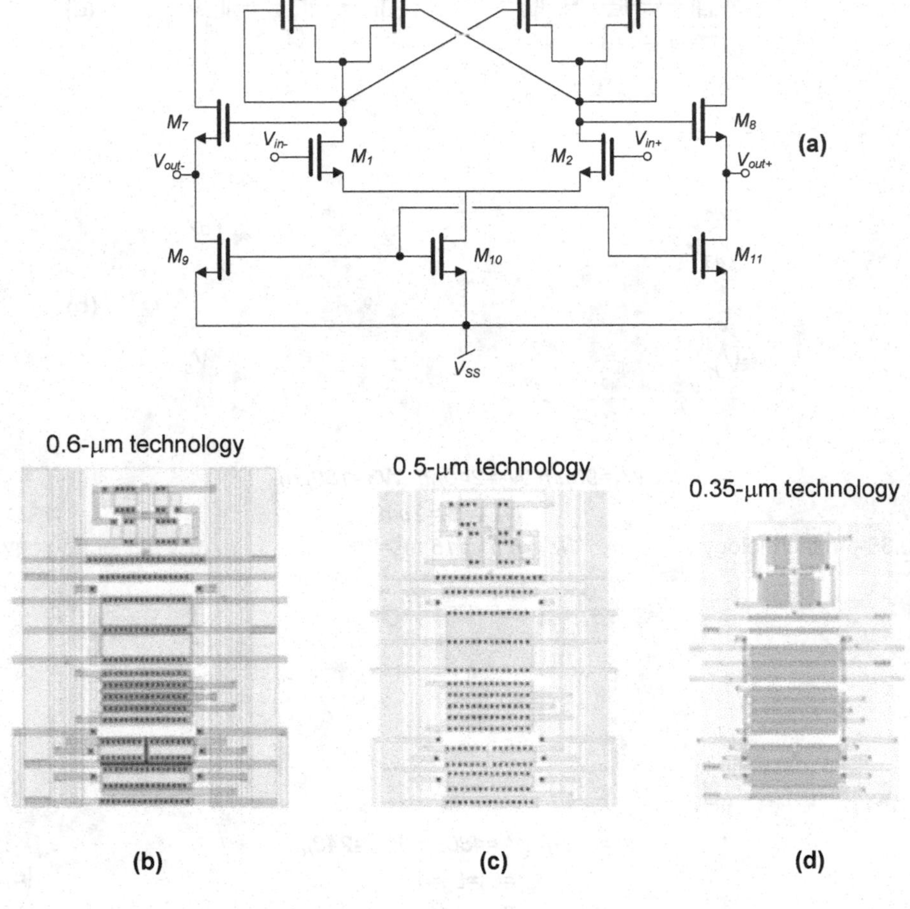

Figure 52. Layout template example: CMOS comparator.

can significantly decrease the generator coding time. For instance, a technician may take 8-10 hours to write the code of the fully differential opamp template in Fig.53 if only built-in pcell SKILL functions are used. Using the *GetLeafCellCS*, *GetPinCS*, *RoutingFunction*, and *MakePin* added functions decreases this time to 4-5 hours.

The three phases of layout template generation are floorplan design, constraint graph generation, and template coding. As floorplan design is common to both template generation and manual full-custom layout design, the difference between these two approaches arises from the time it takes to lay out the shapes manually and the time it takes to generate the graph and develop the

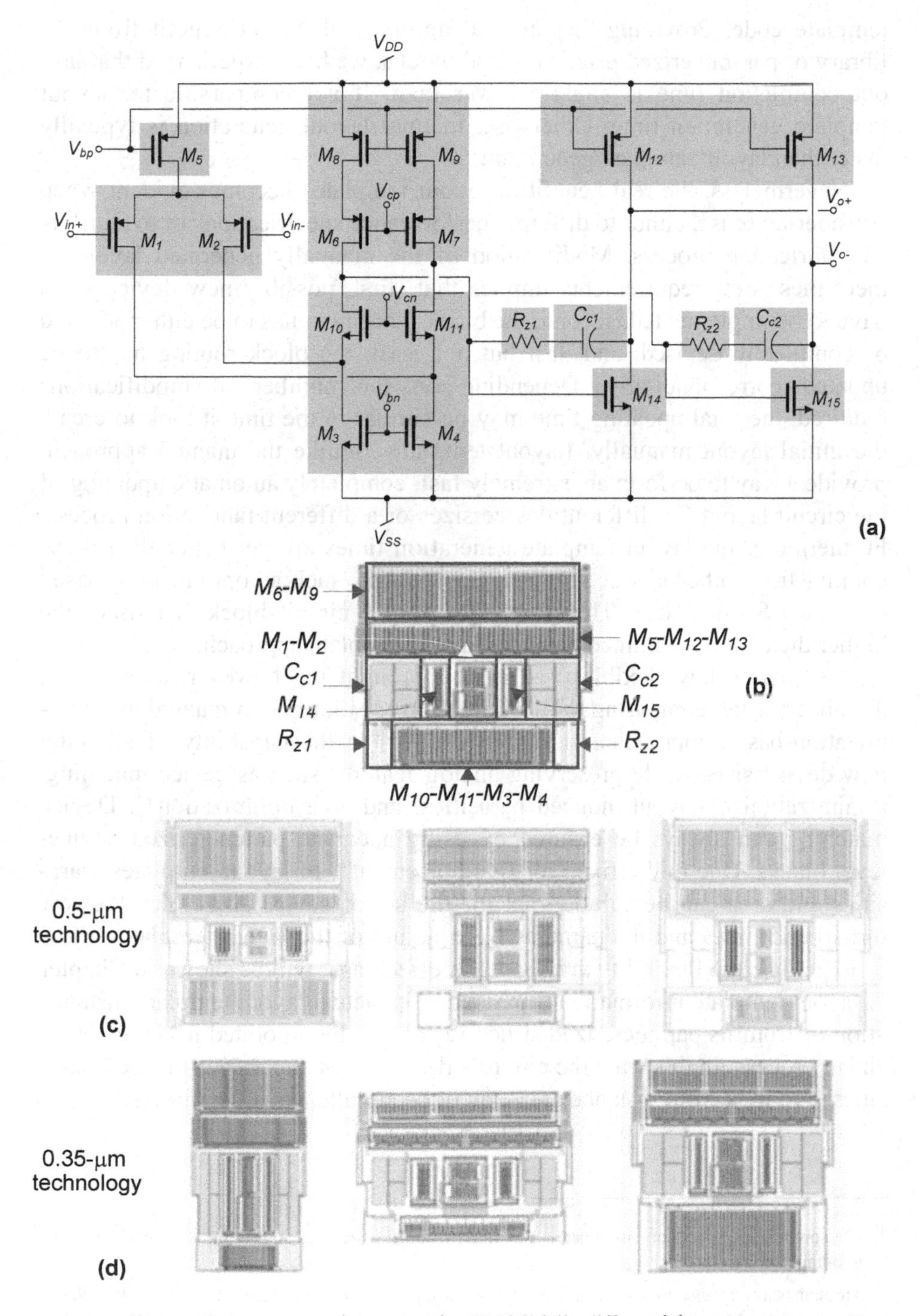

Figure 53. Layout template example: CMOS fully differential opamp core.

template code. Providing that manual approach does not benefit from the library of parameterized primitives and blocks, we have experienced that layout completion time is slightly lower than, if not comparable to, layout template generation time. Otherwise, manual layout generation is typically faster than layout template generation.

Nevertheless, the real benefit of layout templates becomes evident when considering reuse, either to different performance specifications or to a different fabrication process. Modification of the manually generated layout to meet these new requirements implies that, first, possibly new device sizes have to be implemented, second, the block placement has to be either adapted or completely devised, and, last but not least, the block routing has to be updated correspondingly. Depending on the number of modifications required, the total updating time may be similar to the time it took to create the initial layout manually. Layout templates, unlike the manual approach, provide a way to perform an extremely fast, completely automatic updating of the circuit layout for different device sizes or a different fabrication process. Furthermore, the layout template generation times are yet better than those coming from other layout synthesis approaches, such as optimization-based ones (see Section 4.1). Therefore, the more a circuit block is reused, the higher the benefits obtained from the layout template approach.

Template's low flexibility is usually brought up, however, as a severe drawback when comparing the template-based solutions to manual and optimization-based approaches. Flexibility refers to the capability of adapting new device sizes while preserving analog features such as device matching, minimization of layout-induced parasitics, and area optimization[47]. Device matching can always be ensured by using adequate parameterized device-level blocks. Due to the fixed relative placement[48] of layout templates, parasitics and area can be sub-optimal for the set of device sizes. Yet, the low turnaround times and the parameterized nature of the layout templates allow us to palliate the flexibility drawback. In effect, as it will be shown in Chapter 7, layout template information (obtained from actual layout template instantiation or from its parameterized structure) can be incorporated into the circuit sizing process. In this way, the circuit's device sizes, the layout-induced parasitics, and the layout instance area can all be simultaneously optimized.

---

47. Area optimization does not only concern area minimization but maximization of the area density in order to obtain compact layouts.

48. Through heavy parameterization it is always possible to parameterize the placement such that the relative placement can be conveniently modified. Doing so, however, would dramatically increase the complexity of the parameterization task as routing wires should also be parameterized accordingly.

9 SUMMARY

The layout view of the analog reusable block has been explained in this chapter. This layout-reusable analog block has to efficiently deal with the requirements of analog design reuse: performance retargeting and technology migration. From the two analog layout synthesis strategies, namely optimization and knowledge-driven strategies, the latter approaches become more suitable to design reuse as efficient storage and reuse of the analog design knowledge is a major concern. From among the knowledge-driven solutions, the template-based one has been selected to create the layout-reusable analog block.

In this chapter, the problem of creating truly reusable layouts for AMS circuits by means of the template approach has been analyzed. Both retargeting and migration scenarios have been examined and reuse-critical quality features of the AMS layout have been pointed out.

In the light of this study, clear definitions of the layout template properties have been given and a systematic methodology to develop template generators has been defined and thoroughly detailed. This method is composed of three steps: floorplan design, constraint graph generation, and generator coding. Several techniques have been integrated in the methodology that allow the designer to store on the template structure valuable layout knowledge in order to enhance the quality of the reused layout.

Finally, the method has been demonstrated with a complete implementation of a library of layout-reusable blocks in a commercial design environment. Besides, some functions are described that ease the coding of the template generator in this environment.

Chapter 6

Design Examples and Silicon Prototype

In previous chapters, a reuse-based design framework has been described in detail. Chapter 1 set out the motivation for a design paradigm shift and Chapter 2 outlined the solution proposed in this work, based on a renewed concept of circuit reuse. Then, this solution, a combination of design for reusability methodology –to create truly reusable analog and mixed-signal blocks– and a design reusc flow –to incorporate the reusable block into an enhanced hierarchical top-down bottom-up flow– has been developed in Chapters 3, 4, and 5.

The goal of this chapter is to demonstrate the validity of thc reuse-based design framework. This demonstration is carried out in three parts. First, the analog part of an industrial-scale circuit, turned reusable thanks to the design for reusability methodology presented, is hierarchically described. Then, several examples of the design reuse flow applied to this analog circuit are reported. Finally, a silicon prototype that validates the contributions of the work presented in this book is described.

1 INTRODUCTION

The concept of analog reusable block has been developed in the last three chapters. This type of blocks, conceptually closer to the digital firm IP notion, is composed, as shown in Fig. 1, of three separate views –behavioral, structural, and layout– each once embedding valuable, reusable design expertise. Besides, each one of the reusable views is represented by a fully parameterized database, which makes the whole reusable block technology-independent and enables the straightforward retargeting of the analog circuit block. The analog reusable block is attained by a design for reusability methodology, explained in the chapters indicated in Fig. 1.

The analog reusable block can be any circuit block on and above the cell hierarchical level (see Fig. 2). Reusable blocks above the cell level are to be

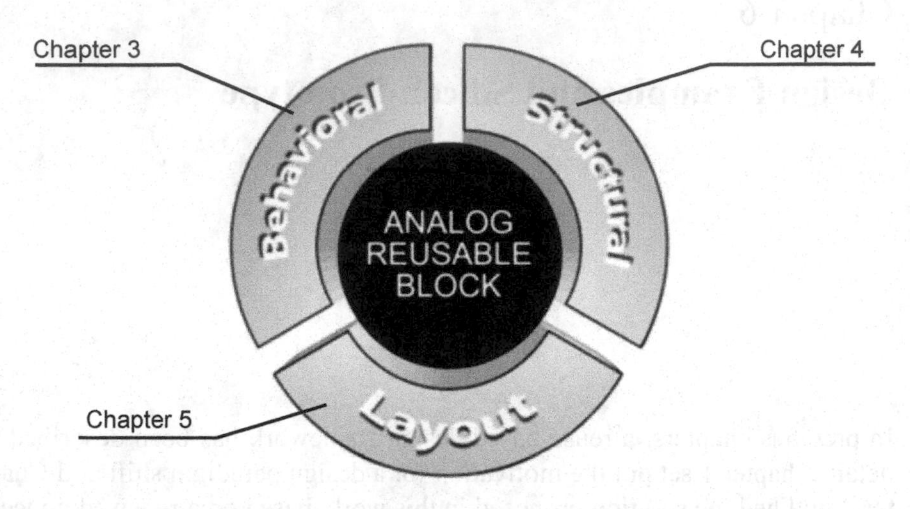

Figure 1. The three-view representation of the analog reusable block.

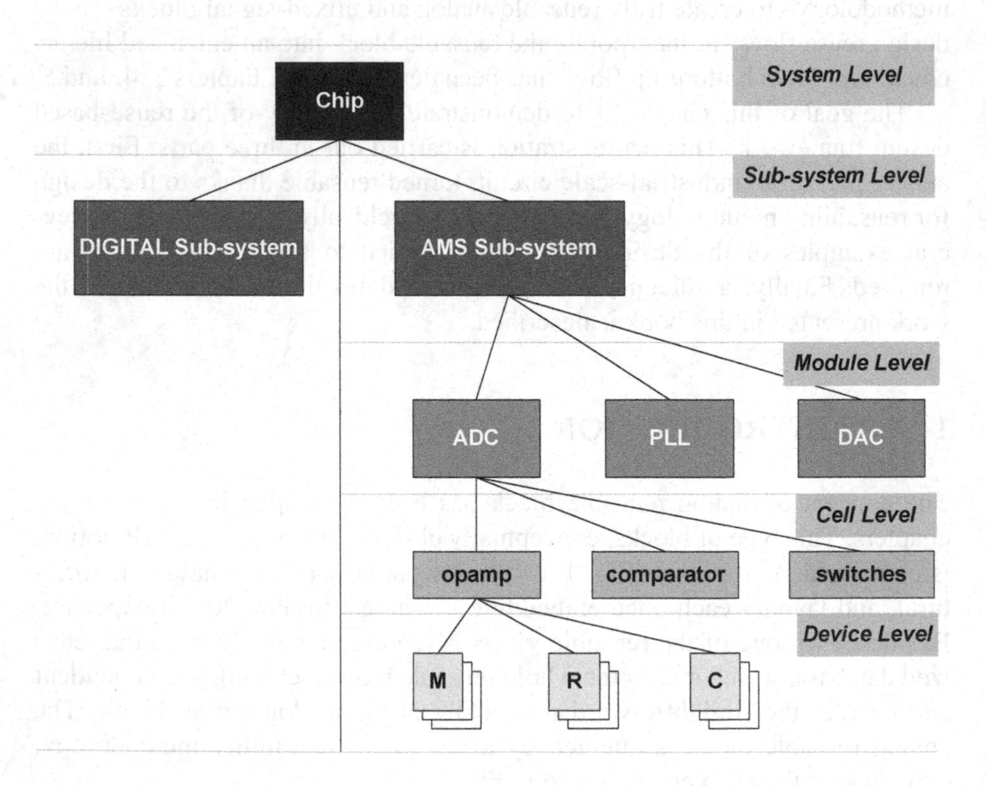

Figure 2. Hierarchical structure of an AMS-SoC design.

composed of lower-level reusable blocks provided that hierarchical decomposition, dividing the overall synthesis problem into separate problems, does simplify it in terms of complexity and required computation time.

The top-down bottom-up design flow for analog and mixed-signal circuits described in this book aims at bridging the nowadays increasing design gap by mixing the concepts of circuit reuse and synthesis. As outlined in Chapter 2, this hierarchical flow, called design reuse flow, benefits from the very nature of the reusable block in order to reduce the overall product-to-market time and to cope with the analog inherent design complexity.

Both the design for reusability methodology and the design reuse flow form the reuse-based design framework, which is the object of the present book. The goal of this chapter is to demonstrate the validity and suitability of such a framework to assist design teams in the today's scenario of analog circuit design. This demonstration uses an industrial-scale mixed-signal circuit, and it is carried out by means of:

- describing the reusable blocks composing the analog portion of the industrial-scale circuit,
- illustrating the design reuse flow by performing several runs for the analog portion of the industrial-scale circuit, and
- presenting and validating the silicon implementation of the industrial-scale circuit.

It is also shown that the whole design flow is susceptible of full automation through appropriate scripting. To that goal, an automation prototype with a graphical user interface (GUI) is presented that illustrates how the design reuse flow can be automated almost to a "push-button" style.

2 THE DEMONSTRATION VEHICLE

This section describes the system selected to demonstrate the reuse-based framework. This system was briefly introduced in Chapter 3. Here, the system is explained in more detail. The system performance specifications are also discussed. Then, the section immediately focuses on the analog part of the system, for which design examples are later reported, by outlining its hierarchy and analyzing its components.

2.1 Application area and rationale for architecture selection

Worldwide semiconductor market trends indicate a rapid increase of chips containing both analog and digital functionality. In order to provide such increased functionality and combined use of analog and digital signals, it is necessary to develop the electronic circuitry that provides the appropriate analog-digital interfacing. The integration of complete systems on a chip will therefore be achieved by assembling a variety of high functionality blocks, from powerful micro-controllers and DSP cores to complex analog and mixed-signal blocks. Such analog and mixed-signal blocks correspond to complete sub-systems that embed all the required functionality to interface analog and digital signals, including data conversion, filtering, and amplification, among other functions. Given the complex multidisciplinary nature of mixed-signal integrated circuits, the design of such analog and mixed-signal blocks is greatly dependent on the application for which the interfacing function is envisaged. One of the fastest growing market segments is the area of wireless communications where baseband modems are in great demand.

Quadrature interfaces are seen in a wide range of wireless communication standards such as GSM, PMR, PHS, CDMA or W-CDMA [Rapa96] (see Fig. 3), and they should be implemented on mainstream digital CMOS technologies for fabrication cost reasons. Essentially, such modems provide the interface between complex digital signals from the baseband digital processors and complex analog signals in quadrature phase (I and Q) from the RF transceiver. Many integrated DA interfacing systems for portable communications have been reported. In [Hasp90] and [Frie96], the digital-to-analog converters (DACs) have a single ended switched-capacitor (SC) implementation, they are followed by a second-order low-pass (LP) filter and, the last stage is an output driver which provides the low impedance to the mixer and performs the single-ended to the fully-differential conversion. In [Laks91], a structure based on a fully differential 2.17-MHz 8-bit current-steering DAC is followed by a single-ended class AB amplifier for driving an external filter. In [Bagg92], the 10-bit DA architecture is based on a two-stage resistor-ladder/

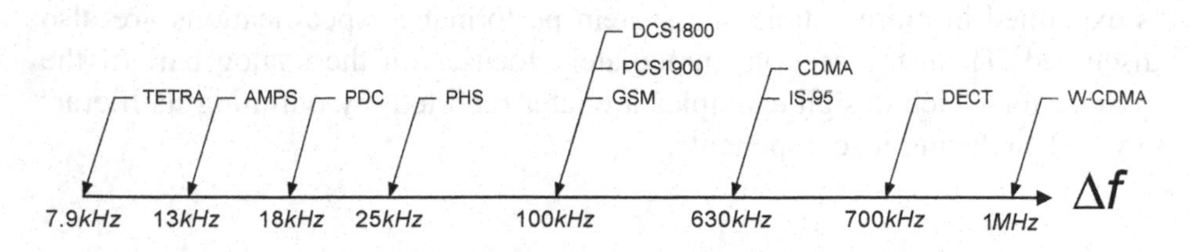

Figure 3. Different wireless communications standards by frequency band.

capacitor-array approach, followed by a fully-differential third-order SC LP filter and then followed by a power buffer. In [Mino95], a 10-bit 4.33-MHz SC DAC is followed by a 4th order Butterworth SC filter and then followed by a 4th-order continuous-time filter. In [Maul95], each channel comprises an interpolation filter followed by a digital second-order delta-sigma modulator followed by a cascade of a first-order SC filter and a second order SC filter, but the continuous-time LP filtering is realized externally and the output offset is about 20 mV.

The IQ DA transmit interface system[1] [Fran99a], whose functional block diagram is shown in Fig. 4, is the AMS system used as demonstration vehicle of the reuse-based framework described in previous chapters. This system provides two fully-differential channels in quadrature phase (90° phase shift) between the input digital ports and the output analog ports. Each channel is formed by a SINC^3 digital processing unit for signal shaping and interpolation, a DAC employing current-steering circuit techniques, a continuous-time second-order low-pass filter (CT-LP Filter), and a first-order programmable-gain amplifier (PGA). An additional calibration unit is employed to adjust the unavoidable offsets and mismatches between both channels. In order to control the functionality and calibration of the complete system, a control unit is also included.

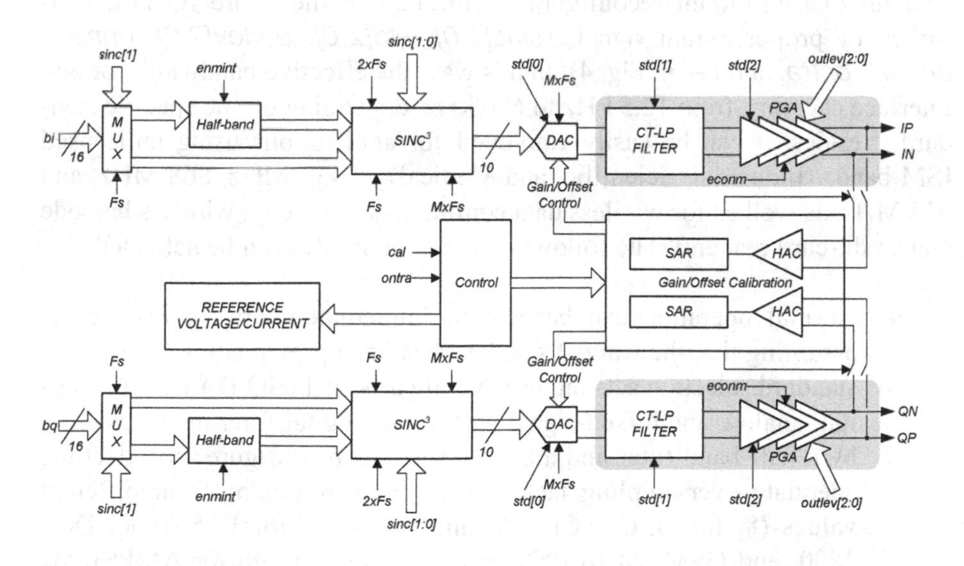

Figure 4. Functional diagram of the IQ DA transmit interface.

---

[1] Hereinafter, the transmit interface will be indifferently referred as the IQ DA transmit interface or, shorter, as the IQ DA.

The system operation is as follows. First, the 16-bit input digital data (signals *bi[15:0]* and *bq[15:0]*) in two's complement at the input bit rate, is oversampled by the interpolating filter, to obtain a good spectral purity, thus relaxing the required resolution of the DAC. Next, the results are applied to the DAC, which has a fully-differential current-steering implementation and employs a segmented structure leading to a good linearity as well as to an optimum area and power dissipation. The purpose of the next block, the CT-LP filter, is to attenuate the image components of the baseband spectrum at multiples of the clock frequency, to smooth the output signals generated by the preceding segmented current-steering DAC, as well as to provide current-to-voltage conversion. The purpose of the PGA, on the other hand, is to provide a digitally-controlled DC amplification of the differential output signal delivered by the CT-LP filter, to enforce the image rejection function of the CT-LP filter, and to support the buffering of its output differential signal to the external load of the chip, in order to directly drive commercially available RF modulators with no need of external components.

The whole IQ DA features a programming capability by which different wireless communication standards, namely DECT, CDMA, IS-95, PCS 1900, DCS 1800, GSM, PDC, AMPS, and Tetrapol, can be supported and several operation modes can be entered. This capability is achieved by means of switching-based circuit reconfiguration throughout the entire IQ DA, controlled by proper digital signals (*sinc[1:0]*, *std[2:0]*, *outlev[2:0]*, *enmint*, *econm*, *ontra*, and *cal* in Fig. 4). In this way, the effective bandwidths of this interface can vary from 12.5 kHz to 600 kHz, depending on the specific standard. Besides, it can be easily employed for applications using unlicensed ISM bands (Industrial, Scientific, and Medical) of 433 MHz, 868 MHz, and 915 MHz, as well as for wireless data communications (e.g., wireless barcode and credit-card readers). The following operation modes can be activated:

- Normal operation can be entered immediately after a reset cycle assuming that the wanted standard has been previously selected. The standard selection acts on the several parts of the IQ DA (i.e., the digital, analog, and mixed-signal parts). The digital filtering, carried out by a half-band filter and a $SINC^3$ filter, is reconfigured to select the adequate oversampling factor, M, which may adopt four different values (8, for DECT, CDMA, and IS-95, 12 for PCS 1900, DCS 1800, and GSM, 24 for PCD and Tetrapol, and 96 for AMPS); the values are selected according to signals *sinc[1:0]* and *enmint*, as shown in Table 1. As illustrated in Table 2, standard selection in the analog and mixed-signal section is done by using the digital word *std[2:0]*. Its three bits reconfigure the blocks to attain a different

Table 1. Half-band and SINC^3 filter interpolation factor control.

Half-band filter oversampling factor	*SINC^3 filter oversampling factor*	*M*	*sinc[1:0]*	*enmint*
--	8	8	[0 0]	0
--	12	12	[0 1]	0
2	12	24	[1 0]	1
2	48	96	[1 1]	1

Table 2. Standard selection in the analog and mixed-signal parts of the IQ DA.

Reconfigured feature	*std[2:0]*	*Standard*
DAC speed	[0 X X]	DECT, CDMA, IS-95
	[1 X X]	GSM, DCS1800, PCS 1900, AMPS, PDC, Tetrapol
CT-LP filter pole frequency	[X 0 X]	DECT, CDMA, IS-95
	[X 1 X]	GSM, DCS1800, PCS 1900, AMPS, PDC, Tetrapol
PGA pole frequency	[X X 0]	DECT, CDMA, IS-95
	[X X 1]	GSM, DCS1800, PCS 1900, AMPS, PDC, Tetrapol

value of the DAC speed, and the pole frequencies of the CT-LP filter and the PGA blocks. It is also possible to control the output level with the selection of the appropriate gain of the PGA, which is done with the digital word *outlev[2:0]*. The reader is referred to Section 2.4 for more details about the reconfiguration capabilities of the analog blocks.

- The *economic* mode is used to minimize current consumption for all standards in case a different load (e.g., because of different RF modulators) is to be driven. This is accomplished by reconfiguring the power dissipation of the PGA block (the block driving the load) by means of the *econm* signal. The reader is referred to Section 2.4 for more details about this reconfiguration capability of the PGA.
- In the *power-down* mode, the biasing currents in the analog blocks are cut, thus reducing the power dissipation. This is done by selecting *ontra* signal to LOW.
- The *offset-calibration* mode (started by having the *cal* signal held HIGH for more than one clock period) is used to make the differential outputs of both channels to converge to less than $\pm 1\,\mathrm{mV}$.

In the following, the development of the demonstration examples is focused on the analog portion of the IQ DA, that is, the analog back-end composed of the CT-LP filter and the PGA blocks.

2.2 System specifications and specifications of the analog back-end

Table 3 shows the performance specifications of the IQ DA transmit interface for several wireless communication standards. In particular, the last five rows give specific requirements for the analog portion of the system, referred here as the analog back-end (the blocks under the shaded area, at the end of the chain, of the IQ DA in Fig. 4).

It is also important to note that some of the specifications of the analog back-end involve the concept of design centering. It implies that the analog back-end has to be carefully designed not only to be optimal with respect to nominal performance, but also with respect to manufacturing process variations. These variations can be local (also known as **intra-die** variations), meaning that they affect every device in the die in a different way, and global (also known as **inter-die** variations), which affect every device in the die exactly in the same way. Eventually, their influence results in a mismatch or difference in how channels I and Q process the signal. Thus, appropriate specifications have to be imposed in order to limit such a mismatch. Accordingly, synthesis of the analog back-end must be carried out taking into account these inter-die and intra-die variations.

2.3 Hierarchy of the analog back-end

Figure 5 illustrates the hierarchical organization of the analog back-end. As said above, the analog back-end (at the sub-system level) is composed of two filtering blocks, a CT-LP filter and a PGA (at the module level). Both blocks are composed of passive devices (for the feedback and feedforward paths) and amplifying blocks A_1 and A_2 (at the CT-LP filter cell level) and A_3 (at the PGA cell level)[2].

The objective of introducing the reuse-based design framework is to support the design of the IQ DA transmit interface, in particular the analog back-end section, by an automatic retargeting or process migration, with minimal changes on their components. This will lead to significant improvements in the accumulated design time when designing such an interface for a variety of wireless communication applications and different fabrication processes.

---

2. Biasing circuitry and other operation-mode selection devices are not shown.

Table 3. IQ DA performance specifications.

	AMPS	*Tetrapol*	*PDC*	*GSM, DCS1800, PCS 1900*	*DECT*	*IS 95, CDMA*
Input bit rate (kbit/s)	40	40	168	1083.3	4608	4915.2
Oversampling factor, M	96	96	24	12	8	8
Sampling frequency, FS (MHz)	3.840	3.840	4.032	13	36.864	39.3216
Signal bandwidth, SB (kHz)	15	3.95	18	100	700	630
DAC attenuation @FS[a] (dB)	48.16	59.75	47.00	42.28	34.43	35.91
Minimum image rejection @FS (dB)	84	74.5	67.5	92	87	71
Minimum attenuation of the analog back-end @FS (dB)	35.84	14.75	20.5	49.72	52.57	35.09
Maximum I/Q gain mismatch (dB)	± 0.7	± 0.7	± 0.7	± 0.7	± 0.7	± 0.7
Maximum group delay deviation (from DC to SB) (ns)	0.45	0.45	0.85	8.3	9.5	9.45
Maximum I/Q group delay mismatch (from DC to SB) (ns)	37.4	37.5	37.2	33.5	6.5	6.5
Maximum I/Q phase mismatch @SB (°)	0.21	0.18	0.245	1.3	1.75	1.7

a. The overall attenuation of the IQ DA is given by $20 \times \log\left[\left(\frac{FS}{F_{p,CTF}}\right)^2\left(\frac{FS}{SB}\right)\left(\frac{FS}{F_{p,PGA}}\right)\right]$ where $F_{p,CTF}$ and $F_{p,PGA}$ are, respectively, the cutoff frequencies of the second-order CT filter and the 1st-order programmable amplifier.

Complete synthesis of the analog back-end involves, first, the mapping of the sub-system performance specifications in Table 3 into specifications for each of the module-level components, i.e., the CT-LP filter and the PGA. Second, these module-level performance specifications are to be translated into values of the passive elements, on the one hand, and performance specifications for each of the three opamps, on the other hand.

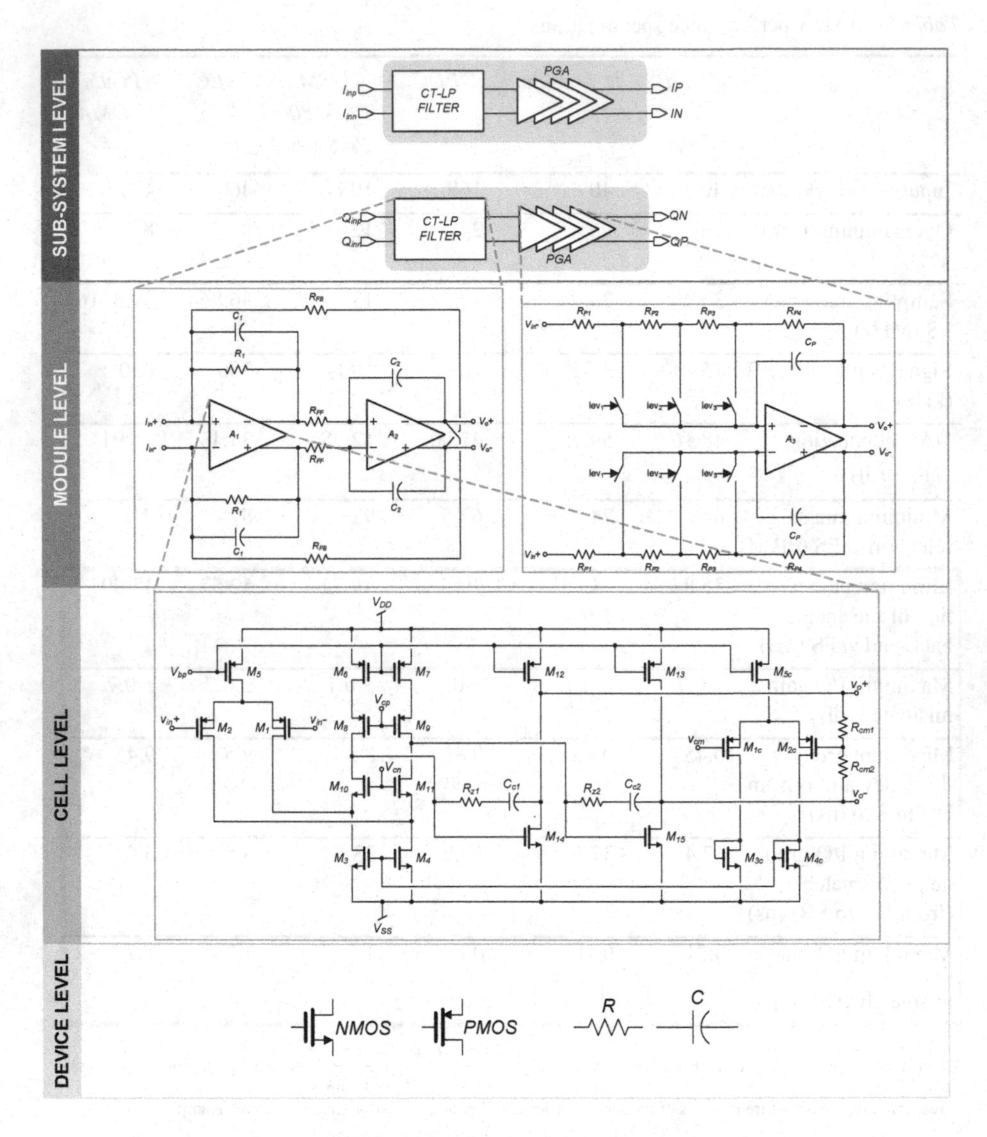

Figure 5. Hierarchical levels of the analog back-end.

2.4 Analysis of the analog back-end

Once the architecture of the analog back-end sub-system is known, it is also important to derive valuable design knowledge about the circuits that are used inside it. This will reveal useful when creating suitable analog reusable blocks, with valuable embedded design knowledge and a proper parameterization[3].

2.4.1 The CT-LP filter

The purpose of the dual CT-LP filter is, as said above, to attenuate the image components of the baseband spectrum at multiples of the clock frequency, to smooth the output signals generated by the preceding segmented current-steering DA converters, as well as to provide current-to-voltage conversion. Taking into account the attenuation inherently provided by the DA converters, as of its sampled data dynamics, and the output spectrum profile required by the two groups of standards (GSM, DCS1800, PCS 1900, AMPS, PDC, and Tetrapol, in one group, and DECT, CDMA, and IS-95 in the other), the necessary selectivity of the CT-LP filter, as well as, its response accuracy may not be high –cut-off frequency variations around $\pm 10\%$ are acceptable. The dynamic range and linearity of the filter, however, must be high to match those of the associated signal processor (10 bits in this application). Further, the filter must have a small DC gain variation and a constant group delay in order not to influence the modulated signal spectrum. Also, excellent gain and phase matching between the I and Q channels are required to avoid additional sideband frequency components when generating the RF signal. Though the Gm-C technique is considered to offer such advantages as high speed and low power, an RC-active structure, based on integrated passive resistors and capacitors, has been adopted for the filtering operation at the transmit interface because of its superior performance regarding dynamic range and linearity[4]. In addition, because the RC-active filter must be programmable to provide multi-standard capabilities to the IQ-DAC interface, it is essential to reduce the distortion contributions of the switch elements, and this can be done with the use of parasitic-insensitive architectures [Durh93]. All the above considerations have been addressed in the topology of Fig. 6 for the implementation of the CT-LP filter. It is a current-driven fully-differential RC-active Tow-Thomas second-order low-pass filter, whose parasitic insensitive feature can be readily seen as all the capacitors are connected between the opamp inputs and outputs –parasitic capacitance will not affect the location of the filter poles directly. This compares favorably to other filter structures as, for instance, many single-amplifier filter topologies like the Sallen-Key filters

---

3. As explained in Chapter 2, the two fundamental principles upon which the design for reusability is based are parameterization and encapsulation of design knowledge.

4. Demanding dynamic range and linearity specifications could be also afforded using SC filters operated at a high sampling rate and followed by less restrictive CT smoothing filters [Hasp90] [Bagg92] [Maul95] [Mino95]. This solution, however, implies a large area and power consumption, and, therefore, a fully continuous-time design has been preferred [Durh93] [Sini01] [Holl01].

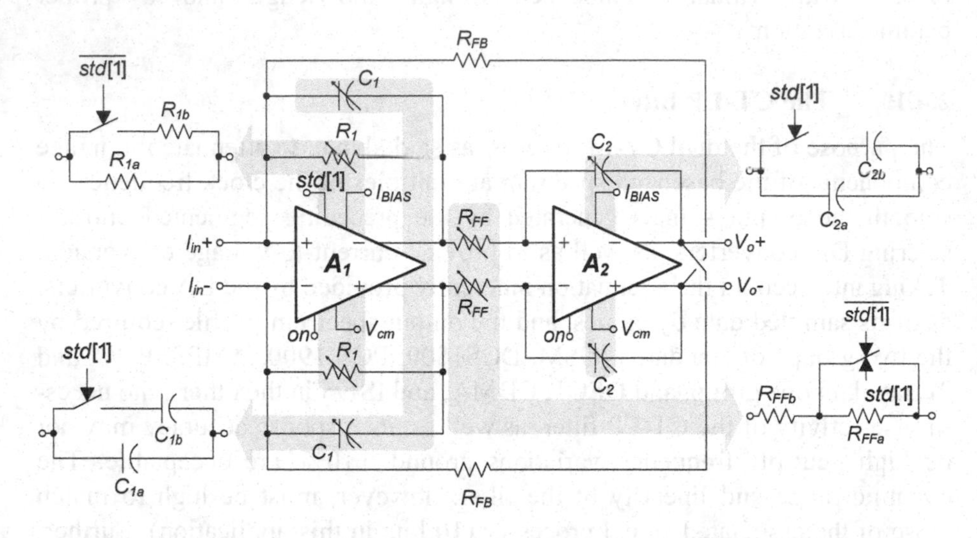

Figure 6. CT-LP filter schematic.

family, which, albeit simpler than the Tow-Thomas topology, exhibit poorer THD performance [Durh93].

In the circuit of Fig. 6, the programmability of the transfer characteristics is accomplished by specifying different settings for its passive elements by means of a digital control signal that selects the wanted standard (*std[1]*). In order to reduce the voltage excursions on the MOS switches used to effect such selection, and hence, to reduce their distortion contributions, the switches must be placed at virtual ground nodes provided by the opamps, as shown in Fig. 6. On the other hand, given the tolerance of the application to deviations on the cut-off frequency, no correction mechanisms will be employed for tuning the RC time constants of the filter. It means that the filter implementation should be based on a precise centering within the design space that account for process variations of the passive elements. This obviously suggests that the behavioral model of the filter should also provide a statistical characterization of its frequency response, because the prescribed image rejection at DAC's sampling frequency must be achieved under worst-case global variations of resistances and capacitances. Further considerations on the design of the CT-LP filter of Fig. 6 can be extracted from a close understanding of the topology. Let us start the analysis from the modified nodal matrix representation of the CT-LP filter, which takes the form:

$$\text{MNA}=\begin{bmatrix} G_1 - G_{FB} + C_1 s & -G_1 - C_1 s & 0 & G_{FB} & 0 & 0 \\ -G_1 - C_1 s & G_1 + G_{FF} + C_1 s & -G_{FF} & 0 & -1 & 0 \\ 0 & -G_{FF} & G_{FF} + C_2 s & -C_2 s & 0 & 0 \\ G_{FB} & 0 & -C_2 s & -G_{FB} + C_2 s & 0 & -1 \\ -A_1 & -1 & 0 & 0 & 0 & 0 \\ 0 & 0 & -A_2 & -1 & 0 & 0 \end{bmatrix} \quad (1)$$

where A_1 and A_2 are, respectively, the open-loop transfer characteristics of the left- and right-most opamps of Fig. 6. Assuming, in first approximation, that they are ideal (infinite gain and bandwidth), the transfer function of the CT-LP filter simplifies as,

$$F(s) = \frac{V_o}{I_{in}}(s) = \frac{G_{FF}}{G_{FB}G_{FF} + G_1 C_2 s + C_1 C_2 s^2} \quad (2)$$

from which

$$K = \frac{1}{G_{FB}} \qquad Q = \frac{1}{G_1}\sqrt{\frac{C_1 G_{FB} G_{FF}}{C_2}} \qquad \omega_n = \sqrt{\frac{G_{FB} G_{FF}}{C_1 C_2}} \quad (3)$$

where K is the DC gain of the filter, Q is the quality factor and ω_n its natural frequency. It should be noted that the DC current-to-voltage factor of the filter is uniquely defined by the overall feedback resistance R_{FB}. Hence, assuming that the full-scale output current of the preceding DA converter takes a fixed value I_{FS}, resistance R_{FB} must be implemented to guarantee a given full-scale output voltage V_{FS}, according to:

$$R_{FB} = V_{FS}/I_{FS} \quad (4)$$

Additional constraints among the passive elements of the filter can be derived by ensuring a maximally-flat transfer of the baseband signal to the filter output (this implies a quality factor of $Q = 1/\sqrt{2}$), by setting the cut-off frequency $f_{p,\,CTF}$ of the filter in accordance to the minimum image rejection imposed by the standard (this defines the filter natural frequency as $\omega_n = 2\pi f_{p,\,CTF}$), and improving matching among similar components (a convenient choice is $C_1 = C_2$).

Design considerations for the opamps of Fig. 6, as well as some additional restrictions among the resistors, can be extracted by performing a sensitivity analysis of the biquad. Clearly, both active and passive sensitivities of Fig. 6 must be as low as possible in order to reduce the variability of the filter characteristics of Eq. (3). This guarantees robust designs against statistical variations of system parameters. This is particularly important in the context of the IQ DA prototype, because no tuning circuitry is used to correct the deviations of ω_n and Q.

It can be easily demonstrated that the sensitivities of the Tow-Thomas biquad to mismatches of the passive elements are the following [Sedra78] [Scha90]:

$$S^{K}_{[R_{FB}, R_{FF}, R_1, C_1, C_2]} = [1, 0, 0, 0, 0] \tag{5}$$

$$S^{Q}_{[R_{FB}, R_{FF}, R_1, C_1, C_2]} = \left[-\frac{1}{2}, -\frac{1}{2}, 1, \frac{1}{2}, -\frac{1}{2}\right] \tag{6}$$

$$S^{\omega_n}_{[R_{FB}, R_{FF}, R_1, C_1, C_2]} = \left[-\frac{1}{2}, -\frac{1}{2}, 0, -\frac{1}{2}, -\frac{1}{2}\right] \tag{7}$$

Regarding active sensitivities, we will separately consider the effects of the finite gain and the finite bandwidth of the amplifiers on the filter characteristics of Eq. (3). With regard to the effect of the opamps' finite gain, it can be shown that the deviations of the filter characteristics defined in Eq. (3) are given by:

$$\zeta_g \equiv \frac{K_g - K}{K} \cong \frac{1}{1 + \dfrac{G_1 K}{A_2}} - 1 \approx -\frac{G_1 K}{A_2} \tag{8}$$

$$\eta_g \equiv \frac{Q_g}{Q} \cong \frac{1}{1 - \dfrac{1}{G_1}\left(\dfrac{G_{FB}}{1 + A_1} - \dfrac{G_{FF}}{1 + A_2}\right)}\left(\frac{\omega_{ng}}{\omega_n}\right) \tag{9}$$

$$\delta_g \equiv \frac{\omega_{ng} - \omega_n}{\omega_n} \cong \sqrt{1 + \frac{G_1 K}{1 + A_2}} - 1 \approx \frac{1}{2}\frac{G_1 K}{1 + A_2} \tag{10}$$

where $\{K_g, Q_g, \omega_{ng}\}$ represent the modified filter characteristics, with respect to the ideal ones $\{K, Q, \omega_n\}$, as a consequence of the finite amplifier gains. The right-hand side approximations of Eq. (8) and Eq. (10) assume that $G_1 K \ll A_2$. Using the above expressions, the gain-sensitivity products of the filter characteristics with respect to A_1 and A_2 take the following form:

$$GS^{[K, Q, \omega_n]}_{A_1} \cong \left[0, \frac{1}{2} - \frac{1}{G_1 K}, \frac{1}{2}\right] \tag{11}$$

$$GS^{[K, Q, \omega_n]}_{A_2} \cong \left[G_1 K, \frac{1}{2} + G_1 K\left(Q^2 - \frac{1}{2}\right), \frac{1}{2}(1 - G_1 K)\right] \tag{12}$$

where parameters K and Q are as defined in Eq. (3). Note that the product $G_1 K$ has a large relevance on the active sensitivities of the filter. For large values of $G_1 K$, the variability of the filter characteristics induced by deviations of amplifier A_2 becomes increasingly important. On the other hand, for low values of G_1K, the Q-sensitivity due to A_1 drastically grows. Two interesting choices for the G_1 value, which overcome the above drawbacks, are $G_1 = 1/K$ and $G_1 = 1/(KQ)$. In both cases, the gain-sensitivity products are small when the maximally flat filter approximation ($Q = 1/\sqrt{2}$) is assumed. Furthermore, both approaches favour matching among resistors, because for $G_1 = 1/K$ we have $R_1 = R_{FB}$ and for $G_1 = 1/(KQ)$, $R_{FF} = R_{FB}$. Unfortunately, these choices lead to large capacitance values in the biquad of Fig. 6 (larger as the cut-off frequency is reduced) and hence, result in large area occupation which may preclude their implementation in area saving contexts. Nevertheless, they represent top implementations regarding sensitivity performance and will be useful as reference benchmarks to other alternative realizations.

Now considering the influence of the frequency limitations of the amplifiers on the filter characteristics, it can be first-order evaluated by using a single-pole model for the opamps and replacing each A_i, $i = 1, 2$, in Eq. (1) by ω_{ti}/s, where ω_{ti} is the unity-gain frequency of the i-*th* amplifier. After solving the transfer function of the filter, the deviations $\Delta\omega_n$ and ΔQ can be readily obtained. An easier and insightful method to evaluate $\Delta\omega_n$ and ΔQ is based on the fact that finite ω_{ti} results in integrators with finite quality factors, or, equivalently, finite phase errors. Fig. 7 shows the signal flow graph associated to the CT-LP filter, in which each integrator is assumed to undergo a real dissipation term, leading to a finite Q factor, and a small deviation on the time constant, i.e., the transfer function of each integrator is transformed according to [Sedra78]:

$$-\frac{a_i}{s} \to -\frac{a_i}{sf_i + \sigma_i} \tag{13}$$

where $a_1 = 1/C_1$ and $a_2 = G_{FF}/C_2$. Using Mason's rule and after some algebra, the actual natural frequency and Q factor of the filter are

$$\omega_{na}^2 = \frac{\omega_n^2}{f_1 f_2}\left(1 + \frac{1}{Q}\frac{\sigma_2}{\omega_n} + \frac{\sigma_1\sigma_2}{\omega_n^2}\right) \tag{14}$$

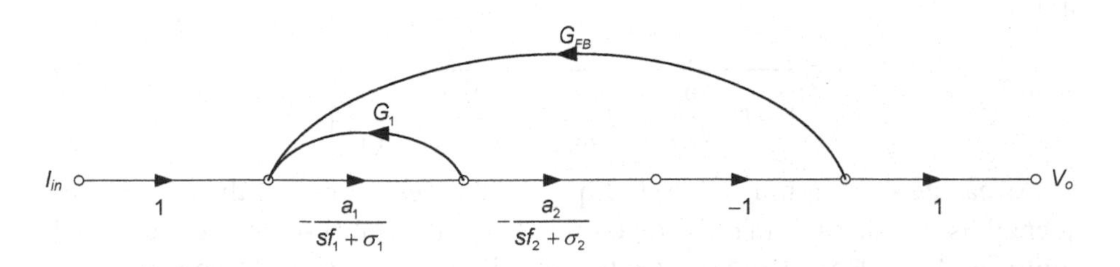

Figure 7. Signal flow graph of the CT-LP filter with the integrators having finite Q-factors.

and

$$Q_a = \frac{\omega_{na}}{\omega_n} \frac{Q}{f_2 + Q\frac{f_2\sigma_1 + f_1\sigma_2}{\omega_n}} \tag{15}$$

respectively, where it can be shown that $f_1 = 1$, $\sigma_1 = -\omega^2/\omega_{t1}$, $f_2 = 1 + a_2/\omega_{t2}$ and $\sigma_2 = -\omega^2/\omega_{t2}$. Replacing these values in Eq. (14) and Eq(15), and after some simplifications based on the assumption that $\omega_n \ll \omega_{ti}$ (for $i = 1, 2$), the relative variation of the natural frequency and the Q factor of the filter can be approximated as

$$\delta \equiv \frac{\omega_{na} - \omega_n}{\omega_n} \cong -\frac{1}{2}\left(\frac{a_2}{\omega_{t2}} + \frac{1}{Q}\frac{\omega_n}{\omega_{t2}} - \frac{\omega_n^2}{\omega_{t1}\omega_{t2}}\right) \tag{16}$$

and

$$\eta \equiv \frac{Q_a}{Q} \cong \frac{1}{1 + \frac{a_2}{\omega_{t2}} + \frac{Q}{\omega_n}\left[-\frac{\omega_n^2}{\omega_{t1}}\left(1 + \frac{\omega_n}{\omega_{t2}}\right) - \frac{\omega_n^2}{\omega_{t2}}\right]} \tag{17}$$

Interestingly enough, the above expressions can be further simplified if we assume the special values for G_1 already obtained in the sensitivity analysis, i.e., $G_1 = 1/K$ and $G_1 = 1/(KQ)$. In the first case, we have $a_2 = Q\omega_n$, from where,

$$\delta \cong -\frac{1}{2}\frac{\omega_n}{\omega_{t2}}\left(Q + \frac{1}{Q}\right) \tag{18}$$

and

$$\eta \cong \frac{1}{1 - Q\left[\frac{\omega_n}{\omega_{t1}}\left(1 + \frac{\omega_n}{\omega_{t2}}\right)\right]} \tag{19}$$

In the second case, $a_2 = \omega_n$, from where

$$\delta \cong -\frac{1}{2}\frac{\omega_n}{\omega_{t2}}\left(1 + \frac{1}{Q}\right) \tag{20}$$

and

$$\eta \cong \frac{1}{1 - Q\left[\frac{\omega_n}{\omega_{t1}}\left(1 + \frac{\omega_n}{\omega_{t2}}\right) + \frac{\omega_n}{\omega_{t2}}\left(1 - \frac{1}{Q}\right)\right]} \tag{21}$$

As can be seen from Eq. (16) to Eq. (21), the deviation δ of the natural frequency is small and mainly determined by the unity-gain frequency of amplifier A_2. Also, it can be noted that the filter suffers from Q enhancement due to the finite bandwidth of the amplifiers. This enhancement is particularly relevant in filters with large Q factors. This, however, is not a serious drawback because, as it has been already mentioned, $Q = 1/\sqrt{2}$. For illustration purposes, Fig. 8 shows the influence of the finite unity-gain frequency of the opamps on the transfer characteristic of the CT-LP filter module. In Fig. 8(a), ω_{t2} is set to a fairly high unity-gain frequency of $2\pi \times 150\,\text{MHz}$ while ω_{t1} is made to swing from $2\pi \times 20\,\text{MHz}$ to $2\pi \times 150\,\text{MHz}$ (i.e., more than a 75 %

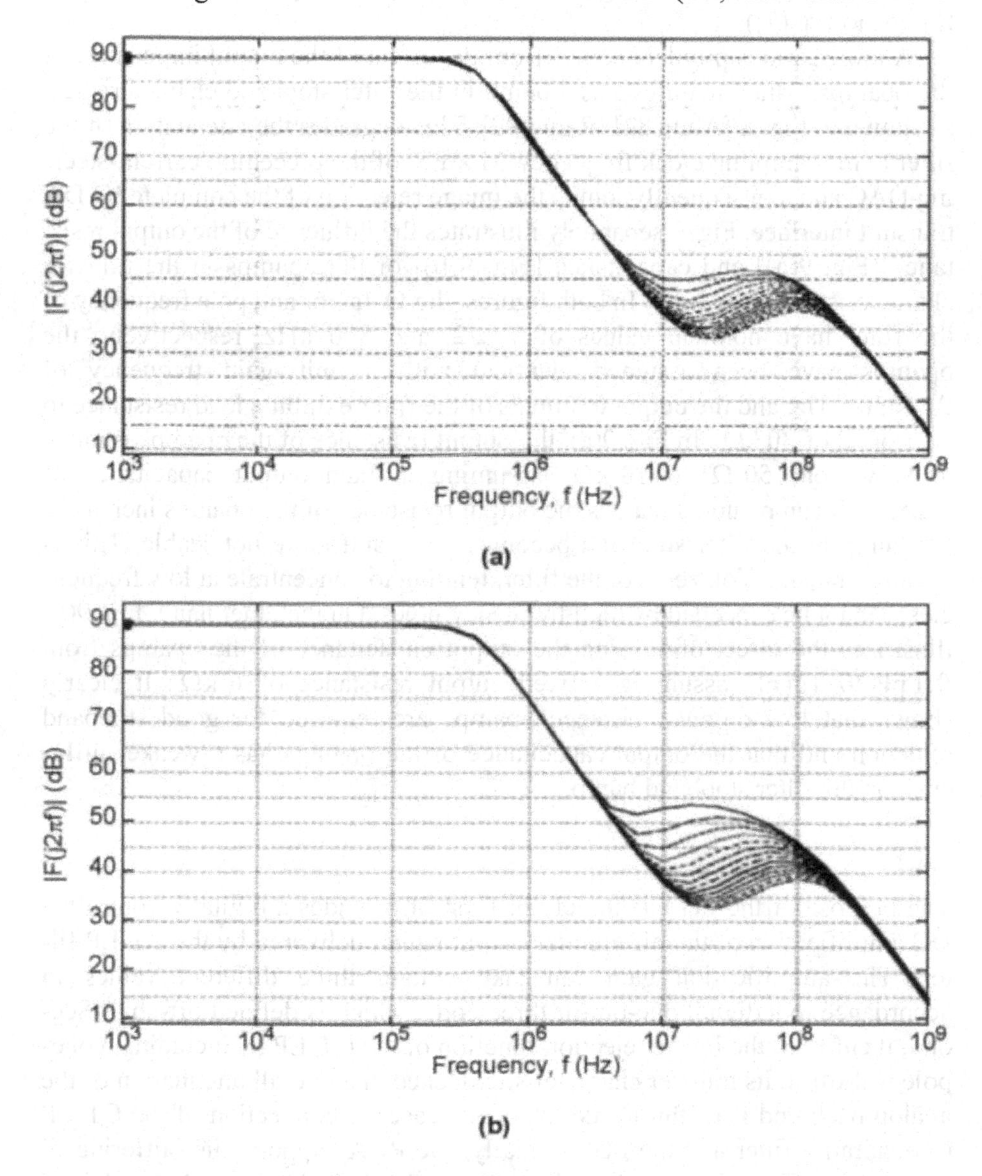

Figure 8. Influence of the bandwidth of the (a) first and (b) second opamps in the CT-LP filter transfer function.

variation around $2\pi \times 85\,\text{MHz}$); whereas in Fig. 8(b), the roles of ω_{t1} and ω_{t2} are swapped. In both cases, the Q factor and pole frequency of the filter have nominal values of $1/\sqrt{2}$ and 300 kHz, respectively. As can be seen, errors in the transfer function at the vicinity of the corner frequency are rather small (δ and η are around 2% in the worst cases) and comparatively below the relative deviations arising from mismatch among passive elements (see Eq. (5) to Eq. (7)).

A more serious problem comes from the effect of the output impedance of the opamps, which manifests as a bump in the filter stopband characteristics, as shown in Fig. 8 [Ahuja82] [Ramet88]. This degrades the attenuation of the filter at the sampling clock frequency $M \times FS$ of the preceding current-steering DAC and, consequently, limits the image rejection of the complete IQ DA transmit interface. Fig. 9 separately illustrates the influence of the output resistance [Fig. 9(a)] and capacitance [Fig. 9(b)] of the opamps in the transfer characteristics of the filter. In both figures, the Q factor and pole frequency of the filter have nominal values of $1/\sqrt{2}$ and 300 kHz, respectively; the opamps have been assumed identical with a unity-gain frequency of $2\pi \times 50\,\text{MHz}$; and the output terminals of the filter exhibit a load resistance to ac ground of $30\,\text{k}\Omega$. In Fig. 9(a), the output resistance of the opamps is made to vary from $50\,\Omega$ to $10\,\text{k}\Omega$, assuming a fixed output capacitance of 0.25pF. It can be noted that, as the output resistance of the opamps increases, the bump in the filter stopband becomes more and more noticeable. This is because transmission zeros of the filter, tending to concentrate at low frequencies, cannot be cancelled by high frequency poles. On the other hand, Fig. 9(b) illustrates the effect of varying the output capacitance of the opamps from $0.1\,\text{pF}$ to $1.4\,\text{pF}$, assuming a fixed output resistance of $6\,\text{k}\Omega$. It clearly shows that low output resistance opamps are required for good stopband rejection and that the output capacitance of the opamps has a weaker influence on the filter stopband bump.

2.4.2 The PGA

The purpose of the PGA is threefold. First, it provides a digitally-controlled DC amplification of the differential output signal delivered by the CT-LP filter. The amplification gain can take among three different values in accordance to a digital one-out-of three code, which is defined off-chip. Second, it enforces the image rejection function of the CT-LP by including a one-pole roll-off in its transfer characteristic. Hence, the overall attenuation of the analog back-end is of third order as of the cascade connection of the CT-LP (biquadratic) filter and the PGA. Finally, the PGA supports the buffering of its output differential signal to the external load of the chip. As explained above, depending on the particular load, an *economic* operation mode can be

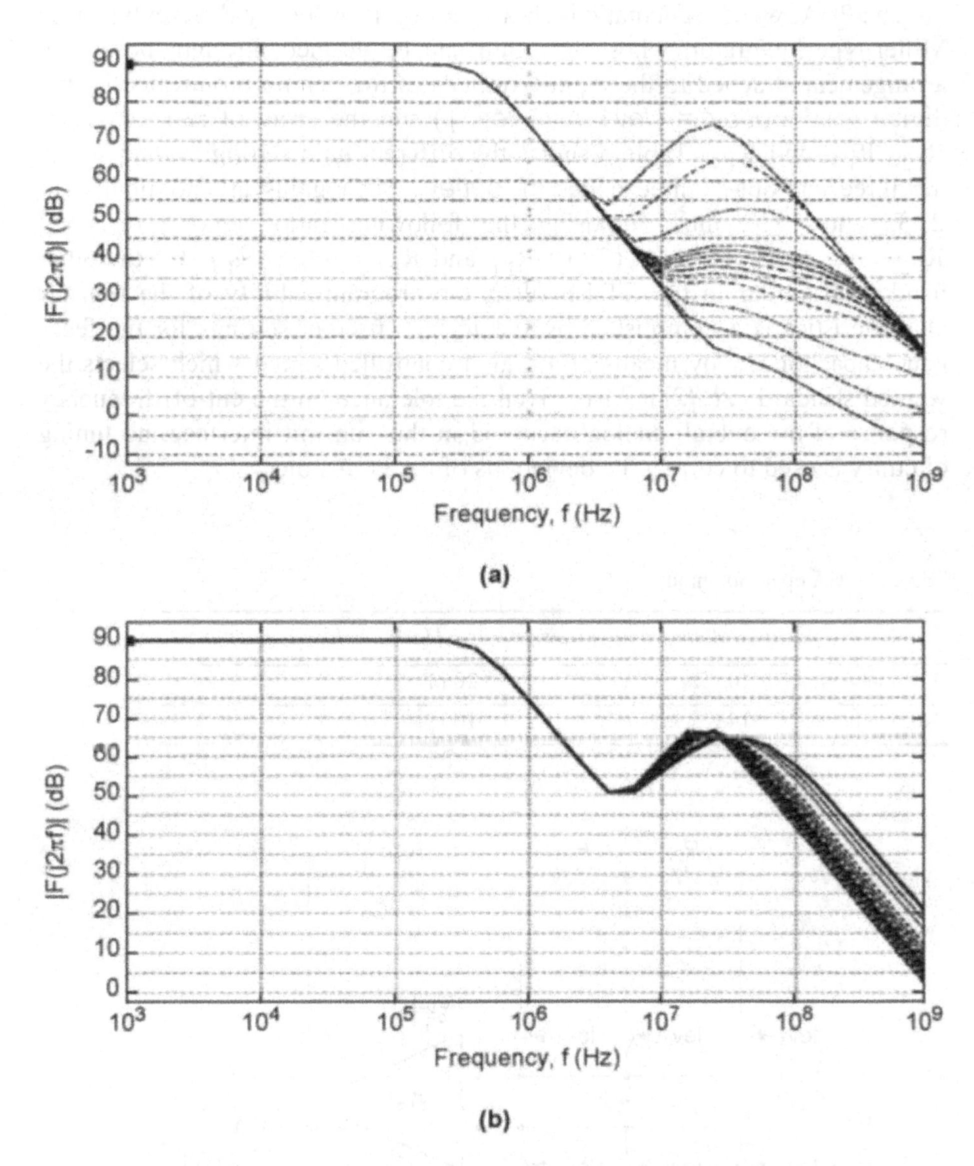

Figure 9. Effects of the output resistance (a) and capacitance (b) of the opamps in the CT-LP filter transfer function.

enabled for the sake of reducing the current consumption of the PGA block. Table 4 shows the worst-case external load conditions for the normal and *economic* PGA operation modes. Of course, all these objectives must be accomplished without degrading the high dynamic range and linearity provided by the previous CT-LP filter.

The PGA, whose schematic is shown in Fig. 10 is a fully-differential lossy Miller-type integrator, whose DC gain can be defined through the switch arrangement attached to the input terminals of the opamp (controlled by the digital word *outlev* = [*lev1, lev2, lev3*]) and the array of resistors R_{P1}, R_{P2}, R_{P3}, and R_{P4}. Table 5 shows the different gain settings associated to the three one-out-of-three codes of *outlev*. Such gains are nominally 1, $2/3$, and $1/3$, thus, imposing the following ratios among resistors: $R_{P2} = (1/5)R_{P1}$, $R_{P3} = (3/10)R_{P1}$ and $R_{P4} = (1/2)R_{P1}$. On the other hand, and similar to the CT-LP filter, the programmability of the transfer characteristics is accomplished by specifying different settings for the feedback capacitor C_P by means of a digital controlled signal which selects the wanted standard (*std[2]*). Also, given the tolerance on the cut-off frequency response of the overall analog back-end in the transmit interface, no tuning circuitry is used to correct the deviations of the PGA pole.

Table 4. PGA operation modes.

	Minimum load resistor	*Maximum load capacitor*
Economic mode	10 kΩ	20 pF
Normal mode	1 kΩ	100 pF

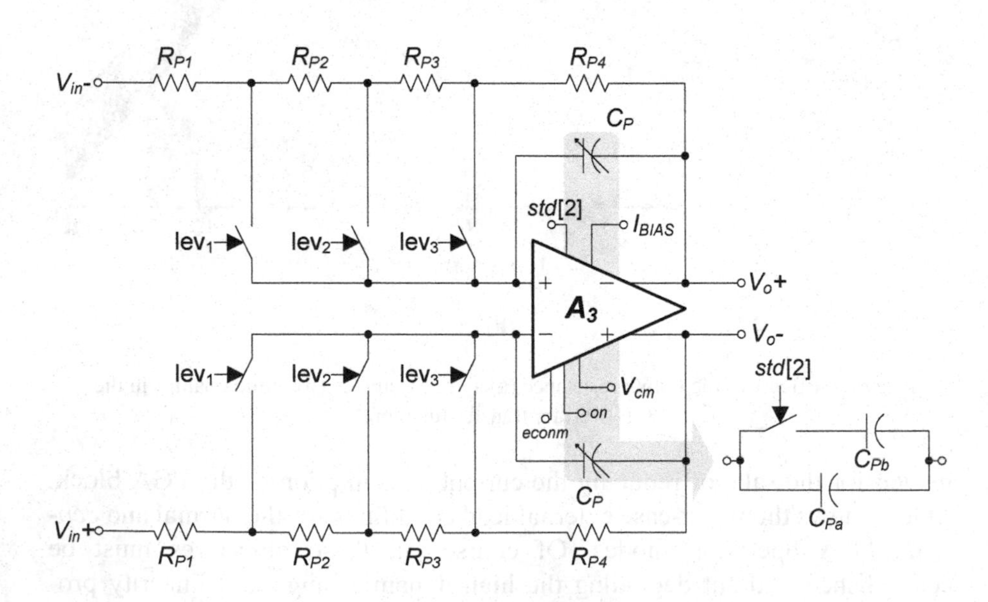

Figure 10. PGA schematic.

Table 5. PGA gain settings.

outlev	*DC gain, K*	*Differential output swing (peak-to-peak)*
[1,0,0]	$(R_{P2}+R_{P3}+R_{P4})/R_{P1}$	1.5 V
[0,1,0]	$(R_{P3}+R_{P4})/(R_{P2}+R_{P1})$	1.0 V
[0,0,1]	$R_{P4}/(R_{P3}+R_{P2}+R_{P1})$	0.5 V

The transfer function of the schematic in Fig. 10 can be obtained by simple inspection as

$$F_P(s) = \frac{V_o}{V_{in}}(s) = \frac{G_{PFF}}{\left(1+\frac{1}{A_3}\right)(G_{PFB}+C_{P1}s)+\frac{G_{PFF}}{A_3}} \quad (22)$$

where A_3 is the open-loop transfer characteristics of the PGA opamp, G_{PFF} and G_{PFB}, are, respectively, the driving and feedback conductances of the structure, whose particular values depend on the scaling factor defined by *outlev*. Assuming, at first instance, that the opamp in Fig. 10 is ideal (infinite gain and bandwidth), it can be seen that the angular frequency of the PGA pole is given by

$$\omega_n = 2\pi f_{p,\,PGA} = G_{PFB}/C_P, \quad (23)$$

and, hence, dependent on the selected DC amplification. Such DC gain is given by $K = G_{PFF}/G_{PFB}$, in accordance to the second column of Table 5.

On the one hand, and like the Tow-Thomas biquad, the PGA sensitivities to passive element mismatches can be easily derived, resulting:

$$S^{K}_{[R_{PFB},\, R_{PFF},\, C_P]} = [1, -1, 0] \quad (24)$$

$$S^{\omega_n}_{[R_{PFB},\, R_{PFF},\, C_P]} = [-1, 0, -1] \quad (25)$$

On the other hand, taking the finite opamp gain into account, the deviations on the PGA amplification and the pole angular frequency have the following form:

$$\zeta_g \equiv \frac{K_g - K}{K} = \frac{1}{1+\frac{(1+K)}{A_3}} - 1 \approx -\frac{(1+K)}{A_3} \quad (26)$$

$$\delta_g \equiv \frac{\omega_{ng} - \omega_n}{\omega_n} = \frac{K}{1+A_3} \quad (27)$$

where $\{K_g, \omega_{ng}\}$ represent the modified DC gain and pole frequency, with respect to the ideal ones $\{K, \omega_n\}$, as a consequence of the finite opamp gain.

Regarding the effect of the finite opamp bandwidth on the PGA transfer characteristic, it can be shown, following a procedure similar to that applied in the analysis of the CT-LP filter, that the deviation of the pole frequency can be approximated as

$$\delta \equiv \frac{\omega_{na} - \omega_n}{\omega_n} \cong -\frac{\omega_n}{\omega_{t3}}(1 + K) \tag{28}$$

where ω_{t3} is the unity-gain frequency of the opamp. The above approximation holds under the assumption that $\omega_n \ll \omega_{t3}$. As can be deduced from Eq. (26) to Eq. (28), there is no need for a high-gain, high-speed opamp to make the deviations on the DC gain and pole frequency of the PGA small as compared to those arising from mismatch among passive elements (see Eq. (24) and Eq. (25)). A major design challenge is the need for a strong driving capability (low resistive and high capacitive loads).

3 REUSABLE BLOCKS

The application of the reuse-based framework to the design of the analog back-end, comprising the characterization of the design reuse flow and the description of the analog reusable blocks employed, is reported in this section.

The design reuse flow presented in Chapter 2 (Fig. 5 on page 55) can be applied to the analog back-end of the IQ DA, by developing appropriate reusable blocks for the components of the hierarchy shown in Fig. 5. These analog reusable blocks (listed from the cell to the system level) are:

- **Opamp** cells. The behavioral facet of these cell-level blocks are used for the sizing of the CT-LP filter (opamps A_1 and A_2) and the PGA (opamp A_3). This facet can also be useful at module verification. The structural facet is used at the opamp-sizing phase and, provide parasitics that have not been included in the sizing process, at the cell verification. Last, the layout facet is employed at the layout generation phase.
- **CT-LP filter** and **PGA** modules. Equivalently, the layout facet is used at the module layout generation. Their structural facet is applied to the module sizing and verification. A reusable behavioral facet is employed for the sub-system (i.e., the analog back-end) sizing and verification.
- **Analog back-end** sub-system. To further improve the design flow, the whole back-end can also be made fully reusable. In such a case, the behavioral facet can be used at system level specification parti-

tioning, the structural facet can be employed for sub-system level sizing and verification, and the layout facet can be utilized for rapid layout generation of the whole analog back-end.

The design reuse flow presented in previous chapters is also flexible enough so as to be modifiable, according to the specific nature of the AMS system to design, gaining efficiency in the process. Thanks to this and due to the very nature of the analog back-end, a few simplifications of the design reuse flow are made that further reduce the overall design time:

1. The module-level components are of relatively low complexity. This implies that module sizing (i.e., obtaining the value of passive devices as well as the performance specifications of the opamps) can be carried out directly from these analog back-end specifications, considering, at the same time, the sizing of the CT-LP filter and the PGA.
2. This also suggests that module verification can be skipped, as verification of the analog back-end using either the behavioral facet of the reusable opamps or, if accuracy is critical, a complete device-level description can be carried out without requiring extremely demanding computational resources.
3. Since the sub-system specifications are directly derived from Table 3, the behavioral model for the whole analog back-end would only be used during verification of the whole transmit interface.

Therefore, only the structural and layout facet of the analog back-end reusable block are required. Regarding the CT-LP filter and PGA modules, there is no need for using their reusable facets, except for the layout one[5]. At the cell-level, complete opamp reusable blocks are required.

At this point, it is also necessary to point out that, in the present case, cell sizing is performed without considering the inclusion of layout knowledge, regarding both layout geometrical quality and parasitic-induced effects. In the following chapter, the description of the reuse-based design framework is completed with discussions on layout-aware synthesis techniques: it will be demonstrated that automation layout-aware analog synthesis is crucial to improve the quality of the solution (in terms of geometry and area) and to avoid time-consuming sizing-layout iterations.

---

5. Nevertheless, the layout templates of these modules can be useful to develop the template of the whole analog back-end. As explained in the precedent chapter, hierarchy facilitates the process of coding the layout knowledge. In our case, corresponding layout templates for the CT-LP filter and the PGA have been developed

While DRC and LVS are not required as the generated layout instances are correct-by-construction, layout formal verification (i.e., extraction) is certainly required to compute the performance degradation induced by layout parasitics. Afterwards, the layout of the complete analog back-end can be generated by using the corresponding layout facet; then, it is extracted (for the same reason as with the cell-level layout instances, no DRC and LVS are required) and verified.

The above-outlined analog back-end design reuse flow is detailed in Section 4, where it is illustrated with several examples. The following two sections describes the analog reusable blocks required for such design reuse flow.

3.1 Reusable blocks: opamps

The three operational amplifiers, A_1, A_2, and A_3 are very similar in topology. The three opamps are fully-differential topologies whose general, conceptual representation is shown in Fig. 11.

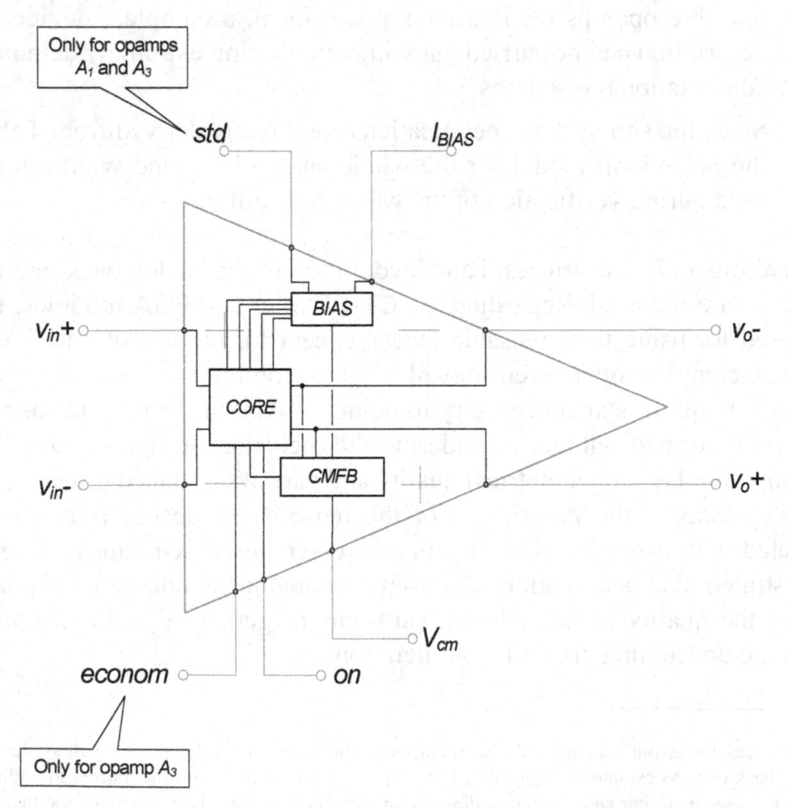

Figure 11. General composition of the opamps.

All three opamps in the analog back-end are implemented as a fully-differential two-stage topology in order to increase the open-loop gain and dynamic range. As shown in Fig. 12(a), this topology consists of a folded-cascode differential stage followed by two common-source amplifiers, one per differential output, and a current-steering common-mode feedback (CMFB) network. The structure uses Miller-type frequency compensation through capacitors C_{c1} and C_{c2} [Gray01]. Only in the case of the first amplifier of the CT-LP filter, a nulling resistor strategy (resistors R_{z1} and R_{z2} are connected in series to capacitors C_{c1} and C_{c2}, respectively) has been adopted to deal with the right half-plane zero arising from the feedforward path through the Miller capacitor[6].

Biasing circuitry is depicted in Fig. 12(b). The devices implementing the *economic* and *power-down* modes, as well as the circuitry enabling different wireless communication standards, are also show in Fig. 12 (inverters generating the $\overline{\texttt{on}}$ and $\overline{\texttt{econm}}$ signals are not displayed). The standard selection is accomplished by varying the input bias current to the core of opamps (see Fig. 12(b)). *Economic* and normal (*not-economic*) modes in the PGA's opamp A_3 are set by using the circuitry in Fig. 12(a) –transistors M_{e1}-M_{e6} and M_{12e}-M_{15e}– that modifies the opamp's output stage in order for the opamp to achieve the same performance features for both loading scenarios (see Table 4).

The behavioral facet of the opamps follows the macromodeling approach described in Chapter 3. According to the analysis in Section 2.4, both finite DC gain and finite bandwidth have a noticeable influence on the CT-LP filter and the PGA performances. In addition, it was shown that the opamp's output resistance needed to be correctly tuned in order to obtain a good stopband rejection, and that the output capacitance of the opamps had a weaker influence on the filter stopband bump. Consequently, the relevant model active parameters for the three opamps to take into account in the synthesis process of the analog back-end are the differential-mode DC open-loop voltage gain, `add_db`, the unity-gain frequency, `ft`, and the output resistance, r_{out}. The phase margin `pm` is also added to ensure correct stability of the feedback circuitries.

Following the approach of experienced designers, and due to the relative independence of the different components of the opamp reusable blocks (i.e.,

---

6. Note that these topologies were considered in the context of a first manual design of the analog back-end. As far as design reuse is concerned, different performance specifications may be targeted (different also meaning 'less demanding'), and the topologies might then result exceedingly suitable; the opposite, obviously, may also occur, i.e., the opamp topologies might result inadequate to address the intended specifications, scenario in which the design reuse flow would flag the specifications as being impossible to meet). The suitability of the topology to the intended performance specifications is not considered in this book, as we did not introduce the problem of topology selection as part of the design reuse flow.

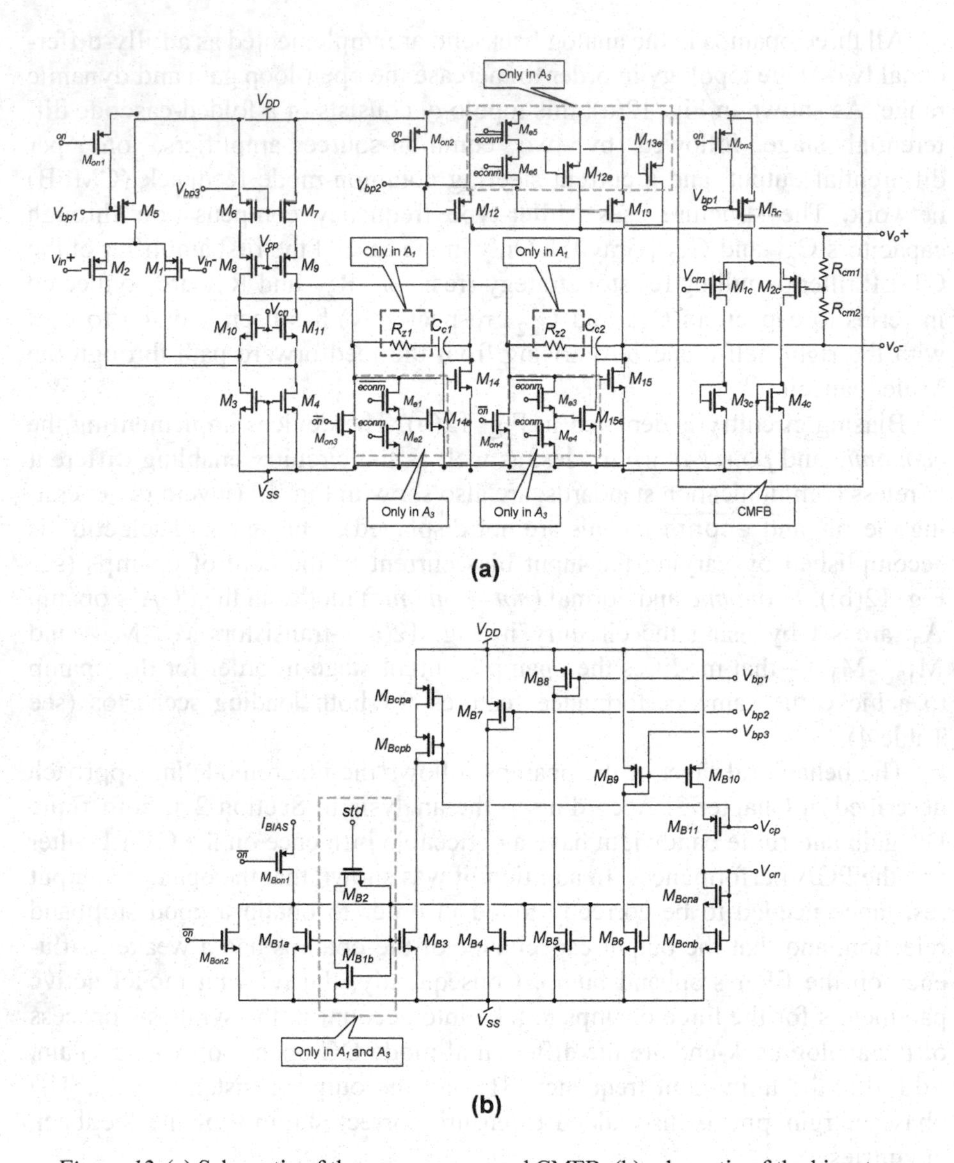

Figure 12. (a) Schematic of the opamps core and CMFB; (b) schematic of the bias stage.

the core, the CMFB, and the biasing circuitry), these are sized consecutively. That is, the core of the opamps is firstly synthesized, then each CMFB circuit is sized while the core is finely tuned to cope with the CMFB circuit loading effects. Finally, the bias devices are sized in accordance with the intended value of biasing current and voltages attained after the core and CMFB sizing phase.

As explained in Chapter 4 and shown in Fig. 13, apart from the parameterized netlist of the opamps, the design knowledge database included at the structural view of the reusable opamps, comprises a set of elements needed for the complete sizing of the opamps themselves.

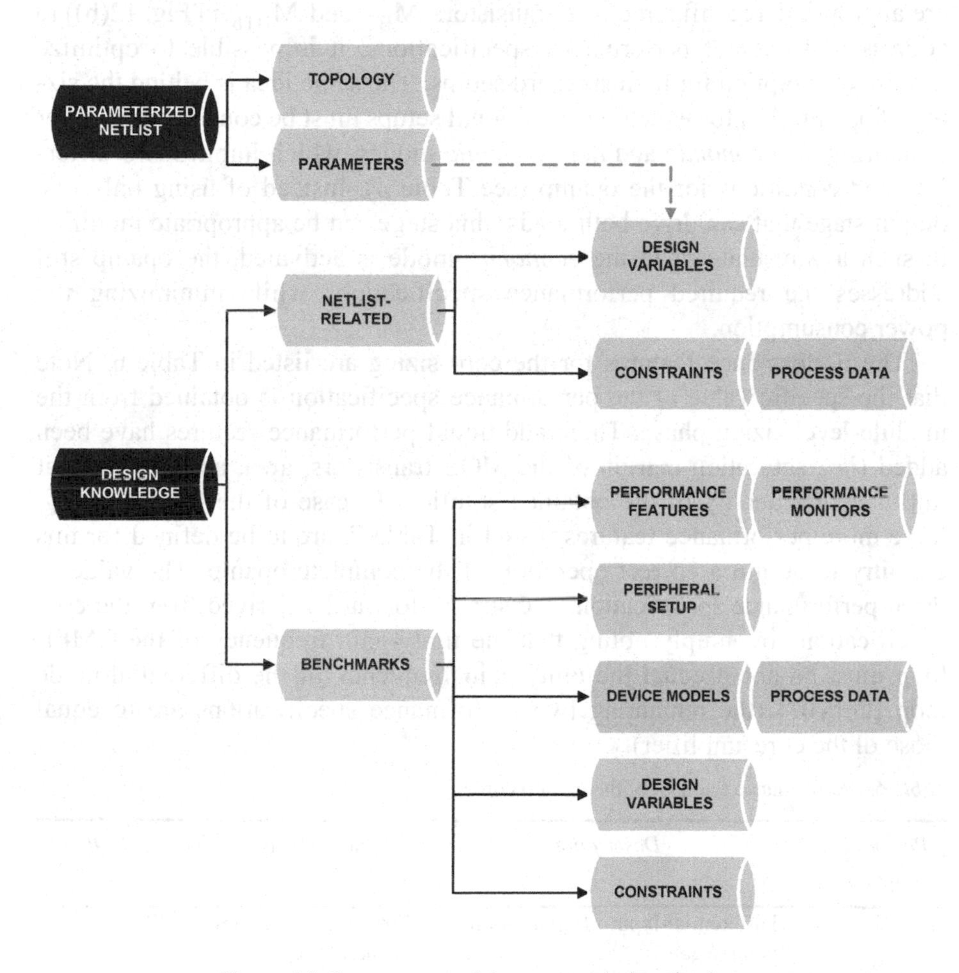

Figure 13. Components of the structural view database.

The netlist-related elements for the sizing of the opamp cores and CMFB circuits have been described in Section 2.1 of Chapter 4 (page 96). For the sizing of the opamp cores, a testbench setup is to be defined for each wireless standard setup the analog back-end has to address. Since there are two standard setups (one for the GSM, DCS1800, PCS 1900, AMPS, PDC, and Tetrapol standards, an another for the DECT, CDMA, and IS-95 standards, as shown in Table 2), two testbench setups must be defined. The purpose of this is to find out two sets of performance specifications for the opamps such that power consumption is optimized over the two standard sets. Obviously, it is also possible to obtain one only performance specification set that provides correct opamp functionality for both standard setups, but this is probably accomplished at the expense of larger power consumption. Instead, if opamps are appropriate reconfigured (see transistors M_{B2} and M_{B1b} in Fig. 12(b)) to address both sets of performance specifications, it is possible to optimize power consumption for both standard setups. The same idea is behind the sizing of opamp A_3, for which two additional setups must be considered in order to address the *economic* and *not-economic* modes, which implies two differing load conditions for the opamp (see Table 4). Instead of using only one output stage that can drive both loads, this stage can be appropriate modified in such a way that, when the *economic* mode is activated, the opamp still addresses the required performance specifications while minimizing the power consumption.

The performance features for the core sizing are listed in Table 6. Note that the specific value of the performance specification is obtained from the module-level sizing phase. Three additional performance features have been added (i.e., saturation margin of the MOS transistors, area, and power) that improves the quality of the obtained solution. In case of the CMFB sizing, three more performance features, listed in Table 7, are to be defined for this circuitry to permit a correct operation of the complete opamp. The value of these performance specifications are straightforwardly derived from the core specifications by simply noting that the unity-gain frequency of the CMFB loop must be about equal the unity-gain frequency of the differential-mode loop [Gray01] (the remaining two performance specifications are to equal those of the core amplifier).

Table 6. Performance features of the opamp cores.

Performance feature	*Description*	*Target/Directive*	*Unit*
add_db	Differential-input differential-output voltage gain	From the module-level sizing phase	dB
ft	Unity-gain frequency		Hz
pm	Phase margin		°
r_{out}	Output resistance		kΩ
DM_i	Saturation margin of transistor M_i	> 110	% of VDS_{sat}
Area		minimize	μm^2
Power		minimize	mW

Table 7. Additional performance features of the opamp CMFB circuitries

Performance feature	*Description*	*Target/Directive*	*Unit*
add_cm_db	Differential-mode DC open-loop voltage gain	From the core performance specifications	dB
ft_cm	Unity-gain frequency		Hz
pm_cm	Phase margin		°

The peripheral setup elements for the core sizing have been described in Section 2.2.2 of Chapter 4 (page 109), with the only difference relying in an ideal CMFB circuitry.

The sizing of the CMFB circuitry is carried out by using the setup illustrated in Fig. 14. The input voltages of the core amplifier ($v_{in}+$ and $v_{in}-$) are set to zero, while an AC input voltage is applied to the CMFB input nodes ($v_{inCMFB}+$ and $v_{inCMFB}-$). In order to correctly set the common-mode input level to the CMFB circuitry, a DC input voltage is also applied; this voltage is the common-mode output voltage which has been obtained from a previous simulation of the complete amplifier (core + CMFB). The output voltages (v_o+ and v_o-) are measured by breaking the loop between the CMFB and the core amplifier as shown in Fig. 14.

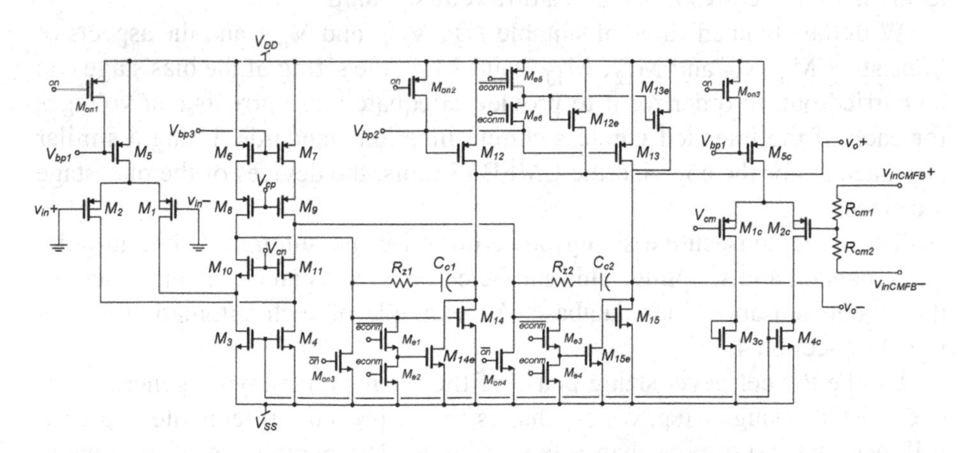

Figure 14. Configuration used to size the CMFB circuitry.

Important elements for the sizing of both the opamp cores and the CMFB circuitries are the loading conditions. The measurement of these loading conditions takes place right after the previous sub-system-level sizing has finished, when all passive devices and opamp required performances have been obtained. Then, separate simulations of the analog back-end, including the passive network, the opamp macromodels (which, as shown in Chapter 3, feature input and output impedances), and the loads listed in Table 4, are carried out so the effective load that each opamp is driving can be ascertained.

Regarding the technological information (i.e., device models and process data), it is important to note that, for the particular case of the analog back-end and even though process variations are involved in the design of this analog sub-system, it is not strictly necessary to consider statistical models (such as Monte Carlo models) for the opamp devices. This is due to the fact that, as shown in Section 2.4, errors in the CT-LP filter and PGA transfer functions due to variations of the opamp characteristics are rather small and well below the deviations arising from mismatch among passive elements. That is, while mismatch in the passive components and non-ideal active components respectively induce second-order deviations of the transfer functions from ideality, mismatch in the active components, resulting in different opamp performances between channels I and Q, leads to third-order deviations.

Additional design variables and constraints are required, as explained in Section 2.2.4 of Chapter 4 (page 113) to complete the peripheral setups. In particular, the bias current IB and the bias voltages V_{cn} and V_{cp} are included. It is important to note that the bias current IB may change from one wireless communication standard to another. For that reason, a separate design variable accounting for the bias current should be considered for each testbench setup corresponding to a different standard.

With the obtained value of suitable IB, V_{cn}, and V_{cp}, and the aspects of transistors M_5-M_7 and M_{12}, M_{13} [7], and M_{5c}, the sizing of the bias stage can be carried out, in order for it to provide adequate bias currents and voltages for each of the intended wireless communication standards. Using a similar approach as for the core and the CMFB circuits, the devices of the bias stage are sized.

The three consecutive sizing procedures can be automated through adequate parsing and scripting. This can be considered as an extra component of the opamp reusable block database. An example of such automation is provided in Section 4.4.

Unlike the cell-level sizing phase of the opamps, the layout generation is executed at a single step, which, thanks to the procedural template approach followed, takes no more than a few seconds. The parameters of the opamp layout templates include the width, length, and multiplicity of all the MOS transistors, the width, length, and number of strips of the resistors R_{z1}, R_{z2}, R_{cm1}, and R_{cm2}, and the width and length of the capacitors C_{c1} and C_{c2}.

[7] Including M_{12e} and M_{13e} for the PGA opamp.

The multiplicities are defined prior the instancing of the layout templates and it is carried out by the user[8], to improve the compactness and area occupation of the resulting designs.

3.2 Reusable blocks: analog back-end

The structural view database for the analog back-end reusable block comprises equivalent elements to those displayed in Fig. 13. This section describes these elements as well as the layout view of the analog back-end reusable block.

The performance features that must be defined to complete the sizing of the analog back-end were listed in Table 3. Table 8 shows an enlarged list of performance features. In addition to the features in Table 3 (outlined in Table 8), a new set of performance features are included that improves the reliability of the obtained sizing solution: performance feature bumpsz controls that the bump in the stopband does not degrades the overall attenuation of the analog back-end; performance feature apeak helps reducing the peaking of the passband characteristic; last but not least, fpolectf_dev, qfactctf_dev, and fpolepga_dev are used to control the deviations from ideal of the CT-LP filter and PGA transfer functions due to mismatch and process variations. Table 8 also displays a sample of the complementary performance monitor for each performance feature.

The specific value of the corresponding performance specifications depends upon the intended wireless communication standard. Note that if the analog back-end is to be synthesized with the *multi-standard* capability (i.e., the capability to deal with more than a single group of wireless communication standard performance requirements thanks to a set of digitally programmed devices), a set of performance specifications for each standard has to be defined.

The performance features in Table 8 entail that statistical variations must be taken into account. Two setups are thus considered: in the first one, only mismatch variations are taken into account to deal with the mismatch specifications (i.e., adcra, gdavgra, and pera), which corresponds to the intra-die variation scenario and will be therefore referred as the *INTRA*-die setup. In the second one, mismatch and process variations are altogether considered, i.e., a combination of the intra and the inter-die scenarios. This last setup, which for the sake of simplicity we will hereafter refer as the *INTER*-die setup, is required to fit all the performance specifications in the worst-case scenario. Note also all performance specifications but apeak and adcra have to be moni-

[8] In the following chapter, a method to automate this step by including the multiplicity variables during the sizing phase is explained.

Table 8. Performance features and monitors of the analog back-end.

Performance feature	*Description*	*Performance monitor*
apeak	Gain peaking	max(vdb(OUT)) - vdb(OUT)@1Hz[a,b]
adcra	I/Q gain mismatch	vdb(OUT)@1Hz
imar	Minimum attenuation of the analog back-end in the frequency range [FS, Fend][c]	min(vdb(OUT)@1Hz - vdb(OUT)@Fi) with Fi ∈ [FS, Fend]
bumpsz	Stopband bump size	Compute the difference between the maximum and minimum of the bump in the stopband of vdb(OUT). Return 0 if there is no bump.
gddev	Group delay deviation (from DC to SB)	abs(max(vt[d](OUT) from 100Hz to SB)-min(vt(OUT) from 100Hz to SB)
gdavgra	I/Q group delay mismatch (from DC to SB)	avg(vt(OUT) from 100Hz to SB)
pera	I/Q phase mismatch @SB	vp[e](OUT) at SB
fpolectf_dev	Deviation of the CT-LP filter pole	CT-LP filter pole calculated through Eq(3)
qfactctf_dev	Deviation of the pole quality factor of the CT-LP filter	Quality factor calculated through Eq(3)
fpolepga_dev	Deviation of the PGA pole	PGA pole calculated through Eq(23)

a. *vdb* stands for voltage in decibels.

b. OUT is the analog back-end differential output node.

c. Fend is the upper limit value of the frequency range defined in the AC simulation setup.

d. *vt* stands for group delay.

e. *vp* stands for phase.

tored differently for every wireless communication standard the IQ DA transmit interface is intended to address, as they use the sampling frequency, FS, or the signal bandwidth values, SB, which have different values for different standards.

The peripheral setup used to evaluate the performance of the analog back-end is illustrated in Fig. 15. This setup is used both for sizing and for verification. Regarding the type of simulation to perform, an AC analysis is required (up to 1GHz) including the *INTRA*-die and *INTER*-die setups. This type of simulations is commonly carried out by using Monte Carlo analysis. For the

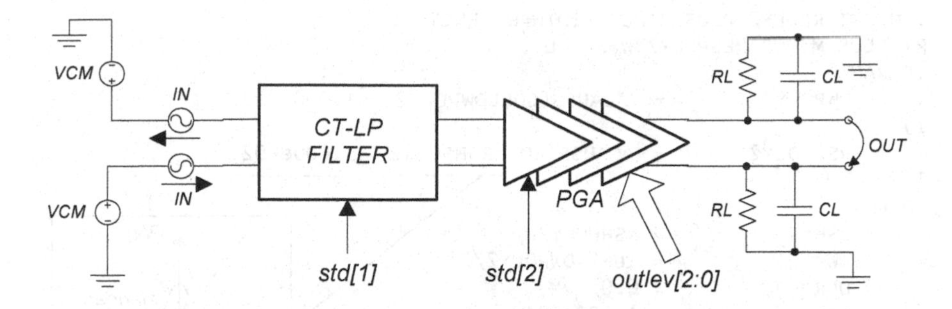

Figure 15. Peripheral setup for the sizing and verification of the analog back-end.

analog back-end, at least 30 Monte Carlo evaluations need to be run for each performance evaluation[9].

As mentioned at the beginning of Section 3, behavioral models are used instead of the full device-level descriptions of the opamps (see Section 3.1). Regarding the passive devices (R_1, R_{FF}, R_{FB}, C_1, and C_2 for the CT-LP filter, R_{P1}, R_{P4}, and C_P in the PGA), a detailed device-level model, including the mismatch and process variations dependences, must be used. For instance, Fig. 16 shows a resistor model for a 0.35-µm CMOS technology, which uses a second polysilicon layer (POLY2). Such a model, written in a HSPICE-like syntax [Hspi04], features a typical model expressed as:

$$R = R_{\square} \cdot \frac{L}{(W - DW)} \tag{29}$$

with L and W being the total length and width of the resistor, $R_{\square}$ the sheet resistance in ohms per square of resistive layer ($\Omega/\square$), and DW the width reduction such that the effective width is $W_{eff} = W - DW$. To model mismatch and process variations, both $R_{\square}$ and DW are determined by Gaussian distributions rather than by a fixed value. The mean value of the Gaussian probability distribution in the *INTRA*-die setup is a fixed-value whereas for the *INTER*-die setup it is represented by an uniform probability distribution.

Similar models are used for the capacitive devices C_1, C_2, and C_P. By way of example, a typical model for the same 0.35-µm CMOS technology is:

---

9. Statistical significance of 30 Monte Carlo runs for synthesis is quite high. If the circuit operates correctly for all 30 runs, there is a 99% probability that over 80% of all possible component values operate correctly, which concerning yield may be regarded as a poor percentage but for synthesis is acceptable. The relative error of a quantity determined by means of Monte Carlo analysis decreases as $1/\sqrt{N}$ with N being the number of runs [Hspi04]. This means that, if the number of iterations specified in the optimization algorithm options is n the total number of simulations for each standard is $n \cdot 30$, which can be very time consuming since n typically hovers around the 1000-2000 iterations. This is why behavioral models of the opamps are essential during the analog back-end design reuse flow.

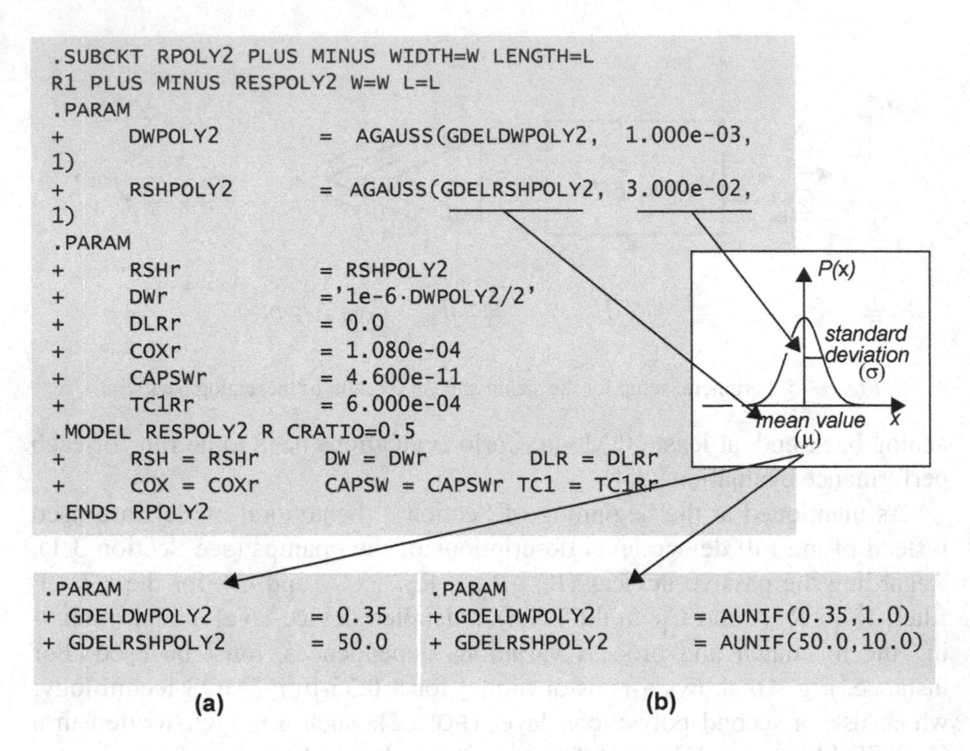

.SUBCKT RPOLY2 PLUS MINUS WIDTH=W LENGTH=L
R1 PLUS MINUS RESPOLY2 W=W L=L
.PARAM
+ DWPOLY2 = AGAUSS(GDELDWPOLY2, 1.000e-03, 1)
+ RSHPOLY2 = AGAUSS(GDELRSHPOLY2, 3.000e-02, 1)
.PARAM
+ RSHr = RSHPOLY2
+ DWr ='1e-6·DWPOLY2/2'
+ DLRr = 0.0
+ COXr = 1.080e-04
+ CAPSWr = 4.600e-11
+ TC1Rr = 6.000e-04
.MODEL RESPOLY2 R CRATIO=0.5
+ RSH = RSHr DW = DWr DLR = DLRr
+ COX = COXr CAPSW = CAPSWr TC1 = TC1Rr
.ENDS RPOLY2
P(x)
standard deviation (σ)
mean value (μ)
x
.PARAM
+ GDELDWPOLY2 = 0.35
+ GDELRSHPOLY2 = 50.0
.PARAM
+ GDELDWPOLY2 = AUNIF(0.35,1.0)
+ GDELRSHPOLY2 = AUNIF(50.0,10.0)
(a)
(b)

Figure 16. Resistor model for a 0.35-μm technology for (a) INTRA-die and (b) INTER-die setups.

$$C = C_A \cdot W \cdot L + C_P \cdot 2 \cdot (W + L) \tag{30}$$

where W and L are the width and length of the rectangular capacitor, and C_A, C_P are the capacitance per unit of area and unit of perimeter, respectively, of the mask layer used. To model mismatch effects and process variations, C_A, C_P are also defined as statistical distributions.

The design variables used in the sizing process are listed in Table 9. As explained in Section 3.2.2 of Chapter 2 (page 56) and in the introductory section of Section 3, among the design variables of this sub-system sizing process there are the performance features of the opamp behavioral models (the final obtained values of these design variables are subsequently transmitted down to the cell-level sizing process). These performance features were discussed in the previous section. Recall also from this section that when multi-standard programmability is addressed, it is required to define a set of opamp design variables for each of the two standard setups controlled by the digital word *std[2:0]*.

The rest of design variables concern the passive devices. As mentioned in Section 2.4, it seems a good design choice to set either $R_1 = K$ or

Table 9. Sizing design variables.

Module-level block	*Cell-level blocks & passive devices*	*Design variable*	*Description*	*Units*
CT-LP filter	OPAMP A_1	add_db	DC gain	dB
		ft	Unity-gain frequency	Hz
		pm	Phase margin	o
		r_{out}	Output resistance	kΩ
	OPAMP A_2	add_db	DC gain	dB
		ft	Unity-gain frequency	Hz
		pm	Phase margin	o
		r_{out}	Output resistance	kΩ
	Passive devices	wctfR1	R_1 resistor width	μm
		wctfRFF	R_{FF} resistor width	μm
		wdtfRFB	R_{FB} resistor width	μm
		fpolectf	CTF pole	Hz
		rr1	Resistance value of R_1	kΩ
PGA	OPAMP A_3	add_db	DC gain	dB
		ft	Unity-gain frequency	Hz
		pm	Phase margin	o
		r_{out}	Output resistance	kΩ
	Passive devices	wpga	Resistor width	μm
		lpga	Resistor length	μm
		fpolepga	PGA pole	Hz

$R_1 = KQ$. in order to reduce the gain-sensitivity products of the CT-LP filter, also favouring matching among resistors. Unfortunately, these choices lead to large capacitance values C_1 and C_2. A different way of sizing the passive devices is then required. In particular, two design variables are defined, namely rr1, the value of resistor R_1, and fpolectf, the cut-off frequency of the CT-LP filter ($\omega_n = 2 \cdot \pi \cdot \text{fpolectf}$). The variation range for rr1 is set such that the it is always larger than K, in order to control the area of the capacitances C_1 and C_2. With rr1 and fpolectf, the values of the rest of passive devices of the CT-LP filter are calculated, as explained in Table 10. Table 10 also indicates the design variables related to the width of resistors R_1, R_{FF}, and R_{FB}, which allows the optimization of the resistor areas.

Table 10. Constraints for the passive devices.

Passive device	*Value*	*Width*	*Length*
R_1	rr1	wctfR1	Eq(29)
R_{FF}	Using $Q = 1/\sqrt{2}$ and $C_1 = C_2$ in Eq(3)	wctfRFF	Eq(29)
R_{FB}	Eq(4)	wctfRFB	Eq(29)
C_1, C_2	Using R_{FF}, R_{FB}, and $2 \cdot \pi \cdot$fpolectf as ω_n in Eq(3)	Using Eq(30) with $W = L$	
R_{P1}	Eq(29)	wpga	$10 \cdot$lpga[a]
R_{P2}	Eq(29)	wpga	$2 \cdot$lpga
R_{P3}	Eq(29)	wpga	$3 \cdot$lpga
R_{P4}	Eq(29)	wpga	$5 \cdot$lpga
C_P	Using fpolega in Eq(23) with $G_{PFB} = R_{P2} + R_{P3} + R_{P4}$ [b]	Using Eq(30) with $W = L$	

a. In this way, the ratios between the PGA resistors ($R_{P2} = (1/5)R_{P1}$, $R_{P3} = (3/10)R_{P1}$, and $R_{P4} = (1/2)R_{P1}$) are ensured, therefore attaining the required nominal gains 1, $2/3$, and $1/3$.

b. This corresponds to outlev[2:0] = [1, 0, 0].

A similar but more direct approach is followed to size the PGA's passive devices. The design variables are fpolepga, the cut-off frequency of the PGA, and the width and length, wpga and lpga, of a unitary resistor by means of which each resistor in the PGA is made up to correctly achieve the different gain settings (see Table 5).

As depicted in Figures 6 and 10, passive elements are actually made up of various passive component and switches to implement the multi-standard capability (e.g., R_1 is composed of R_{1a}, R_{1b}, and the switch controlled by signal **std***[1]*). This means that when performing the analog back-end sizing, two different values of R_1, R_{FF}, C_1, C_2, and C_P have to be obtained, one for each standard setup. Eventually, these two values are implemented by assigning proper values to the passive components of each passive device (e.g., for the standard setup *std[1,1,1]*[10], the value obtained of R_1 corresponds to R_{1a}, while the value obtained for the standard setup *std[0,0,0]*[11]

10. Corresponding to communication standards GSM, DCS1800, PCS 1900, AMPS, PDC, and Tetrapol, which, for the sake of simplicity, will be hereinafter referred as the GSM standard setup.

11. Corresponding to communication standards DECT, CDMA, and IS-95, which, for the sake of simplicity, will be hereinafter referred as the DECT standard setup.

corresponds to the combination $R_{1a} \| R_{1b}$, from which the value of R_{1b} can be worked out). In addition, the passive components of each passive device should be made up of unitary devices at the layout level, mainly to improve matching and area occupation. Furthermore, the passive devices of the two differential paths (e.g., R_1 above and below amplifier A_1 in Fig. 6) should be also laid out such that matching between them is also improved. For instance, to implement C_1, composed of C_{1a} and C_{1b}, it has been used a common-centroid capacitor array in which both passive component is divided into unitary capacitors (see Fig. 49 on page 198 of Chapter 5); similar layout techniques have been used for the remaining passive devices. This all implies that the layout phase might become increasingly complex, since it has to deal with varying passive values, divided into several components, in their turn divided into unitary devices, which should be laid out in complex arrangements to improve device matching and area occupation.

To ease such layout implementation, a set of constraints have been defined that relate the value of passive devices from one standard setup to the other. These constraints are collected in Table 11. By using these constraints, the values of the passive devices for each standard setup are bound to fulfill a certain ratio (and so their passive components), which ultimately simplifies the layout phase as the arrangement can be easily specified beforehand in the layout template. As noted in Table 11, a direct consequence is that design

Table 11. Multi-standard constraints for the passive devices.

Passive device	*Standard setup*		*Constrainedratio (GSM to DECT)*	*Which implies that...*
	std[2:0] = [1,1,1] (GSM)	*std[2:0] = [0,0,0] (DECT)*		
R_1	R_{1a} [a]	$R_{1a} \| R_{1b}$	3	$R_{1b}=R_{1a}/2$
R_{FF} [b]	$R_{FFa}+R_{FFb}$	R_{FFa}	9	$R_{FFb}=8 \cdot R_{FFa}$
C_1, C_2 [c]	$C_{1a}+C_{1b}$	C_{1a}	2	$C_{1a}=C_{1b}$
C_P [d]	$C_{Pa}+C_{Pb}$	C_{Pa}	4	$C_{Pb}=3 \cdot C_{Pa}$

a. R_{1a}, R_{1b}, R_{FFa}, R_{FFb}, C_{1a}, C_{1b}, C_{Pa}, and C_{Pb} are the components of passive devices R_1, R_{FF}, C_1, and C_P, respectively (see Fig. 6 and Fig. 10). Switches controlled by the digital word *std[2:0]* connect or disconnect these components in order to compose appropriate passive devices for each standard setup, thereby attaining appropriate values of the CT-LP filter and PGA nominal characteristics (gain, pole frequency, and, additionally for the CT-LP filter, the pole quality factor).

b. This constraint comes up as a consequence of constraint on R_1 and the pursued value of the pole quality factor ($Q = 1/\sqrt{2}$) of the CT-LP filter.

c. This constraint implies that $\mathrm{fpolectf}_{DECT} = 6 \cdot \mathrm{fpolectf}_{GSM}$.

d. This constraint implies that $\mathrm{fpolega}_{DECT} = 4 \cdot \mathrm{fpolega}_{GSM}$.

variables fpolectf and fpolega have also their values constrained from one standard to the other.

The layout template of the analog back-end comprises the layout templates for the CT-LP filter and the PGA. Although the latter blocks were not defined as reusable blocks, it is also very useful to construct the analog back-end layout template in a fully hierarchical way, as explained in the precedent chapter.

Figure 17 illustrates the layout template of the CT-LP filter. The floorplanning style suggested in Chapter 5 has been followed. Likewise, the layout template for the PGA, shown in Fig. 18, follows the same floorplanning style. By way of illustration, Fig. 19 shows several examples of layout retargeting (i.e., automatic updating of the layout template parameters) and layout migration, for the PGA case. The floorplanning of the whole analog back-end is shown in Fig. 20

4 DESIGN EXAMPLES

This section illustrates the design reuse flow of the analog back-end with several design examples. The first, introductory example is the cell-level sizing

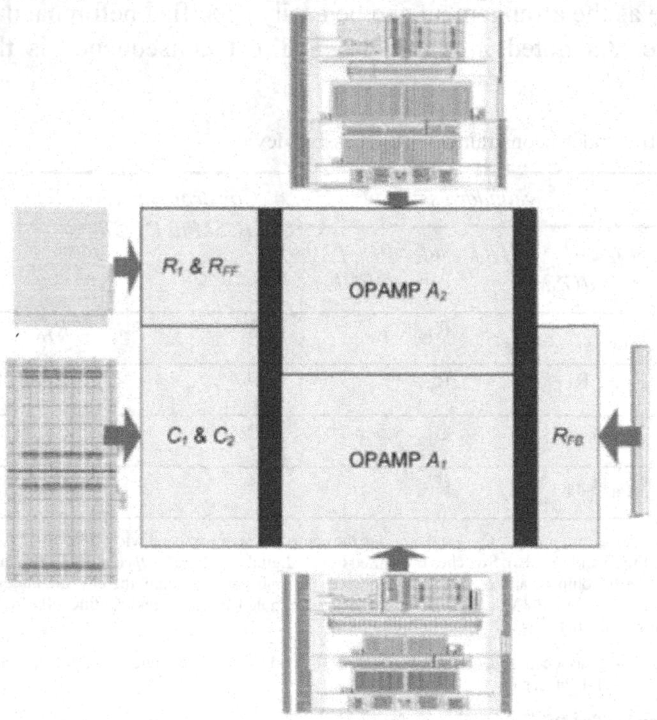

Figure 17. CT-LP filter layout template.

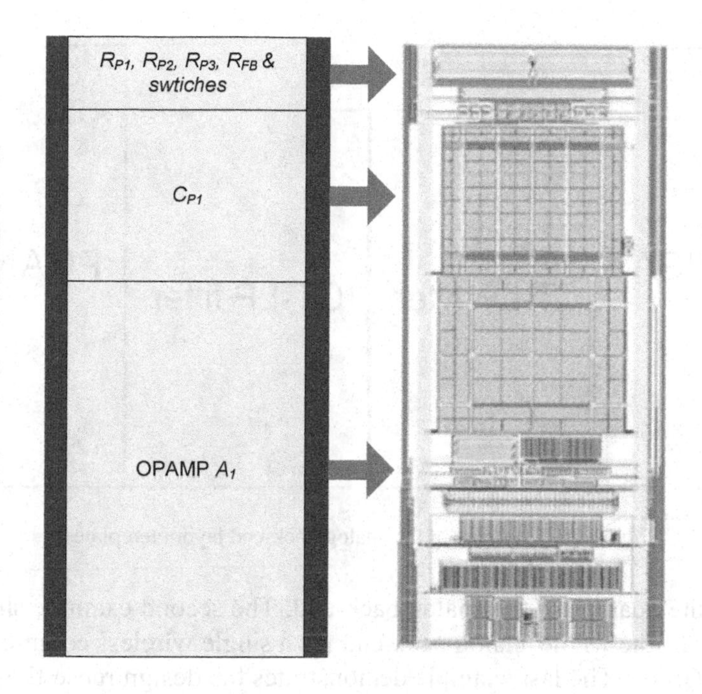

Figure 18. PGA layout template.

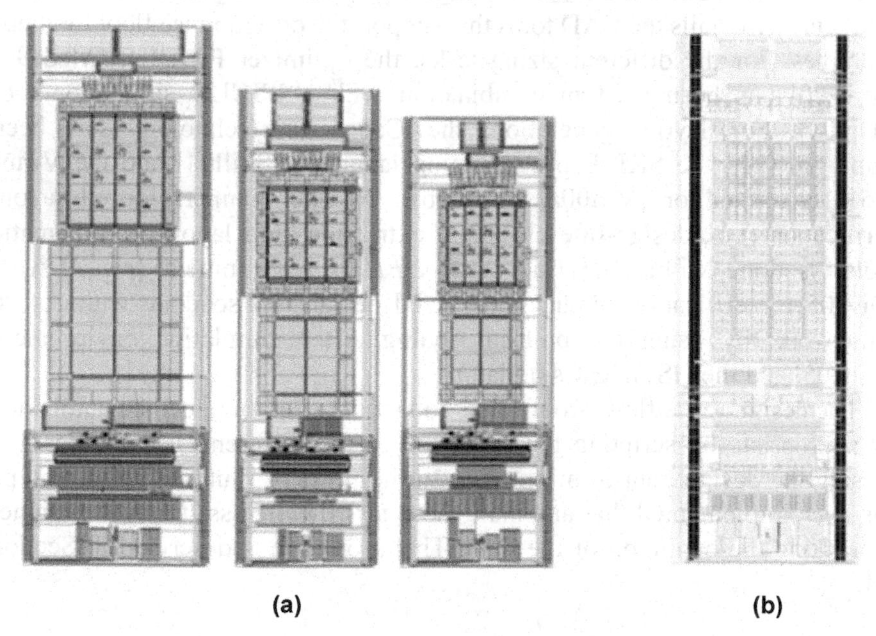

Figure 19. (a) Several examples of layout retargeting (0.35-μm technology); (b) example of layout migration (0.5-μm technology).

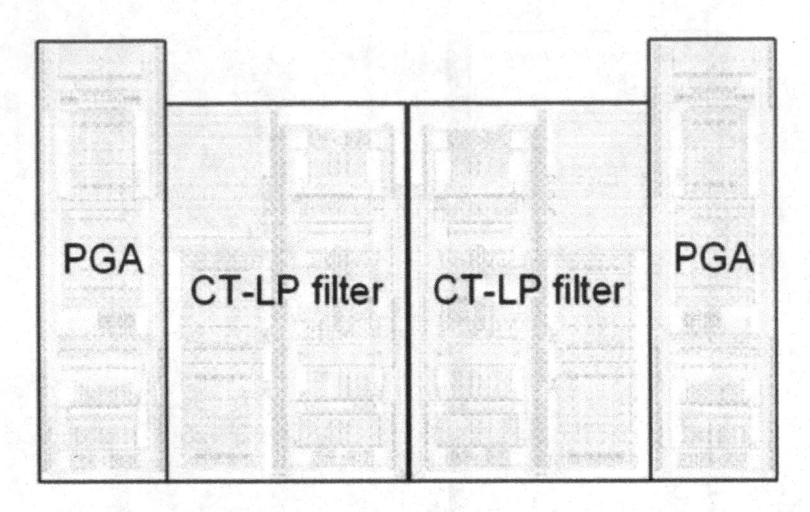

Figure 20. Floorplan of the analog back-end layout template.

of one of the opamps of the analog back-end. The second example shows the retargeting of the whole analog back-end for a single wireless communication standard (GSM). The last example demonstrates the design reuse flow for the multi-standard (GSM and DECT) retargeting of the analog back-end.

Figure 21 details the CAD tools that support the design reuse flow for these examples. For the different sizing tasks, the optimizer FRIDGE [Mede99] [Perez00] has been used in combination with HSPICE® simulation tool [Hspi04]. For layout generation, the Cadence® technology has been employed (i.e., the SKILL programming language [Skill04] and the Virtuoso® layout editor [Virt00a] [Virt00b]). In like manner, formal layout verification (i.e., design-rule checking, extraction, and layout-*vs.*-schematic tools) is done within the Cadence® *DFII II* environment (e.g., Diva® [Diva05]). Verification of the synthesized design is also done within such framework, by using the built-in Analog Artist simulation environment [Cade05] and the HSPICE® simulator.

The design reuse flow, from sizing to layout generation, can be automated by appropriate by scripting the execution of the different design steps and parsing the design data from one step to the next. An automation prototype has been implemented that automate these tasks while assisting the designer in tracking the evolution of the flow. This prototype is described in Section 4.4.

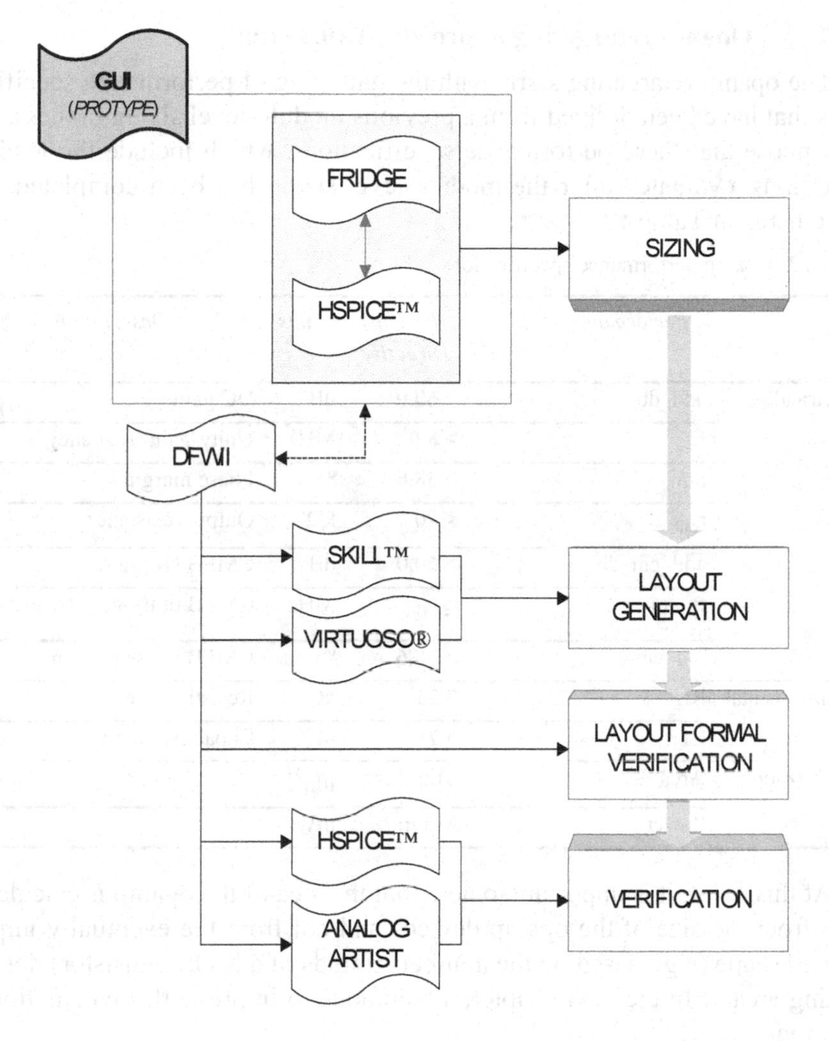

Figure 21. Supporting CAD tools of the design reuse flow.

4.1 Design example (I): design retargeting and migration of the opamp

In this introductory design example, the second operational amplifier of the CT-LP filter (A_2) is synthesized using the analog reusable block and its databases for sizing and layout generation, explained in Section 3.1. First, the opamp is synthesized in a 0.35-µm CMOS process and then this design is ported to a different-foundry 0.5-µm CMOS fabrication process[12].

[12] Both processes feature a 3.3V supply voltage range, three metal layers, and two polysilicon layers.

4.1.1 Opamp retargeting in process A (0.35 μm)

The opamp retargeting starts with the gathering of performance specifications that have been derived from a previous module-level sizing process. Let us suppose that these performance specifications, which include the loading conditions, evaluated after the module-level sizing has been completed, are those listed in Table 12.

Table 12. Opamp performance specifications.

	Performance feature	*Target/ Directive*	*Units*	*Description*
Electrical	add_db	> 60.0	dB	DC gain
	ft	> 8.0	MHz	Unity-gain frequency
	pm	> 38.6	°	Phase margin
	r_{out}	< 70	kΩ	Output resistance
	add_cm_db	> 60.0	dB	CMFB DC gain
	ft_cm	≈ ft	MHz	CMFB unity-gain frequency
	pm_cm	> 38.6	°	CMFB phase margin
Environmental	RL	7.23	kΩ	Resistive load
	CL	5.73	pF	Capacitive load
Optimization	Area	Minimize	μm^2	
	Power	Minimize	mW	

At this point, it is important to note that the area of the opamp is calculated only from the size of the opamp devices and not from the eventually implemented shape (e.g., based on the number of folds of a MOS transistor) nor the routing area[13]. In the next chapter, a technique to improve this calculation is reported.

With such set of performance specifications and the sizing database of the opamp reusable block (i.e., its structural facet), the sizing can be carried out. Sizing is done sequentially: first the opamp core is sized, followed by the CMFB circuit and, last, the biasing devices. The structural view database has been adapted to the specific optimization tool used, FRIDGE. Basically, the data are organized in the following separate sections:

1. One or more **configurations** sections, where most testbench setups are stored. In particular these sections contain the parameterized

[13] Nevertheless, such area figure provides a valid way to roughly compare design solutions from the sizing iterations.

netlist, performance monitors, the peripheral setups (loading conditions included), as well as the device models and process data.

2. **Variables** section, where all the design variables (i.e., device sizes and biasing conditions), whose variation ranges define the explorable design space, are stored.
3. **Constraints** section, where the netlist-related and the testbench setup constraints are defined.

For each sizing process, it is also necessary to define the specific value of the performance target/directive in the **targets** section. Optionally, the user can specify several control settings of the optimization algorithms employed (explained in Section 3.2.1 of Chapter 2). The reader is referred to [Mede99] for further details on the optimization tool input.

The evolution of the cost functions for the opamp core and CMFB circuit sizings are shown in Fig. 22. As it can be noticed in both figures, good results are achieved at early iterations (iterations *I1* and *I2*), when the cost function is below zero [Mede99], but the optimizer still looks for a better solution. When the optimization reaches the iteration *I3*, there is a process of fine tuning of the design that continues until the end of the sizing. In the final solution, not also the targets are also achieved but also the objectives (minimization of area and power) have been improved.

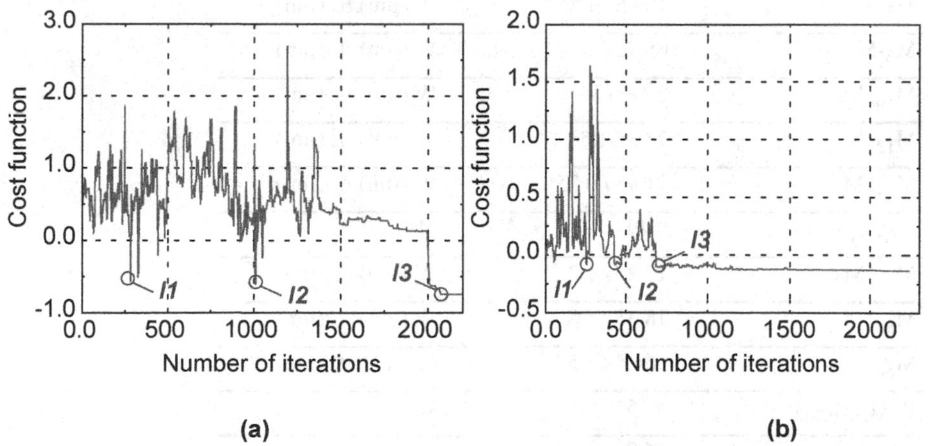

Figure 22. Evolution of the cost function of the (a) opamp core and (b) CMFB circuitry.

The results of the opamp core and CMFB sizing are shown in Table 13.

The sizes of all the devices in the opamp core and the CMFB circuitry, as well as the biasing conditions resulting from the sizing process are displayed in Table 14.

Table 13. Obtained opamp performance.

Performance feature	*Target/Directive*	*Sizing result*	*Units*
add_db	> 60.0	62.70	dB
ft	> 8.0	8.4	MHz
pm	> 38.6	40.0	°
r_{out}	< 70	9.5	kΩ
add_cm_db	> 60.0	63.2	dB
ft_cm	≈ ft	8.51	MHz
pm_cm	> 38.6	40.3	°
Area	Minimize	1528.845	μm^2
Power	Minimize	0.27	mW

Table 14. Obtained device sizes of the opamp core and CMFB and biasing conditions.

Device or Bias current/voltage	*Size/Value*	*Unit*
M_1, M_2	20.0 / 1.2	W (μm) /L (μm)
M_3, M_4	16.0 / 1.5	W (μm) /L (μm)
M_5	109.6 / 1.5	W (μm) /L (μm)
M_6-M_9	65.76 / 1.5	W (μm) /L (μm)
M_{10}, M_{11}	2.4 / 0.55	W (μm) /L (μm)
M_{12}, M_{13}	75.3 / 1.5	W (μm) /L (μm)
M_{14}, M_{15}	11.95 / 0.35	W (μm) /L (μm)
C_{c1}, C_{c2}	0.49	pF
M_{1c}, M_{2c}	12.25 / 1.2	W (μm) /L (μm)
M_{3c}, M_{4c}	18.35 / 1.8	W (μm) /L (μm)
M_{5c}	164.4 / 1.5	W (μm) /L (μm)
R_{CM1}, R_{CM2}	9.75	kΩ
I_{BIAS}	16.8	μA
V_{cn}, V_{cp}	-0.5, -0.5	V

If a solution is not achieved, a change in the design space definition (i.e., the range variation of the design variables) is required to search for solutions in unexplored parts of the entire design space. If good solutions were found, the next step in the flow is the tuning of the biasing circuitry. After completing the sizing of the opamp reusable block, the size/value of each device is obtained. Table 15 displays the full value list of the optimized design variables.

Table 15. Obtained device sizes/values.

Device	*Size/Value*	*Units*
M_1, M_2	20.0 / 1.2	W (μm) /L (μm)
M_3, M_4	16.0 / 1.5	W (μm) /L (μm)
M_5	109.6 / 1.5	W (μm) /L (μm)
M_6-M_9	65.8 / 1.5	W (μm) /L (μm)
M_{10}, M_{11}	2.4 / 0.55	W (μm) /L (μm)
M_{12}, M_{13}	75.3 / 1.5	W (μm) /L (μm)
M_{14}, M_{15}	12.0 / 0.35	W (μm) /L (μm)
M_{B7}	27.4 / 1.5	W (μm) /L (μm)
M_{B8}	164.45 / 1.5	W (μm) /L (μm)
M_{Bcab}	65.8 / 1.5	W (μm) /L (μm)
M_{Bcpa}	1.65 / 13.55	W (μm) /L (μm)
M_{Bcna}	13.4 / 0.55	W (μm) /L (μm)
M_{Bcnb}	2.35 / 7.85	W (μm) /L (μm)
M_{1c}, M_{2c}	12.25 / 1.2	W (μm) /L (μm)
M_{3c}, M_{4c}	18.35 / 1.8	W (μm) /L (μm)
M_{5c}	164.4 / 1.5	W (μm) /L (μm)
M_{B9}-M_{B11}	27.4 / 1.5	W (μm) /L (μm)
R_{CM1}, R_{CM2}	9750.0 kΩ	kΩ
C_{c1}, C_{c2}	0.49 pF	pF
M_{B1a}	8.0 / 6.0	W (μm) /L (μm)
M_{B3}	8.0 / 6.0	W (μm) /L (μm)
M_{B4}	8.0 / 6.0	W (μm) /L (μm)
M_{B5}	48.0 / 6.0	W (μm) /L (μm)
M_{B6}	8.0 / 6.0	W (μm) /L (μm)
M_{Bon1},M_{Bon2}	3.5 / 0.5	W (μm) /L (μm)

The next task in the design reuse flow is to transform the obtained design sizes into useful data to generate the layout. First, it is necessary to calculate device multiplicities (i.e., the number of transistor finger or resistor strips) and actual capacitor sizes[14]. Second, it is necessary to correct off-grid errors that may arise as consequence of multiplicities different from the unity. Table 16 shows the detailed opamp sizing.

Table 16. Final opamp sizing.

Device	*Multiplicity*	*Unitary size*	*Implemented*[a]
M_1, M_2	8	2.5	20.0 / 1.2
M_3, M_4	8	2.0	16.0 / 1.5
M_5	8	13.7	109.6 / 1.5
M_6-M_9	4	16.45	65.8 / 1.5
M_{10}, M_{11}	1	2.4	2.4 / 0.55
M_{12}, M_{13}	4	18.85	75.4 / 1.5
M_{14}, M_{15}	4	3.0	12.0 / 0.35
M_{B7}	2	13.7	27.4 / 1.5
M_{B8}	10	16.45	164.5 / 1.5
M_{Bcab}	4	16.45	65.8 / 1.5
M_{Bcpa}	1	1.65	1.65 / 13.55
M_{Bcna}	1	2.4	2.4 / 0.55
M_{Bcnb}	1	2.35	2.35 / 7.85
M_{1c}, M_{2c}	4	3.05	12.2 / 1.2
M_{3c}, M_{4c}	8	2.3	18.4 / 1.8
M_{5c}	12	13.7	164.4 / 1.5
M_{B9}-M_{B11}	2	13.7	27.4 / 1.5
R_{CM1}, R_{CM2}[b]	5	25.35[c]	9750.0 kΩ[d]
C_{c1}, C_{c2}[e]	2	16.60[f]	0.4862 pF[g]
M_{B1a}	4	2.0	8.0 / 6.0
M_{B3}	4	2.0	8.0 / 6.0
M_{B4}	4	2.0	8.0 / 6.0
M_{B5}	24	2.0	48.0 / 6.0

14. This is done manually. In the next chapter, a technique that calculates the device multiplicities and capacitor sizes to improve the layout geometrical quality early in the sizing process, is reported.

Table 16. Final opamp sizing.(*cont.*)

Device	*Multiplicity*	*Unitary size*	*Implemented*[a]
M_{B6}	4	2.0	8.0 / 6.0
M_{Bon1},M_{Bon2}	1	3.5	3.5 / 0.5

a. The column represents W/L for transistors and resistance (capacitance) value for passive elements.

b. In this case, the finger width is the unitary resistor length, and W/L is the total resistance with 1.0 μm of width.

c. $R = R_{\square} \cdot \frac{L}{(W - 0.35)}$, where $R_{\square}$ is 50 Ω/□ .

d. R_{unit}=1950.0kΩ.

e. In this case, unitary size is the width of the unitary capacitor and W/L is the total capacitance.

f. The expression to calculate the width (L_{side}) of the capacitors is:, $L_{side} = \frac{-4 \cdot C_P + \sqrt{16 \cdot C_P^2 + 4 \cdot C_A \cdot C}}{2 \cdot C_A}$ since, $C = C_A \cdot L^2_{side} + C_P \cdot 4 \cdot L_{side}$ where C is the capacitance, C_A is the capacitance per unit of area, whose typical value is 0.86fF/μm^2, and C_P is the capacitance per unit of perimeter, with a typical value of 0.092fF/μm.

g. C_{unit}=0.2431 pF.

The layout instancing is a process in which the sizes obtained at the previous stages of the flow are used as input parameters of the layout templates to obtain a physical representation of the analog circuit. This is done automatically once the circuit sizing is available and it takes from 1 to 10 seconds (depending on the complexity of the template) to generate the new layout. Fig. 23 depicts the layout instance resulting from the circuit sizing in Table 16 and using the layout facet of the opamp reusable block. Total occupied area is $232.9 \times 244.45\ \mu m^2$.

The opamp is finally extracted and verified using device-level simulation. The result of this verification is summarized in Table 17. For the sake of clarity and comparison purposes, simulation results include two scenarios: a first one, labeled 'LAY' (for layout), where all layout-induced parasitics have been considered, on a second one, labeled 'SCH' (for schematic), where no parasitics but those associated to the diffusion areas of MOS transistors have been included. This same comparison and labeling holds for the ensuing examples.

Running times were as follows: sizing of the complete opamp (i.e., core + CMFB + bias stages) took 170.5s of CPU time[15], layout generation and extraction took 10s, and opamp verification took 10s. That is, retargeting the opamp takes around 3.1 minutes of CPU time.

15. All examples here reported have been performed on a PentiumIV@1.3GHz PC with 512Mb RAM.

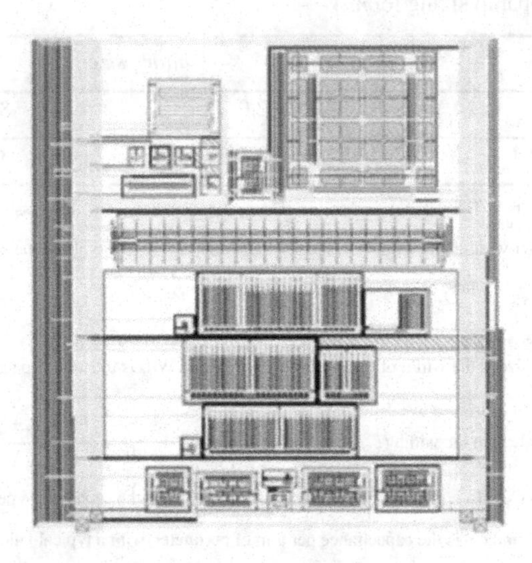

Figure 23. Opamp layout instance in process A (0.35μm).

Table 17. Final opamp verification.

Performance feature	*Specified*	*Type*	*Value*	*Units*
add_db	> 60.0	SCH	63.79	dB
		LAY	63.79	
ft	> 8.09	SCH	10.58	MHz
		LAY	8.13	
pm	> 38.67	SCH	42.58	o
		LAY	40.55	
add_cm_db	> 60.0	SCH	64.08	dB
		LAY	64.08	
ft_cm	≈ft	SCH	10.95	MHz
		LAY	8.19	
pm_cm	> 38.67	SCH	41.25	o
		LAY	39.07	

4.1.2 Opamp migration to process B (0.5 μm)

To illustrate how the design reuse flow deals with design migration, the above-described opamp design is ported to a different fabrication process.

First, a simple migration is carried out. This process consists in keeping the transistor aspects as well as capacitor and resistor values, while avoiding violations of the minimum dimension errors. The obtained design by simple migration is the same than the one shown in Table 15, but with a change in the group M_{14}-M_{15} to fit minimum dimensions in process B, where the minimum gate length is 0.5 µm. Keeping this group's aspect yields the new aspect $17.5/0.5$.

Figure 24 shows the simulated gain of the designed opamp in process A and that of the design in process B, which has been ported as described above. As it can be noticed, this ported design is not appropriate since it does not meet the required specs (see Table 18).

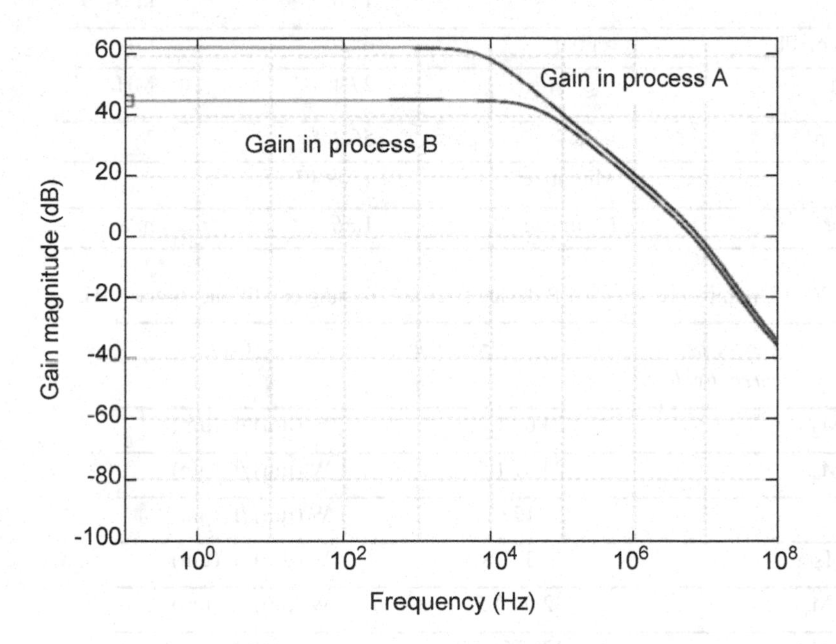

Figure 24. Gain in process A vs. gain in process B (simple migration of the design).

Table 18. Simulation results from the opamp simple migration.

Performance feature	*Specification*	*Simple migration results*	*Units*
add_db	> 60.0	45.24	dB
ft	> 8.09	6.75	MHz
pm	> 38.67	46.12	o

Therefore, design migration may imply, as shown above, a redesign of the analog circuit. In the following example, a complete design migration (i.e., database migration and design retargeting) is illustrated.

From the same performance specifications (see Table 12), a sizing process is carried out in process B, yielding the results shown in Table 19.

The sizes of all the devices in the opamp core and the CMFB circuits are displayed in Table 20.

Table 19. Obtained opamp performance in process B.

Performance feature	*Target/Directive*	*Sizing result*	*Units*
add_db	> 60.0	70.2	dB
ft	> 8.09	26.2	MHz
pm	> 38.7	46.5	o
r_{out}	< 70	13.0	kΩ
add_cm_db	> 60.0	67.0	dB
ft_cm	≈ ft	27.4	MHz
pm_cm	> 38.7	56.3	o
Area	Minimize	1509.17	μm^2
Power	Minimize	1.36	mW

Table 20. Opamp core and CMFB device sizes and biasing conditions (process B).

Device or Bias current/voltage	*Size/Value*	*Units*
M_1, M_2	24.0 / 1.2	W (μm) /L (μm)
M_3, M_4	24.0 / 1.5	W (μm) /L (μm)
M_5	57.15 / 1.5	W (μm) /L (μm)
M_6-M_9	34.3/ 1.5	W (μm) /L (μm)
M_{10}, M_{11}	12.0 / 0.55	W (μm) /L (μm)
M_{12}, M_{13}	178.55 / 1.5	W (μm) /L (μm)
M_{14}, M_{15}	66.4 / 0.5	W (μm) /L (μm)
C_{c1}, C_{c2}	0.61	pF
M_{1c}, M_{2c}	32.0 / 1.2	W (μm) /L (μm)
M_{3c}, M_{4c}	43.7 / 1.35	W (μm) /L (μm)
M_{5c}	85.75 / 1.5	W (μm) /L (μm)
R_{CM1}, R_{CM2}	21.38	kΩ
I_{BIAS}	43.2	μA
V_{cn}, V_{cp}	-0.5, -0.5	V

Full-detail sizing is shown in Table 21.

Table 21. Final opamp sizing in process B.

Device	*Multiplicity*	*Unitary size*	*Implemented*[a]
M_1, M_2	8	3.0	24.0 / 1.2
M_3, M_4	8	3.0	24.0 / 1.5
M_5	8	7.15	57.2 / 1.5
M_6-M_9	4	8.6	34.4 / 1.5
M_{10}, M_{11}	4	3.0	12.0 / 0.55
M_{12}, M_{13}	6	29.8	178.8 / 1.5
M_{14}, M_{15}	4	16.6	66.4 / 0.5
M_{B7}	2	7.15	14.3 / 1.5
M_{B8}	4	21.45	85.8 / 1.5
M_{Bcab}	1	34.4	34.4 / 1.5
M_{Bcpa}	1	2.75	2.75 / 6.0
M_{Bcna}	4	3.0	12.0 / 0.55
M_{Bcnb}	1	4.0	4.0 / 2.85
M_{1c}, M_{2c}	4	8.0	32.0 / 1.2
M_{3c}, M_{4c}	8	5.45	43.6 / 1.35
M_{5c}	12	7.15	85.8 / 1.5
M_{B9}-M_{B11}	2	7.15	14.3 / 1.5
R_{CM1}, R_{CM2}[b]	5	12.25[c]	21.44[d] kΩ
C_{c1}, C_{c2}[e]	2	16.7[f]	0.61[g] pF
M_{B1a}	4	2.0	8.0 / 6.0
M_{B3}	4	2.0	8.0 / 6.0
M_{B4}	4	2.0	8.0 / 6.0
M_{B5}	24	2.0	48.0 / 6.0
M_{B6}	4	2.0	8.0 / 6.0
M_{oni}, M_{Boni}	1	3.5	3.5 / 0.5

a. The column represents W/L for transistors and resistance (capacitance) value for passive elements.

b. In this case, the finger width is the unitary resistor length, and W/L is the total resistance with width equal to 1.0 μm.

c. $L = R_{\square} \cdot \frac{L}{W}$, where $R_{\square}$ is 350 Ω/square.

d. R_{unit}=4.2875kΩ.

e. In this case, unitary size is the width of the unitary capacitor and W/L is the total capacitance.

f. The expression to calculate the width (L_{side}) of the capacitors is: $L_{side} = \sqrt{\frac{C}{C_A}}$, since, $C = C_A \cdot L^2_{side}$, where C is the capacitance, C_A is the capacitance per unit of area, whose typical value is 1.1fF/μm^2.

g. C_{unit}= 0.3067 pF.

As with the retargeted design in process A, the ported opamp design is finally verified using electrical simulation at the schematic and layout levels. The results are listed in Table 22.

Table 22. Verification of the ported opamp design.

Performance feature	*Specified*	*Type*	*Value*	*Units*
add_db	> 60.0	SCH	70.7	dB
		LAY	70.7	
ft	> 8.0	SCH	28.5	MHz
		LAY	24.9	
pm	> 38.7	SCH	48.0	°
		LAY	43.0	
add_cm_db	> 60.0	SCH	66.1	dB
		LAY	66.1	
ft_cm	≈ ft	SCH	29.0	MHz
		LAY	25.9	
pm_cm	> 38.7	SCH	63.4	°
		LAY	52.6	

The layout implementation of the opamp in process B technology is depicted in Fig. 25. The instancing of this cell was automatic thanks to the technological portability property of the layout templates. Total occupied area is $309.6 \times 301.55\ \mu m^2$.

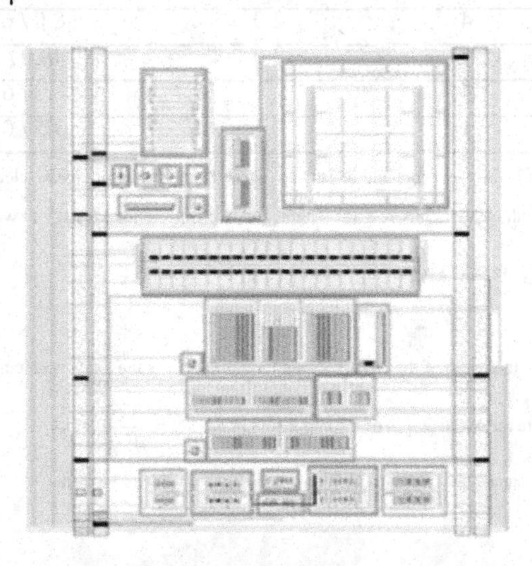

Figure 25. Layout of the ported opamp.

4.2 Design example (II): GSM retargeting of the analog back-end

In this example, the design reuse flow is executed to retarget the whole analog back-end to fit the set of GSM performance specifications. This example and the following one have been carried out in the fabrication process A (0.35μm) mentioned in the opamp design example in the precedent section. Likewise, in the current and the following examples the full-scale output current of the preceding DA converter is $I_{FS} = 350\,\mu A$ and the full-scale output voltage is $V_{FS} = 1.5\,V$.

As said above, during the design of the analog back-end it is necessary to define a number of testbench setups that account for all possible scenarios this circuit has to comply with. That is, sizing the analog back-end implies to use both the *INTRA*-die and the *INTER*-die setups, as well as the three gain configurations of the PGA (controlled by the word outlev[2:0]), in order to completely evaluate the analog back-end performance at each iteration of the sizing process (if multi-standard capability is considered, the number of testbench setups is doubled to take into account the two standard setups controlled by the digital word std[2:0]). Similarly, the sizing of the opamp A_3 requires two testbench setups to take into account the *economic* and *not-economic* modes. For the sake of clarity, however, only the worst-case values from the sizing results of the analog back-end and the results from the not-*economic* mode opamp A_3 are shown here.

The design retargeting process starts with the sizing of the module-level components, i.e., the tuning of the passive devices and the performance specifications of opamps A_1, A_2 and A_3, for the analog back-end to meet the GSM performance specifications shown in Table 23.

The obtained worst-case analog back-end performance after this initial sizing process is detailed in Table 24. Likewise, the obtained passive components of the CT-LP filter and the PGA are listed in Table 25[16].

The requested opamp performance specifications for correct operation of the analog back-end are shown in Table 26. Note that these specifications are worst-case values, obtained from the Monte Carlo simulations of the analog back-end performed during the sizing process.

Following the analog back-end sizing, a cell-level sizing takes place for each opamp, similar to the one explained in the precedent section. The results of these processes are shown in Table 27. Table 28 contains the resulting sizes and values for the opamp cores and CMFB circuits.

---

[16]. The actual physical device sizes will be provided below.

Table 23. GSM performance specifications for the analog back-end.

Performance feature	*Specification*	*Units*
apeak	< 1.5	dB
adcra	< 0.7	dB
imar	> 50	dB
bumpsz	< 1	dB
gddev	< 8.3e-9	s
gdavgra	< 33.5e-9	s
pera	< 1.3	o
fpolectf_dev	< 3.0e5	Hz
qfactctf_dev	< 0.15	
fpolepga_dev	< 1.7e5	Hz

Table 24. Obtained analog back-end performance after initial sizing.

Performance feature	*Specification*	*Obtained*	*Units*
apeak	< 1.5	0.78	dB
adcra	<0.7	0.038	dB
imar	> 50	57.65	dB
bumpsz	< 1	0	dB
gddev	< 8.3e-9	2.72e-9	s
gdavgra	< 33.5e-9	0.64e-9	s
pera	<1.3	0.02	o
fpolectf_dev	< 3.0e5	1.70e5	Hz
qfactctf_dev	<0.15	0.02	
fpolepga_dev	< 1.7e5	1.69e5	Hz

The detailed sizing is shown in Table 29. In this table, the multiplicities and sizes of the unitary devices are also displayed. Note that the total dimensions are slightly changed because of the need to adjust the unitary device sizes to the process grid. In this table, *m* stands for multiplicity, W_{finger} is the width of the unitary element in μm, and W is the total width of the device, also in μm. Note as well that the devices of opamp A_3 enabling the *economic* and *not-economic* modes are also displayed. Table 30 shows detailed sizing of the passive devices, where the unitary components of each passive device are also listed.

Table 25. Passive device values.

Module-level block	*Device*	*Value*	*Units*
CT-LP filter	R_1	21.975	kΩ
	R_{FF}	225.353	kΩ
	R_{FB}	4.286	kΩ
	C_1	4.267	pF
	C_2	4.267	pF
PGA	R_{P1}	31.73	kΩ
	R_{P2}	6.340	kΩ
	R_{P3}	9.520	kΩ
	R_{P4}	15.87	kΩ
	C_P	2.850	pF

Table 26. Requested performance of the opamps.

	Performance feature	*Target/Directive*			*Units*
		Opamp A_1	Opamp A_2	Opamp A_3	
Electrical	add_db	> 45.22	> 32.03	> 15.69	dB
	ft	> 15.67	> 8.09	> 1.96	MHz
	pm	> 41.05	> 38.67	> 56.53	°
	r_{out}	< 40	< 70	< 70	kΩ
	add_cm_db	> 45.22	> 32.03	> 15.69	dB
	ft_cm	≈ft	≈ft	≈ft	MHz
	pm_cm	> 41.05	> 38.67	> 56.53	°
Environmental	RL	41.0	7.23	0.98	kΩ
	CL	2.77	5.73	100.12	pF
Optimization	Area	Minimize	Minimize	Minimize	μm^2
	Power	Minimize	Minimize	Minimize	mW

Table 27. Optimized performance after cell-level sizing.

Performance feature	Opamp A_1	Opamp A_2	Opamp A_3	*Units*
add_db	75.19	62.70	79.21	dB
ft	18.12	8.431	2.04	MHz
pm	50.98	40.03	58.08	o
r_{out}	22.06	13.60	27.41	kΩ
add_cm_db	80.30	63.21	77.52	dB
ft_cm	18.76	8.50	2.00	MHz
pm_cm	62.54	40.27	63.46	o
Area	1419.2	1613.684	2002.94	μm^2
Power	0.53	0.27	0.75	mW

Table 28. Core and CMFB device sizes.

Device or Bias current/voltage	Opamp A_1	Opamp A_2	Opamp A_3	*Units*
M_1, M_2	16.0 / 1.2	20.0 / 1.2	16.25 / 1.2	W (μm) /L (μm)
M_3, M_4	16.0 / 1.5	16.0 / 1.5	17.25 / 1.5	W (μm) /L (μm)
M_5	69.45 / 1.5	109.6 / 1.5	43.25 / 1.5	W (μm) /L (μm)
M_6-M_9	41.65 / 1.5	65.75 / 1.5	25.95 / 1.5	W (μm) /L (μm)
M_{10}, M_{11}	3.2 / 0.55	2.4 / 0.55	9.8 / 0.55	W (μm) /L (μm)
M_{12}, M_{13}	131.25 / 1.5	75.3 / 1.5	252.45 / 1.5	W (μm) /L (μm)
M_{14}, M_{15}	9.75 / 0.35	11.95 / 0.35	23.85 / 0.35	W (μm) /L (μm)
C_{c1}, C_{c2}	0.54	0.48	3.04	pF
M_{1c}, M_{2c}	17.20 / 1.2	22.6 / 1.2	27.1 / 1.2	W (μm) /L (μm)
M_{3c}, M_{4c}	11.65 / 1.5	21.44 / 2.8	21.19 / 3.55	W (μm) /L (μm)
M_{5c}	92.6 /1.5	164.4 / 1.5	54.05 / 1.5	W (μm) /L (μm)
R_{CM1}, R_{CM2}	10.77	9.75	18.92	kΩ
R_{z1}, R_{z2}	2.07			kΩ
I_{BIAS}	22.65	16.8	16.05	μA
V_{cn}, V_{cp}	-0.5, -0.5	-0.5, -0.5	-0.5, -0.5	V

Table 29. Detailed sizing of the opamps.

Device	Opamp A_1			Opamp A_2			Opamp A_3		
	m	W_{finger}	*W/L*	*m*	W_{finger}	*W/L*	*m*	W_{finger}	*W/L*
M_1,M_2	8	2.0	16.0/1.2	8	2.5	20.0/1.2	16	1.05	16.8/1.2
M_3,M_4	8	2.0	16.0/1.5	8	2.0	16.0/1.5	8	2.15	17.2/1.5
M_5	12	5.8	69.6/1.5	8	13.7	109.6/1.5	16	2.75	44.0/1.5
M_6-M_9	4	10.45	41.8/1.5	4	16.45	65.8/1.5	4	6.5	26.0/1.5
M_{10},M_{11}	1	3.2	3.2/0.55	1	2.4	2.4/0.55	2	4.9	9.8/0.55
M_{12},M_{13}	8	16.4	131.2/1.5	4	18.85	75.4/1.5	2	25.25	50.5/1.5
M_{14},M_{15}	4	2.45	9.8/0.35	4	3.0	12.0/0.35	4	1.2	4.8/0.35
M_{B7}	6	5.8	34.8/1.5	2	13.7	27.4/1.5	4	2.75	11.0/1.5
M_{B8}	4	17.35	69.4/1.5	10	16.45	164.5/1.5	2	27.05	54.1/1.5
M_{Bcab}	4	10.45	41.8/1.5	4	16.45	65.8/1.5	4	6.5	26.0/1.5
M_{Bcpa}	1	1.65	1.65/15	1	1.65	1.65/13.55	1	1.35	1.35/22.6
M_{Bcna}	1	3.2	3.2/0.55	1	2.4	2.4/0.55	2	4.9	9.8/0.55
M_{Bcnb}	1	2.3	2.3/9.15	1	2.35	2.35/7.85	1	1.2	1.2/9.2
M_{1c},M_{2c}	4	4.3	17.2/1.2	4	11.3	22.6/1.2	8	3.35	26.8/1.2
M_{3c},M_{4c}	8	1.45	11.6/1.0	8	2.7	21.6/2.8	8	2.6	20.8/1.35
M_{5c}	16	5.8	92.8/1.5	12	13.7	164.4/1.5	20	2.75	55.0/1.5
M_{B9}-M_{B11}	2	5.8	11.6/1.5	2	13.7	27.4/1.5	2	2.75	55.0/1.5
R_{CM1},R_{CM2}[a]	5	28.0	10.76 kΩ	5	25.35	9750.0 kΩ	5	49.15	18.90 kΩ
R_{z1},R_{z2}[b]	2	9.35	2.077 kΩ						
C_{c1},C_{c2}[c]	2	17.6	0.54 pF	2	16.60	0.48 pF	2	41.9	3.05 pF
M_{B1a}	4	2.5	10.0/4.0	4	2.0	8.0/6.0	4	2.5	10.0/4.0
M_{B1b}	4	2.5	10.0/4.0				4	2.5	10.0/4.0
M_{B2}	1	30.0	30.0/0.5				1	30.0	30.0/0.5
M_{B3}	4	2.5	10.0/4.0	4	2.0	8.0/6.0	4	2.5	10.0/4.0
M_{B4}	12	2.5	30.0/4.0	4	2.0	8.0/6.0	8	2.5	20.0/4.0
M_{B5}	24	2.5	60.0/4.0	24	2.0	48.0/6.0	40	2.5	100.0/4.0
M_{B6}	4	2.5	10.0/4.0	4	2.0	8.0/6.0	4	2.5	10.0/4.0
M_{12e},M_{13e}							m_{M12}·4	8	202.0/1.5
M_{14e},M_{15e}							16	1.2	19.2/0.35
M_{oni},M_{Boni}	1	3.5	3.5/1.5	1	3.5	3.5/1.5	1	3.5	3.5/1.5
M_{e1}-M_{e4}							1	30.0	30.0/0.35
M_{e5},M_{e6}							1	15.0	15.0/0.35

a. In this case, W_{finger} is the unitary resistor length (1.0μm width) and column W/L is the total resistance.

b. In this case, W_{finger} is the unitary resistor length (0.8μm width) and column W/L is the total resistance.

c. In this case, W_{finger} is the width of the unitary, square capacitor and column W/L is the total capacitance.

Table 30. Detailed sizing of the passive devices.

Module-level block	*Device*	*Value*	*Width*	*Length*
CT-LP filter	$R_{1,unit}$	10.987 kΩ	1.15 μm	175.8 μm
	$R_{FF,unit}$	5.007 kΩ	2.05 μm	170.25 μm
	$C_{1,2,unit}$	0.533 pF	24.7 μm	24.7 μm
	R_1	21.975 kΩ	$2 \cdot R_{1,unit}$	
	R_{FF}	225.331 kΩ	$45 \cdot R_{FF,unit}$	
	R_{FB}	4.284 kΩ	1.15 μm	68.55 μm
	C_1, C_2	4.270 pF	$8 \cdot C_{1,2,unit}$	
PGA	$R_{1,2,3,unit}$	3.173 kΩ	1.0 μm	41.25 μm
	$R_{4,unit}$	7.934 kΩ	1.0 μm	103.15 μm
	$C_{P,unit}$	0.356 pF	20.15 μm	20.15 μm
	R_{P1}	31.730 kΩ	$10 \cdot R_{1,2,3unit}$	
	R_{P2}	6.340 kΩ	$2 \cdot R_{1,2,3unit}$	
	R_{P3}	9.519 kΩ	$3 \cdot R_{1,2,3unit}$	
	R_{P4}	15.869 kΩ	$2 \cdot R_{4,unit}$	
	C_P	2.850 pF	$8 \cdot C_{P,unit}$	

As in the previous example, the implemented cell-level designs must be verified at the layout level. Table 31 collects the verification results and compares them with the schematic verification.

The resulting layout instance of the analog back-end is depicted in Fig. 26. Total occupied area is 762.0×1338.8 μm^2.

Finally, the verification of the analog back-end retargeted for GSM standard requirements is resumed in Table 32. This verification has been done by using device-level descriptions of the opamps. Total retargeting time was 34.5 minutes CPU time, which includes 10 minutes for analog back-end sizing, 9.5 minutes for cell-level sizing, layout generation, extraction, and verification, and about 15 minutes for layout generation (10s), extraction (30s), and verification (14 minutes) of the analog back-end.

Table 31. Verification of the cell-level blocks.

Cell-level block	*Performance feature*	*Specified*	*Type*	*Value*	*Units*
Opamp A_1	add_db	>45.22	SCH	75.50	dB
			LAY	75.56	
	ft	>15.67	SCH	20.33	MHz
			LAY	16.5	
	pm	>41.05	SCH	60.3	°
			LAY	50.29	
	add_cm_db	>45.22	SCH	80.27	dB
			LAY	80.44	
	ft_cm	≈ ft	SCH	20.14	MHz
			LAY	17.07	
	pm_cm	>41.05	SCH	68.13	°
			LAY	63.06	
Opamp A_2	add_db	>32.03	SCH	63.79	dB
			LAY	63.79	
	ft	>8.09	SCH	10.58	MHz
			LAY	8.13	
	pm	>38.67	SCH	42.58	°
			LAY	40.55	
	add_cm_db	>32.03	SCH	64.08	dB
			LAY	64.08	
	ft_cm	≈ ft	SCH	9.95	MHz
			LAY	7.85	
	pm_cm	>38.67	SCH	41.25	°
			LAY	39.07	
Opamp A_3	add_db	>15.69	SCH	79.38	dB
			LAY	79.39	
	ft	>1.96	SCH	2.14	MHz
			LAY	1.98	
	pm	>56.53	SCH	59.87	°
			LAY	59.52	
	add_cm_db	>15.69	SCH	77.28	dB
			LAY	77.29	
	ft_cm	≈ ft	SCH	2.69	MHz
			LAY	1.92	
	pm_cm	>56.53	SCH	65.18	°
			LAY	65.28	

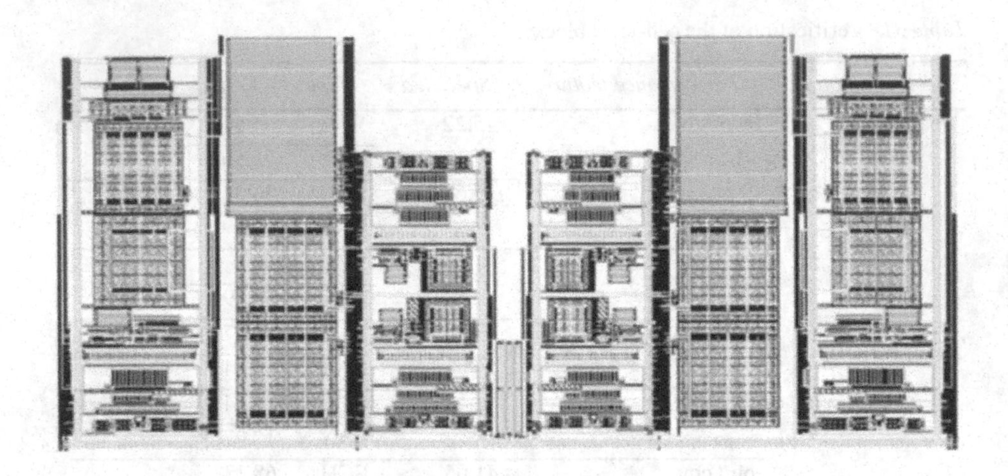

Figure 26. Layout implementation result of the analog back-end for the single-standard retargeting example.

Table 32. Final analog back-end verification.

Performance feature	*Specification*	*Optimized*[a]	*Type*	*Value*	*Units*
adcra	< 0.7	0.038	SCH	0.04	dB
			LAY	0.02	
gdavgra	< 33.5e-9	0.64e-9	SCH	2.04e-9	s
			LAY	2.48e-9	
pera	< 1.3	0.02	SCH	0.07	°
			LAY	0.08	
apeak	< 1.5	0.78	SCH	0.37	dB
			LAY	1.29	
imar	> 50.0	57.65	SCH	56.99	dB
			LAY	54.57	
bumpsz	< 1.0	0	SCH	0	dB
			LAY	0	
gddev	< 8.3e-9	2.72e-9	SCH	1.99e-9	s
			LAY	2.35e-9	

a. This column resumes the obtained values from the analog back-end sizing process.

4.3 Design example (III): multi-standard retargeting of the analog back-end

The present example illustrates the multi-standard retargeting of the analog back-end. For the sake of simplicity, only results for the GSM and DECT wireless communication standards (one for each of the two standard setups set by the digital word std[2:0]) are collected here. As in the previous example, only worst-case results from the analog back-end sizing and the *not-economic* sizing results of the opamp A_3 are reported.

As in the single-standard example, the results shown here are the worst-case ones. The process starts at the analog back-end sizing, continues to the cell-level sizing problem, down to the layout generation, and ends with the verification of the obtained design.

The high-level specifications for both standards are displayed in Table 33.

Table 33. GSM and DECT performance specifications for the analog back-end.

	Performance feature	*Specification*	*Units*
GSM	apeak	< 1.5	dB
	adcra	< 0.7	dB
	imar	> 50	dB
	bumpsz	< 1	dB
	gddev	< 8.3e-9	s
	gdavgra	< 33.5e-9	s
	pera	< 1.3	o
	fpolectf_dev	< 3.0e5	Hz
	qfactctf_dev	< 0.15	
	fpolepga_dev	< 1.7e5	Hz
DECT	apeak	< 1.5	dB
	adcra	< 0.7	dB
	imar	> 55	dB
	bumpsz	< 1	dB
	gddev	< 9.5e-9	s
	gdavgra	< 6.5e-9	s
	pera	< 1.75	o
	fpolectf_dev	< 1.4e6	Hz
	qfactctf_dev	< 0.15	
	fpolepga_dev	< 8.0e5	Hz

The attained values of performance features after the analog back-end sizing process are listed below, in Table 34.

Another outcome of the analog back-end sizing process is the set of required resulting passive devices, which are shown in Table 35.

Likewise, the performance specifications for the cell-level sizing of the reusable opamps are another result of the analog back-end sizing. These specifications are provided in Table 36. As said above, two sets of specifications for each opamp are obtained for each standard setup (note that required performance specifications for the opamp A_2 are the same for both standard setups) in order to optimize power consumption in each standard setup. Recall from Section 3.1 that it is by changing the biasing current (performed by transistors M_{B2}, M_{B1b}, and the digital signals *std[1]* –for A_1– and *std[2]* –for A_3–, as shown in Fig. 12), that both performance specifications sets are addressed. Consequently, sizing A_1 or A_3 has to be carried out by making

Table 34. Obtained analog back-end multi-standard performance after initial sizing.

	Performance feature	*Specification*	*Obtained*	*Units*
GSM	apeak	< 1.5	0.2	dB
	adcra	< 0.7	0.48	dB
	imar	> 50	57.37	dB
	bumpsz	< 1	0	dB
	gddev	< 8.3e-9	8.29e-9	s
	gdavgra	< 33.5e-9	8.45e-9	s
	pera	< 1.3	0.30	°
	fpolectf_dev	< 3.0e5	0.92e5	Hz
	qfactctf_dev	< 0.15	0.02	
	fpolepga_dev	< 1.7e5	1.41e5	Hz
DECT	apeak	< 1.5	1.49	dB
	adcra	< 0.7	0.50	dB
	imar	> 55	59.94	dB
	bumpsz	< 1	0	dB
	gddev	< 9.5e-9	2.23e-9	s
	gdavgra	< 6.5e-9	1.43e-9	s
	pera	< 1.75	0.36	°
	fpolectf_dev	< 1.4e6	5.58e5	Hz
	qfactctf_dev	< 0.15	0.02	
	fpolepga_dev	< 8.0e5	5.63e5	Hz

Table 35. Required passive devices.

Module-level block	*Standard*	*Device*	*Value*	*Units*
CTF	GSM	R_1	41.355	kΩ
		R_{FF}	798.110	kΩ
		R_{FB}	4.286	kΩ
		C_1	3.728	pF
		C_2	3.728	pF
	DECT	R_1	13.785	kΩ
		R_{FF}	88.675	kΩ
		R_{FB}	4.286	kΩ
		C_1	1.864	pF
		C_2	1.864	pF
PGA	GSM/DECT	R_{P1}	25.0	kΩ
		R_{P2}	5.0	kΩ
		R_{P3}	7.5	kΩ
		R_{P4}	12.5	kΩ
	GSM	C_P	3.984	pF
	DECT	C_P	0.995	pF

the opamp to fulfill both sets of specifications using the same sizing scheme, the only difference relying in using distinct biasing currents. Eventually, both values of the biasing current are implemented by properly sizing the biasing stage (in particular M_{B2} and M_{B1b}).

With the performance targets and directives in Table 36, the cell-level sizing yields the solution summarized in Table 37. In like manner, the resulting device sizes and values of the opamp core and CMFB circuits after the corresponding cell-level sizing processes are listed in Table 38.

The detailed sizing of the opamps is shown in Table 39.

In this table, the multiplicities and sizes of the unitary devices are also displayed. Note that the total dimensions are slightly changed because of the need to adjust the unitary device sizes to the process grid. In this table, *m* stands for multiplicity, W_{finger} is the width of the unitary element in μm, and W is the total width of the device, also in μm. Note as well that the devices of opamp A_3 enabling the *economic* and *not-economic* modes are also displayed.

Table 36. Requested performance of the opamps.

		Performance feature	Target/Directive			Units
			Opamp A_1	Opamp A_2	Opamp A_3	
GSM	Electrical	add_db	> 50	> 60	> 50	dB
		ft	> 26.43	> 51.41	> 21.47	MHz
		pm	> 72.83	> 47.78	> 45.42	°
		r_{out}	< 40	< 70	< 70	kΩ
		add_cm_db	> 50	> 60	> 50	dB
		ft_cm	≈ ft	≈ ft	≈ ft	MHz
		pm_cm	> 72.83	> 47.78	> 45.42	°
	Environmental	RL	36.5	6.0	0.97	kΩ
		CL	3.9	3.8	100.35	pF
	Optimization	Power	Minimize	Minimize	Minimize	mW
DECT	Electrical	add_db	> 60	> 60	> 60	dB
		ft	> 43.92	> 51.41	> 80.92	MHz
		pm	> 39.05	> 47.78	> 51.13	°
		r_{out}	< 40	< 70	< 70	kΩ
		add_cm_db	> 60	> 60	> 60	dB
		ft_cm	≈ ft	≈ ft	≈ ft	MHz
		pm_cm	> 39.05	> 47.78	> 51.13	°
	Environmental	RL	10.4	6.0	0.97	kΩ
		CL	1.90	3.8	100.08	pF
	Optimization	Power	Minimize	Minimize	Minimize	mW
GSM/ DECT	Optimization	Area	Minimize	Minimize	Minimize	μm^2

Table 40 shows detailed sizing of the passive devices, where the unitary components of each passive device is also specified.

After layout generation of the cell-level blocks, their performance must be verified at the layout level. Table 41 and Table 42 collect the verification results and compares them with the schematic verification (i.e., without layout-induced parasitics).

The resulting layout instance of the analog back-end for this multi-standard retargeting is depicted in Fig. 27. Total occupied area is $750.4 \times 1250.6\ \mu m^2$.

Table 37. Obtained opamp performances after cell-level sizing.

	Performance feature	Opamp A_1	Opamp A_2	*Opamp A_3*	*Units*
GSM	add_db	102.5	94.59	98.99	dB
	ft	30.40	55.69	56.35	MHz
	pm	85.19	62.49	69.86	°
	r_{out}	36.26	67.94	69.98	kΩ
	add_cm_db	112.65	108.03	109.32	dB
	ft_cm	30.87	56.52	55.91	MHz
	pm_cm	75.78	65.50	64.44	°
	Power	1.07	0.72	0.86	mW
DECT	add_db	95.08	94.595	89.12	dB
	ft	44.24	55.69	81.72	MHz
	pm	88.37	62.49	68.88	°
	r_{out}	21.56	67.94	40.60	kΩ
	add_cm_db	100.87	108.0	93.77	dB
	ft_cm	44.31	56.52	78.45	MHz
	pm_cm	77.53	65.50	72.14	°
	Power	2.15	0.72	1.74	mW
GSM/ DECT	Area	977.337	984.39	1648.67	μm^2

Table 38. Core and CMFB device sizes.

Device or Bias current/voltage	Opamp A_1	Opamp A_2	Opamp A_3	*Units*
M_1, M_2	25.45 / 1.2	96.7 / 1.2	48.2 / 1.2	W (μm) /L (μm)
M_3, M_4	14.9 / 1.5	11.85 / 1.5	32.6 / 1.5	W (μm) /L (μm)
M_5	17.3 / 1.5	18.55 / 1.5	58.4 / 1.5	W (μm) /L (μm)
M_6-M_9	10.35 / 1.5	11.1 / 1.5	35.0 / 1.5	W (μm) /L (μm)
M_{10}, M_{11}	24.6 / 0.55	13.4 / 0.55	18.5 / 0.55	W (μm) /L (μm)
M_{12}, M_{13}	96.3 / 1.5	47.85 / 1.5	174.6 / 1.5	W (μm) /L (μm)
M_{14}, M_{15}	72.25 / 0.35	45.95 / 0.35	106.2 / 0.35	W (μm) /L (μm)
C_{c1}, C_{c2}	0.62	0.52	0.59	pF
M_{1c}, M_{2c}	43.85 / 1.2	71.7 / 1.2	79.5 / 1.2	W (μm) /L (μm)
M_{3c}, M_{4c}	21.91 / 1.35	29.75 / 1.25	36.80 / 0.75	W (μm) /L (μm)
M_{5c}	23.05	27.8 / 1.5	73.0 / 1.5	W (μm) /L (μm)
R_{CM1}, R_{CM2}	9.85	7.45	7.75	kΩ
R_{z1}, R_{z2}	828.52			kΩ
I_{BIAS}	46.3	25.5	83.0	μA
V_{cn}, V_{cp}	-0.5, -0.5	-0.5, -0.5	-0.5, -0.5	V

Table 39. Detailed sizing of the opamps.

Device	Opamp A_1			Opamp A_2			Opamp A_3		
	m	W_{finger}	*W/L*	*m*	W_{finger}	*W/L*	*m*	W_{finger}	*W/L*
M_1, M_2	8	3.20	25.6 / 1.2	8	12.1	96.8 / 1.2	16	3.0	48 / 1.2
M_3, M_4	8	1.85	14.8 / 1.5	8	1.5	12.0 / 1.5	8	4.1	32.8 / 1.5
M_5	12	1.45	17.4 / 1.5	8	2.35	18.8 / 1.5	16	3.65	58.4 / 1.5
M_6-M_9	4	2.6	10.4 / 1.5	4	2.8	11.2 / 1.5	4	8.75	35 / 1.5
M_{10}, M_{11}	4	6.15	24.6 / 0.55	4	3.35	13.4 / 0.55	4	4.65	18.6 / 0.55
M_{12}, M_{13}	10	9.65	96.5 / 1.5	4	11.95	47.8 / 1.5	2	17.5	35 / 1.5
M_{14}, M_{15}	4	18.05	72.2 / 0.35	4	11.5	46.0 / 0.35	4	5.4	21.6 / 0.35
M_{B7}	6	1.45	8.7 / 1.5	2	2.35	4.7 / 1.5	4	3.65	14.6 / 1.5
M_{B8}	2	8.65	17.3 / 1.5	2	13.95	27.9 / 1.5	8	9.15	73.2 / 1.5
M_{Bcab}	1	10.4	10.4 / 1.5	1	11.2	11.2 / 1.5	4	8.75	35 / 1.5
M_{Bcpa}	1	1.4	1.4 / 12.0	1	1.9	1.9 / 10.0	1	1.55	1.55 / 10.0
M_{Bcna}	4	6.15	24.6 / 0.55	4	3.35	13.4 / 0.55	4	4.65	18.6 / 0.55
M_{Bcnb}	1	2.4	2.4 / 10.0	1	2.6	2.6 / 6.0	1	2.7	2.7 / 8.0
M_{1c}, M_{2c}	4	10.95	43.8 / 1.2	4	17.95	71.8 / 1.2	8	9.95	79.6 / 1.2
M_{3c}, M_{4c}	8	2.75	22.0 / 1.35	8	3.75	30.0 / 1.25	8	4.6	36.8 / 0.75
M_{5c}	16	1.45	23.2 / 1.5	12	2.35	28.2 / 1.5	20	3.65	73.0 / 1.5
M_{B9}-M_{B11}	2	1.45	2.9 / 1.5	2	2.35	4.7 / 1.5	2	3.65	7.3 / 1.5
R_{CM1}, R_{CM2}[a]	5	17.8	9.88 kΩ	5	13.45	7.47 kΩ	5	14.0	7.77 kΩ
R_{z1}, R_{z2}[b]	2	3.75	833.33 Ω						
C_{c1}, C_{c2}[c]	2	18.9	0.62 pF	2	17.3	0.52 pF	2	18.4	0.59 pF
M_{B1a}	4	2.5	10.0 / 4.0	4	2.0	8.0 / 6.0	4	2.5	10.0 / 4.0
M_{B1b}	4	2.5	10.0 / 4.0				4	2.5	10.0 / 4.0
M_{B2}	1	30.0	30.0 / 0.5				1	30.0	30.0 / 0.5
M_{B3}	4	2.5	10.0 / 4.0	4	2.0	8.0 / 6.0	4	2.5	10.0 / 4.0
M_{B4}	12	2.5	30.0 / 4.0	4	2.0	8.0 / 6.0	8	2.5	20.0 / 4.0
M_{B5}	24	2.5	60.0 / 4.0	24	2.0	48.0 / 6.0	40	2.5	100.0 / 4.0
M_{B6}	4	2.5	10.0 / 4.0	4	2.0	8.0 / 6.0	4	2.5	10.0 / 4.0
M_{12e}, M_{13e}							m_{M12}·4	17.5	140.0 / 1.5
M_{14e}, M_{15e}							16	5.4	86.4 / 0.35
M_{oni}, M_{Boni}	1	3.5	3.5 / 1.5	1	3.5	3.5 / 1.5	1	3.5	3.5 / 1.5
M_{e1}-M_{e4}							1	30.0	30.0 / 0.35
M_{e5}, M_{e6}							1	15.0	15.0 / 0.35

a. In this case, W_{finger} is the unitary resistor length (0.8μm width) and column W/L is the total resistance.

b. In this case, W_{finger} is the unitary resistor length (0.8μm width) and column W/L is the total resistance.

c. In this case, W_{finger} is the width of the unitary, square capacitor and column W/L is the total capacitance.

Table 40. Detailed sizing of the passive devices.

Module-level block	*Standard*	*Device*	*Value*	*Width*	*Length*
CT-LP filter		$R_{1,unit}$	20.675 kΩ	0.75 μm	165.4 μm
		$R_{FF,unit}$	17.733 kΩ	0.8 μm	159.6 μm
		$C_{1,2,unit}$	0.465 pF	23.05 μm	23.05 μm
	GSM	R_1	41.35 kΩ	$2 \cdot R_{1,unit}$	
		R_{FF}	797.999 kΩ	$45 \cdot R_{FF,unit}$	
		R_{FB}	4.287 kΩ	0.75 μm	34.3 μm
		C_1, C_2	3.723 pF	$8 \cdot C_{1,2,unit}$	
	DECT	R_1	13.783 kΩ	$(2/3) \cdot R_{1,unit}$	
		R_{FF}	88.666 kΩ	$5 \cdot R_{FF,unit}$	
		R_{FB}	4.287 kΩ	0.75 μm	34.3 μm
		C_1, C_2	1.861 pF	$4 \cdot C_{1,2,unit}$	
PGA		$R_{1,2,3,unit}$	2.5 kΩ	1.5 μm	57.5 μm
		$R_{4,unit}$	6.25 kΩ	1.5 μm	143.75 μm
		$C_{P,unit}$	0.497 pF	23.85 μm	23.85 μm
	GSM/DECT	R_{P1}	25.0 kΩ	$10 \cdot R_{1,2,3unit}$	
		R_{P2}	5.0 kΩ	$2 \cdot R_{1,2,3unit}$	
		R_{P3}	7.5 kΩ	$3 \cdot R_{1,2,3unit}$	
		R_{P4}	12.5 kΩ	$2 \cdot R_{4,unit}$	
	GSM	C_P	3.984 pF	$8 \cdot C_{P,unit}$	
	DECT	C_P	0.994 pF	$2 \cdot C_{P,unit}$	

The last step in the design reuse flow is the performance verification of the analog back-end, which, in this case, is done using device-level simulation. Table 43 and Table 44 summarize the verification results for the GSM and the DECT wireless communication standard, respectively.

Total retargeting time was 62.2 minutes CPU time, which includes 15 minutes for analog back-end sizing[17], 14.6 minutes for cell-level sizing, layout generation, extraction, and verification, and about 31 minutes for layout generation (10s), extraction (30s), and verification (30 minutes) of the analog back-end.

---

17. Recall that there are two wireless communication standards to address.

Table 41. CT-LP filter opamps (A_1 and A_2) verification.

Cell-level block	*Standard*	*Performance feature*	*Specified*	*Type*	*Value*	*Units*
Opamp A_1	GSM	add_db	> 50	SCH	101.8	dB
				LAY	101.7	
		ft	> 26.43	SCH	30.6	MHz
				LAY	29.26	
		pm	> 72.83	SCH	88.06	°
				LAY	83.01	
		add_cm_db	> 50	SCH	111.5	dB
				LAY	111.4	
		ft_cm	≈ ft	SCH	30.7	MHz
				LAY	28.66	
		pm_cm	> 72.83	SCH	78.34	°
				LAY	74.67	
	DECT	add_db	> 60	SCH	100.3	dB
				LAY	99.5	
		ft	> 43.92	SCH	44.9	MHz
				LAY	43.93	
		pm	> 39.05	SCH	90.93	°
				LAY	86.21	
		add_cm_db	> 60	SCH	107.1	dB
				LAY	106.8	
		ft_cm	≈ ft	SCH	44.36	MHz
				LAY	42.86	
		pm_cm	> 39.05	SCH	79.44	°
				LAY	74.23	
Opamp A_2	GSM/ DECT	add_db	> 60	SCH	93.94	dB
				LAY	93.87	
		ft	>51.41	SCH	57.39	MHz
				LAY	52.36	
		pm	> 47.78	SCH	70.53	°
				LAY	57.03	
		add_cm_db	> 60	SCH	107.6	dB
				LAY	107.5	
		ft_cm	≈ ft	SCH	58.02	MHz
				LAY	52.96	
		pm_cm	> 47.78	SCH	67.4	°
				LAY	57.77	

Table 42. PGA opamp (A_3) verification.

Standard	*Performance feature*	*Specified*	*Type*	*Value*	*Units*
GSM	add_db	> 50	SCH	98.49	dB
			LAY	98.04	
	ft	> 21.47	SCH	57.66	MHz
			LAY	52.16	
	pm	> 45.42	SCH	73.98	o
			LAY	66.08	
	add_cm_db	> 50	SCH	109.2	dB
			LAY	108.5	
	ft_cm	≈ ft	SCH	55.48	MHz
			LAY	52.41	
	pm_cm	> 45.42	SCH	54.62	o
			LAY	47.43	
DECT	add_db	> 60	SCH	96.06	dB
			LAY	95.7	
	ft	> 80.92	SCH	86.51	MHz
			LAY	78.91	
	pm	> 51.13	SCH	74.49	o
			LAY	66.49	
	add_cm_db	> 60	SCH	103.2	dB
			LAY	102.6	
	ft_cm	≈ ft	SCH	78.65	MHz
			LAY	74.58	
	pm_cm	> 51.13	SCH	57.96	o
			LAY	48.65	

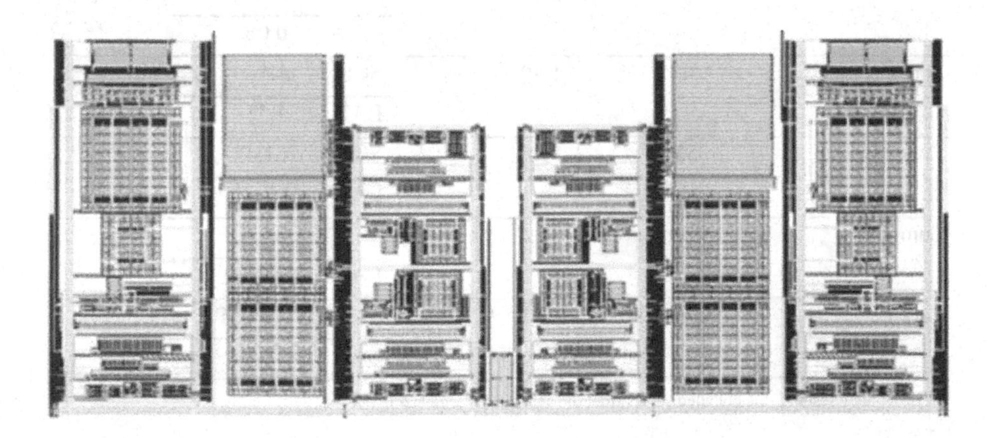

Figure 27. Analog back-end layout instance for the multi-standard retargeting example.

Table 43. Final analog back-end verification for the GSM standard.

Performance feature	*Specification*	*Optimized*[a]	*Type*	*Value*	*Units*
adcra	< 0.7	0.48	SCH	0.015	dB
			LAY	0.011	
gdavgra	< 33.5e-9	8.45e-9	SCH	1.59e-9	s
			LAY	1.43e-9	
pera	< 1.3	0.3	SCH	0.05	°
			LAY	0.05	
apeak	< 1.0	0.20	SCH	0.24	dB
			LAY	0.43	
imar	> 50.0	57.37	SCH	64.57	dB
			LAY	66.64	
bumpsz	< 1.0	0	SCH	0	dB
			LAY	0	
gddev	< 8.3e-9	8.29e-9	SCH	7.20e-9	s
			LAY	7.54e-9	

a. This column resumes the obtained values from the analog back-end sizing process.

Table 44. Final analog back-end verification for the DECT standard.

Performance feature	*Specification*	*Optimized*[a]	*Type*	*Value*	*Units*
adcra	< 0.7	0.5	SCH	0.01	dB
			LAY	0.01	
gdavgra	< 6.5e-9	0.14e-9	SCH	0.26e-9	s
			LAY	0.24e-9	
pera	< 1.75	0.36	SCH	0.06	°
			LAY	0.06	
apeak	< 1.5	1.49	SCH	1.17	dB
			LAY	1.55	
imar	> 55	59.94	SCH	61.48	dB
			LAY	67.51	
bumpsz	< 1.0	0	SCH	0	dB
			LAY	0	
gddev	< 9.5e-9	2.23e-9	SCH	2.36e-9	s
			LAY	2.27e-9	

a. This column resumes the obtained values from the analog back-end sizing process.

4.4 Automation prototype

To show how the execution of the design reuse flow illustrated above can be fully automated, a graphical user interface prototype has been built. Such interface represents a homogeneous environment that provides the required data management and translation (which can be costly and time-consuming for complex analog designs) between the different sizing steps of the design reuse flow: the system, module, and cell-level sizing steps. It also facilitates the backtracking of the design process by dynamically watching the evolution of the sizing process, by reporting on the resulting solutions and, therefore, giving the user critical information in case any of the sizing steps fails to accomplish its objectives. It is important here to clearly state that this prototype is but a practical demonstration of the benefits of automating the analog design flow (especially when, to the knowledge of the authors, there is a commercial shortage of analog CAD tools providing the same level of automation and flow monitoring), thanks to a great extent to the underlying principles of reuse-based design, where storing design knowledge via parameterization is the key. Note also that the example shown here is suited to the analog back-end used as demonstration vehicle of the work proposed in this book, but that all scripts and parsers can be easily transformed to work with any other analog system and hierarchy.

The automation prototype consists, on the one hand, of a number of scripts that collect the output data of the precedent sizing step, transform it, and handle it over for the execution of the following sizing task. For instance, after the analog back-end sizing step, the prototype collects the required cell-level performance specifications, transforms the corresponding optimization input files, and feeds them to the optimization tool (FRIDGE) for it to complete the cell-level sizing step. The scripts have been written in PERL.

On the other hand, a user-friendly interface, written in TCL-TK [Welch03], provides a visual monitoring of the evolution of the design reuse flow. This interface is illustrated in Fig. 28. First, the user is prompted to specify the database for the higher-level sizing, its required performance specifications, and an empty file where the values of the layout template parameters, necessary to automatically instance the circuit layout, are saved.

The design reuse flow is then tracked with a dedicated interface. Three windows form this interface: the flow navigator, the status window, and the datasheet window. In the flow navigator, the evolution of the flow is monitored by highlighting the ongoing process, the already accomplished, and the failed ones. It is also possible to retrieve the output information from each sizing step. The corresponding datasheet is displayed in the datasheet window. Several examples of the type of documentation provided by the user interface are shown in Fig. 29.The status window reports, in text format, what sizing step is being carried out. It also provides information on sizing failures (e.g., if any design variable has been exhausted without attaining the optimization goals).

With this automation prototype, the resulting design reuse flow becomes fully traceable and thoroughly documented.

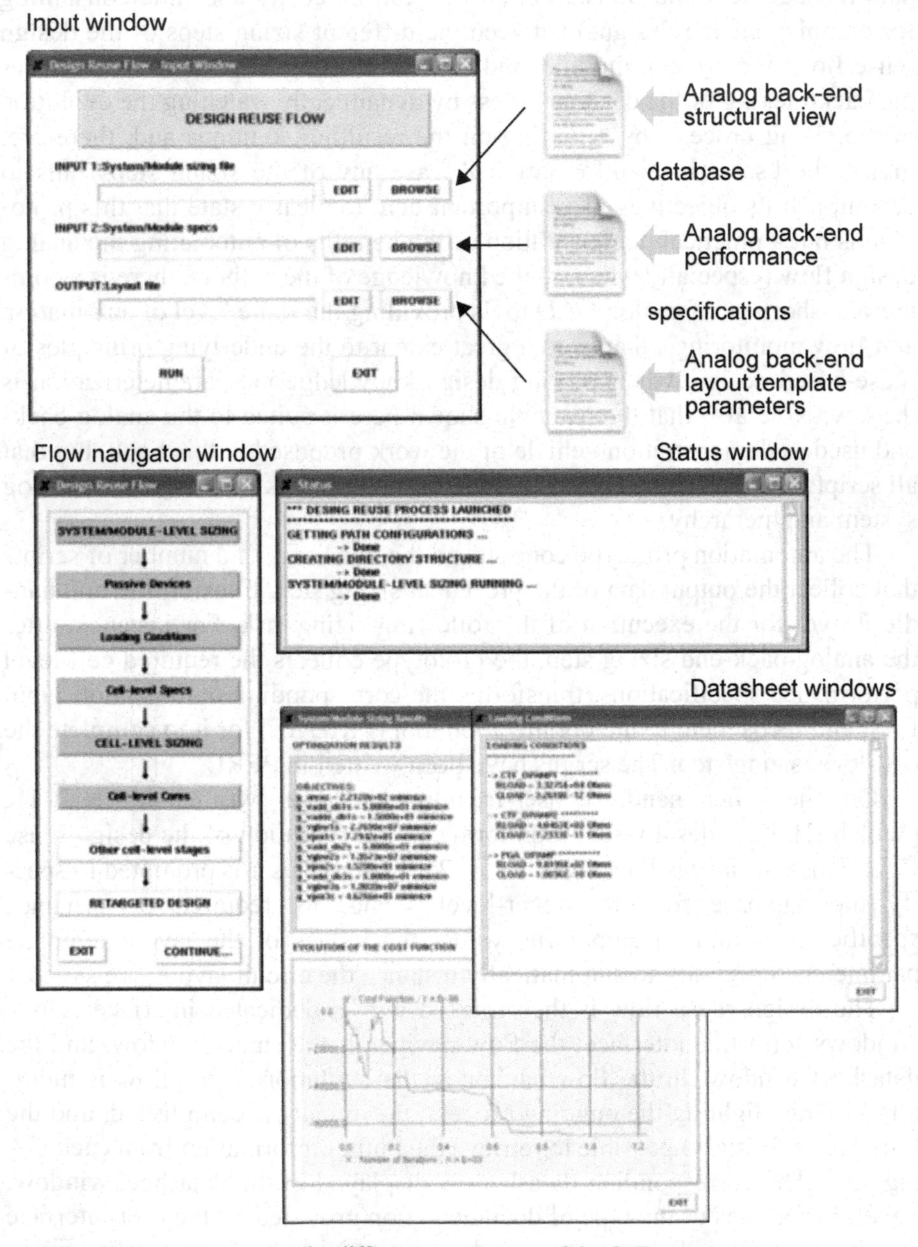

Figure 28. The different components of the GUI prototype.

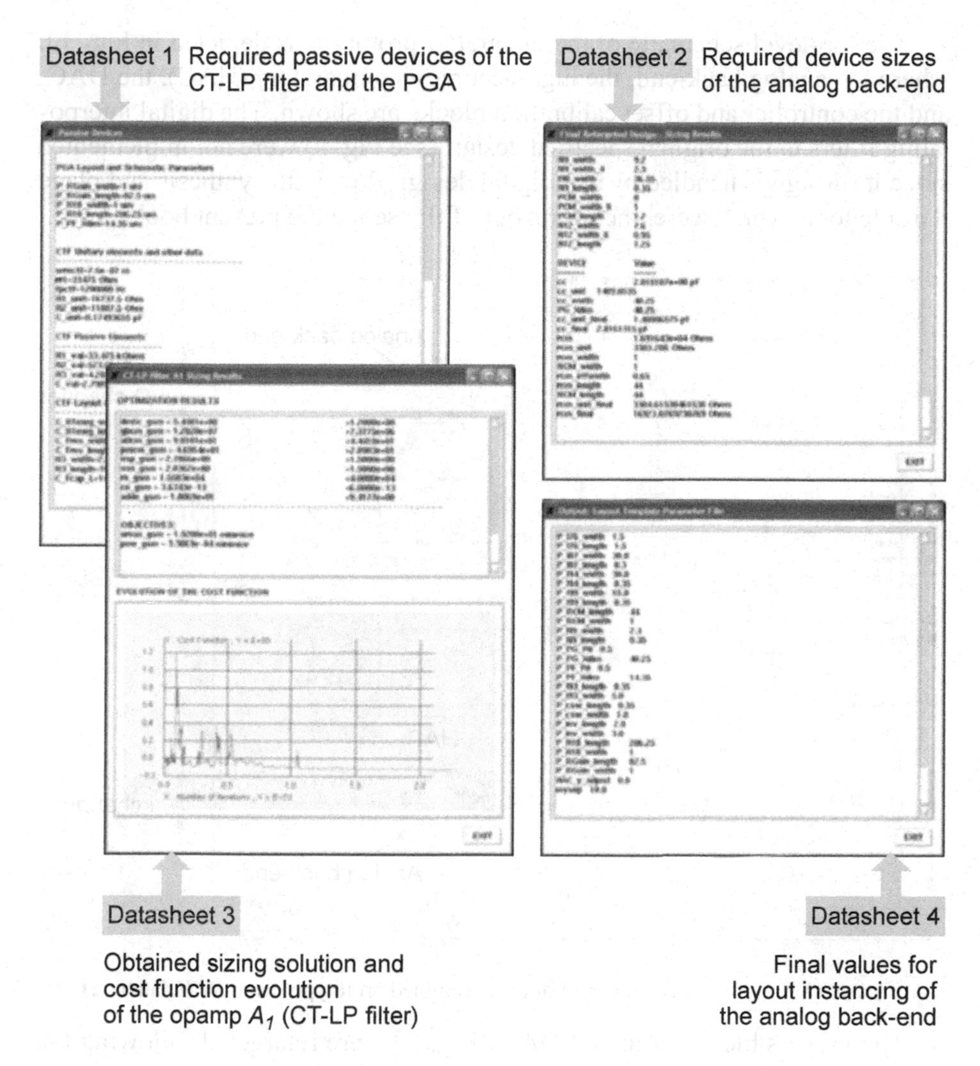

Figure 29. Illustration of the documentation provided by the automation prototype.

5 SILICON PROTOTYPE

The objective of this section is to describe the prototype of the IQ DA transmit interface, whose design has been completed by following the design reuse methodology presented in this book, and fabricated in a double-poly, 3.3V supply voltage 0.35-μm CMOS technology. This integrated silicon prototype serves as a demonstrator of the reuse-based design framework.

The top-level schematic of the integrated prototype is depicted in Fig. 30, where the analog back-end, the high-accuracy comparators (HAC), the DAC, and the controller and offset calibration blocks are shown. The digital interpolating filters in the original specified design (see Fig. 4) were not implemented since its design is handled by any digital design flow with synthesis and place & route tools which were, therefore, out of the scope the present book.

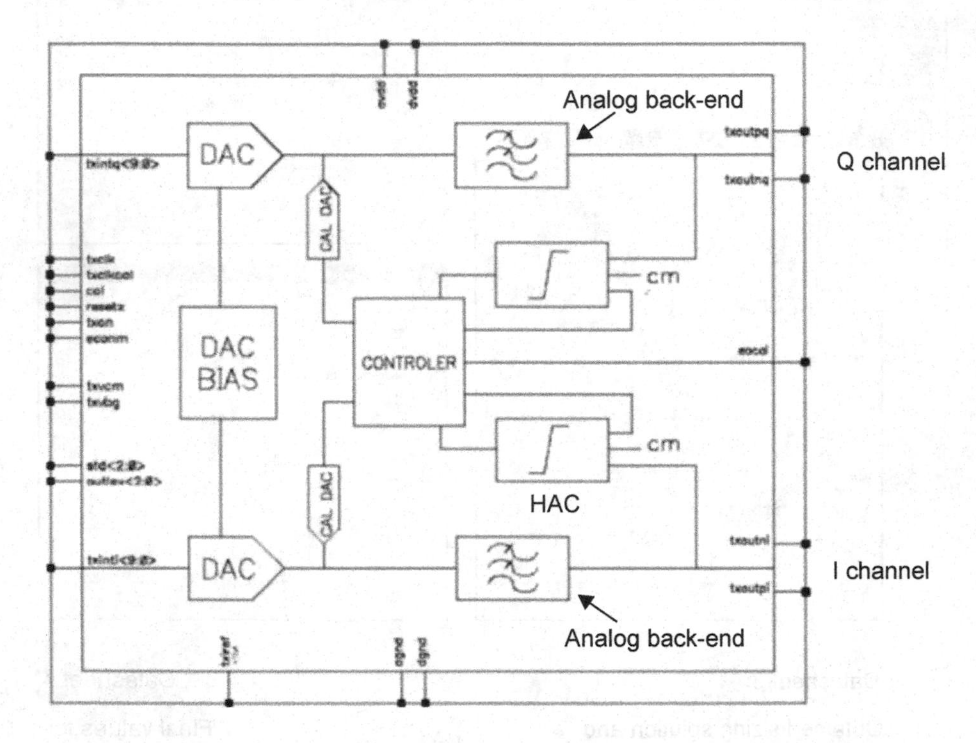

Figure 30. Top-level block diagram of the assembled prototype (from *DFII* database).

The various blocks of the IQ DA in Fig. 30 were retargeted following the design reuse methodology presented in this book[18], verified, and taped-out for fabrication. The prototype floorplan is illustrated in Fig. 31 and the corresponding top-level layout assembly of the retargeted IQ DA is shown in Fig. 32.

The pad distribution was carefully selected to palliate any coupling between analog and digital signals. All the analog signals are present at the bottom of Fig. 32, together with the power cuts between the analog and digital supply rings. The 10-bit I and Q signal inputs are located, respectively, at the right and left sides of Fig. 32. Control pins are located at the top of Fig. 32. A CLCC-44pin package was selected to accommodate the prototype.

[18] In the case of the DAC blocks, a similar methodology, presented in [Jing01], was used.

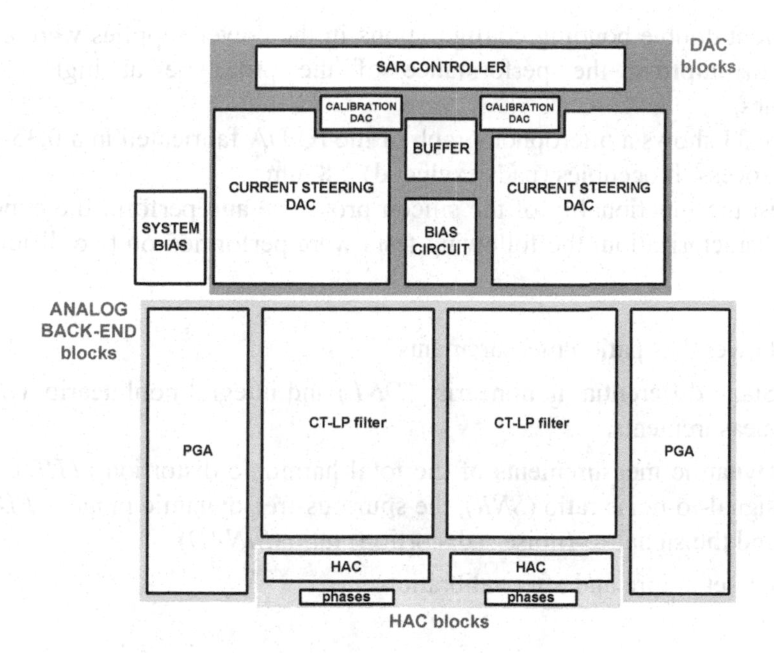

Figure 31. Illustration of the prototype floorplan.

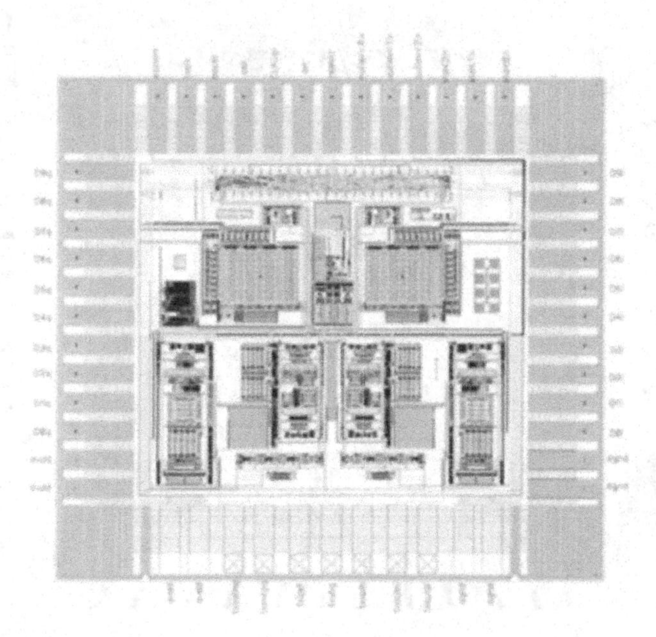

Figure 32. Top-level layout with pads of the assembled prototype (courtesy of Chipidea® Microelectrónica).

Different double bonding configurations in the power supplies were considered to improve the performance of the prototype at high clock frequencies.

Figure 33 shows a microphotograph of the IQ DA fabricated in a 0.35-μm CMOS process. It occupies (pads excluded) 2.8 mm^2.

To test the functionality of the silicon prototype and perform the experimental characterization, the following tests were performed on two different samples:

1. Power dissipation measurements.
2. Static differential nonlinearity (*DNL*) and integral nonlinearity (*INL*) measurements.
3. Dynamic measurements of the total harmonic distortion (*THD*), the signal-to-noise ratio (*SNR*), the spurious-free dynamic range (*SFDR*), and the signal-to-(noise + distortion) ratio (*SINAD*).
4. Offset before and after calibration.

Figure 33. Microphotograph of the prototype in the 0.35-μm CMOS technology (courtesy of Chipidea® Microelectrónica).

The default test conditions are summarized in Table 45. A DC power supply system (HP E3632A) is used to generate the power supply signals and the logic input data are acquired from a logic analyzer (HP 16702A) and a data generator (HP 80000). An audio-precision system (System One, SYS-222G) is used to analyze the output data. The test set-up is controlled with C routines from a workstation. MATLAB is used to process the DAC output.

The experimental results are summarized in Table 46. Fig. 34, Fig. 35, and Fig. 36 illustrate some of the experimental measurements.

Table 45. Default test conditions.

Test condition	*Value*
Analog power supply	3.3V
Digital power supply	3.3V
Master clock	14MHz
Standard mode std[2:0]	[1 1 1][a]
Economic mode	YES

a. It corresponds to the GSM wireless communication standard.

Table 46. Experimental results summary.

Characteristic	*Specified*			*Measured*			*Units*
	MIN	*TYP*	*MAX*	*MIN*	*AVERAGE*	*MAX*	
Current consumption (AVDD), normal mode		13.8		13.33	13.68	14.03	mA
Current consumption (DVDD), normal mode				0.149	0.15	0.152	mA
Current consumption (AVDD), *power-down* mode				24	24.5	25	nA
Current consumption (DVDD), *power-down* mode				0.149	0.15	0.152	nA
SINAD		57		60.91	65.42	71.28	dB
SNR		60		63.29	70.50	75.28	dB
THD		-60		-62.23	-67.44	-73.89	dB
SFDR		65		66.29	70.84	97.72	dB
DNL		± 1.0		-1.29	-0.21	0.87	LSB
INL		± 1.0		-1.42	-0.28	0.86	LSB

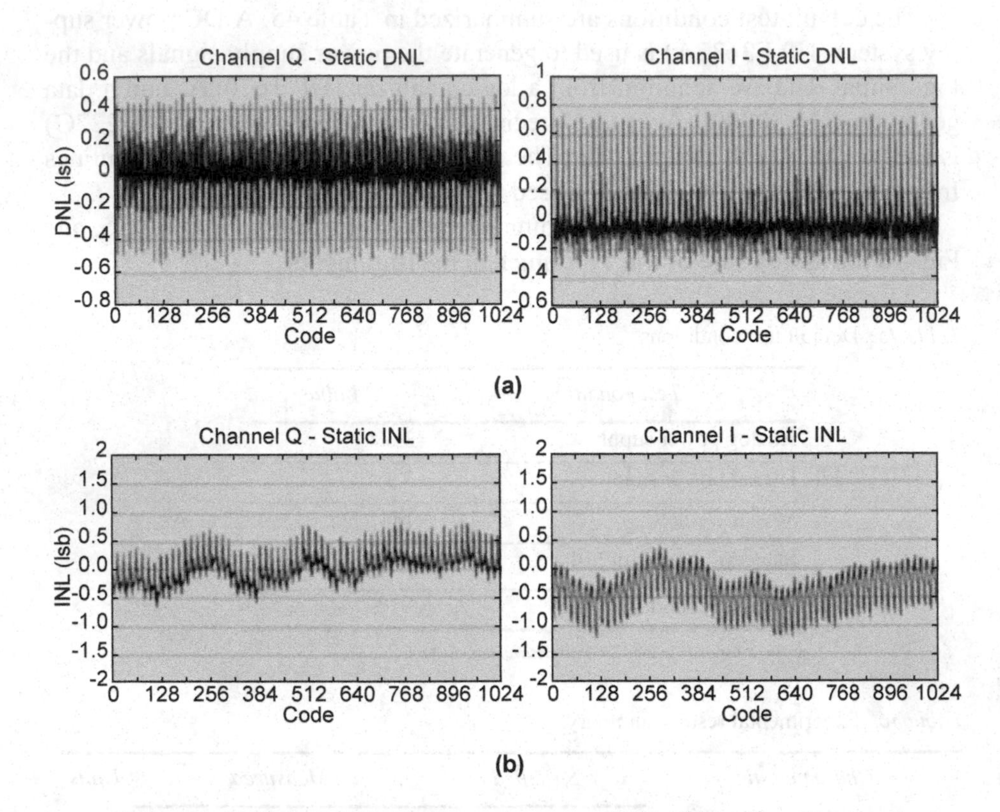

Figure 34. (a) DNL and (b) INL figures (courtesy of Chipidea® Microelectrónica).

For example, Fig. 36 shows typical waveforms at the output of one channel, both single-ended centered in the common-mode voltage as well as the differential waveform obtained by a difference function in the oscilloscope. From top to bottom, outlev[2:0] was varied from [0, 0, 1] to [1, 0, 0].

Globally, the achieved results are better than the required specifications for the GSM standard. Specifically:

1. By design, current consumption should typically be 13.8 mA, and the measured value is between 13.33 mA and 14.02 mA; this variation is within ±2.5 % and can be justified by accuracy of the current mirrors for biasing purposes.
2. *INL* was set by design to be around 1.0 LSB with a yield of 90%; this means that 10% of the samples can exhibit *INL* values above 1.0 LSB. According to a Gaussian distribution, the limits imposed by 90% of the area are defined by ±1.65 sigma. This means that

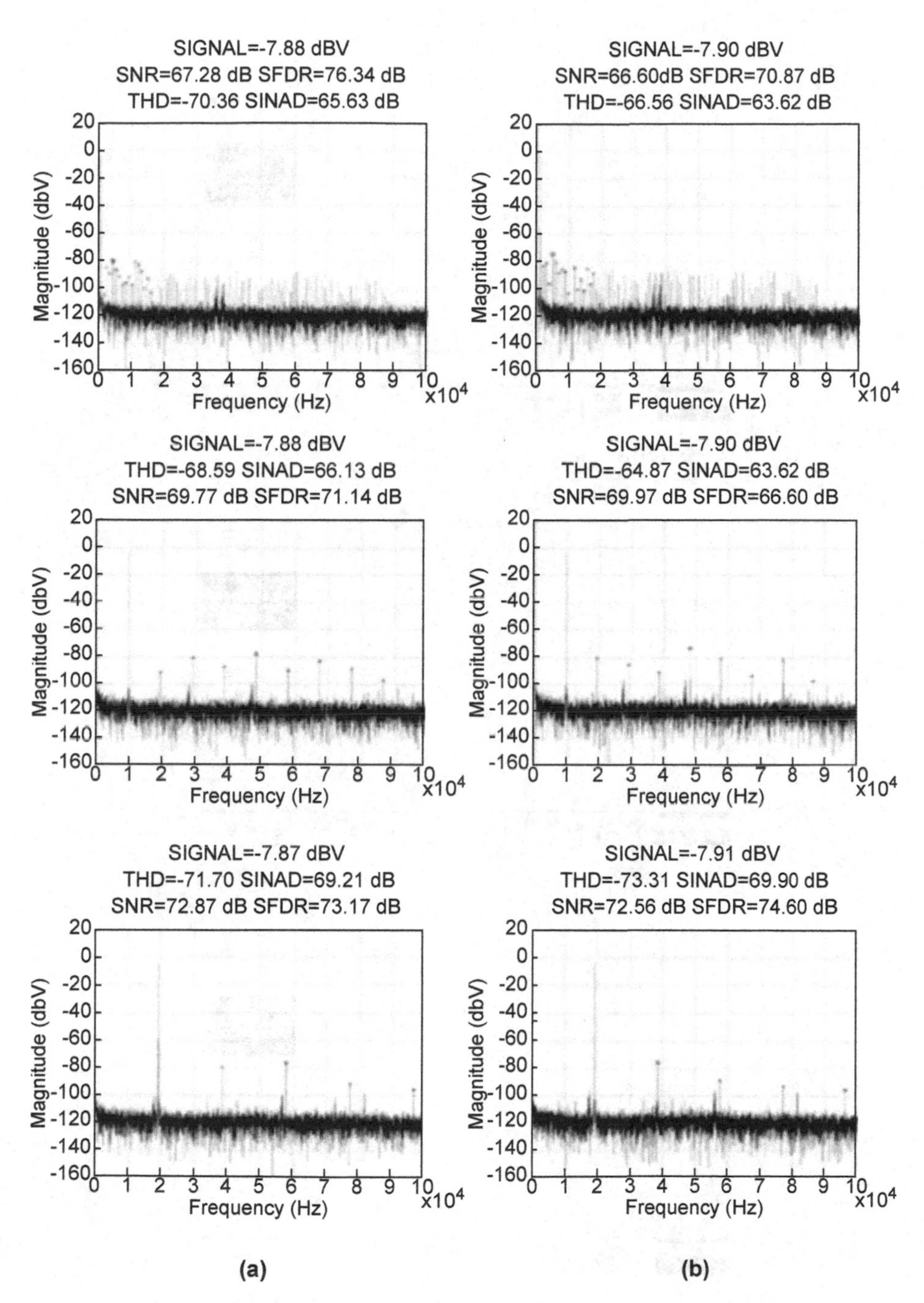

Figure 35. Spectrum samples: (a) Q channel; (b) I channel (courtesy of Chipidea® Microelectrónica).

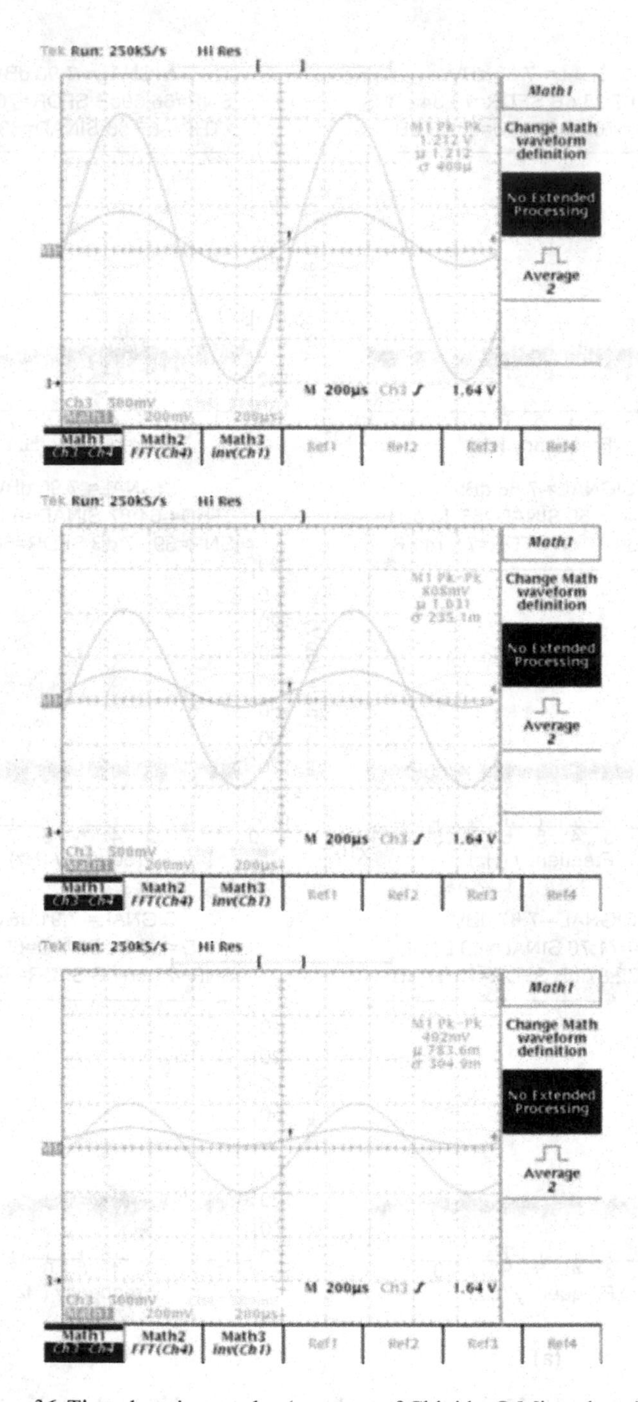

Figure 36. Time domain samples (courtesy of Chipidea® Microelectrónica).

sigma=1.0LSB/1.65, and the usual limits of ±3 sigma are around 1.81 LSB. The observed values are all within these limits.

- The measured values for the dynamic characteristics are all better than the requirements for GSM. *THD* is always below -62 dB, having an average value of -67 dB over all tested conditions, whereas the requirement for this system was -60 dB. *SNR* is always above 63 dB, with an average value in all tested conditions of 70 dB, whereas the requirement was 60 dB. *SFDR* is always above 66 dB, having an average value of 70 dB, whereas the requirement was 65 dB. *SINAD* is always above 60 dB, having an average value of 65 dB, whereas the requirement was 57 dB.

Therefore, the results from the silicon characterization presented here, and the design examples provided in the previous section, altogether validate the reuse-based framework describe in this book, since, first, the system operates correctly, exceeding specifications, and, second, the examples prove that working designs can be attained with the framework in a considerable short amount of time.

6 COSTS AND BENEFITS

At this point, it is necessary to make a few comments on the development costs of design for reusability and the benefits derived from adopting a design reuse flow.

Development cost is the necessary effort to develop the reusable block with the design for reusability methodology explained in previous chapters, which it is not only direct engineering cost but also the reusability overhead cost involving the design knowledge encapsulation and parameterization of the databases. The development cost of design for reusability of the analog back-end has been evaluated in nearly 6 $\text{persons} \cdot \text{month}$, from which around 50% is devoted to developing the reusable opamps and roughly 50% of the cost for each reusable block was devoted to the development of its layout facet[19].

Prior to the design for reusability of the analog back-end, it took about 4 $\text{persons} \cdot \text{month}$ to complete its design following a traditional design flow. The investment that the extra development cost on design for reusability entails is, on the other hand, to be recovered in the lifetime reuses of the

19. Providing that a library of layout templates (i.e., macro-cells) for basic building blocks (e.g., differential pairs, mirrors, capacitor arrays, resistor arrays, etc.) is already available.

design. Besides, as stated explicitly in the design examples section (Section 4), the running times for the complete reuse of the analog back-end circuit demonstrator (once it is completely reusable) takes, from sizing to final verification, around 30-60 minutes of CPU time, if no layout-sizing spins occur. Consequently, this means that the application of the reuse-based design framework to this or other analog and mixed-signal circuits will certainly dilute the initial investment, rendering the difference in efforts increasingly irrelevant.

7 SUMMARY

In this chapter, the reuse-based design framework has been illustrated with several reuse examples for the analog part of an industrial-scale mixed-signal system. This analog sub-system has been thoroughly described in all its reusable facets, as explained in previous chapters. Furthermore, a silicon prototype of this system has been fabricated and experimental results demonstrates the capability of this design approach in coping with analog inherent complexity and drastically reducing the time to market.

However, there are two aspects of the presented reuse-based framework that have not been demonstrated yet[20]. These aspects are the geometrically constrained sizing and the parasitic-aware sizing; they have been briefly outlined in Chapter 2. The first is important to automatically ensure that the reused design achieves an effective use of silicon area because, in volume production, the area of the chip is paramount for determining the final production cost. This can be assured by establishing careful floorplanning, maximizing the layout regularity, and optimizing the component aspect ratios. The second is important to avoid time-consuming iterations between layout generation and cell-level sizing by including the impact of layout-induced degradation early in the sizing step. Both aspects are discussed in the following chapter.

---

[20]. Despite this, the lack of this demonstration does not erode the validity of the presented design examples as these aspects are planned to enhance the reuse-based design framework (to further automate and improve the overall design time), rather than to correct unexpected faults of the framework.

Chapter 7

Layout-Aware Circuit Sizing

This chapter completes the description and demonstration of the reuse-based design framework presented here. It basically delves into the why and how of including layout geometric information and layout-induced parasitics within the optimization loop of the cell-level sizing process. Outlined in Chapter 2, these sizing techniques, namely geometrically constrained sizing and parasitic-aware sizing, are now the object of the present chapter.

1 INTRODUCTION

The required features for design quality and productivity are multiple. One of them is to ensure that the reused design achieves an effective use of silicon area because, in volume production, the area of the chip is paramount for determining the final production cost. This can be assured by maximizing the layout regularity, optimizing the component aspect ratios, or even by establishing careful floorplanning for the circuit. Another important concern is related to the degradation of performance due to layout-induced parasitics. Such degradation may require time-consuming, unsystematic iterations between layout generation and circuit sizing to improve a design that fails to meet the intended performance specifications.

In this sense, the design reuse flow described in Chapter 2 envisages the application of appropriate techniques to deal with both concerns (see Fig. 1). **Parasitic-aware** sizing, to include layout-induced parasitics in the circuit sizing process, and **geometrically constrained** sizing, to include layout geometric information for the layout to meet geometric objectives, both compose the **layout-aware** circuit sizing methodology that aims at successfully overcoming such concerns. In this chapter, automation of such a methodology is described.

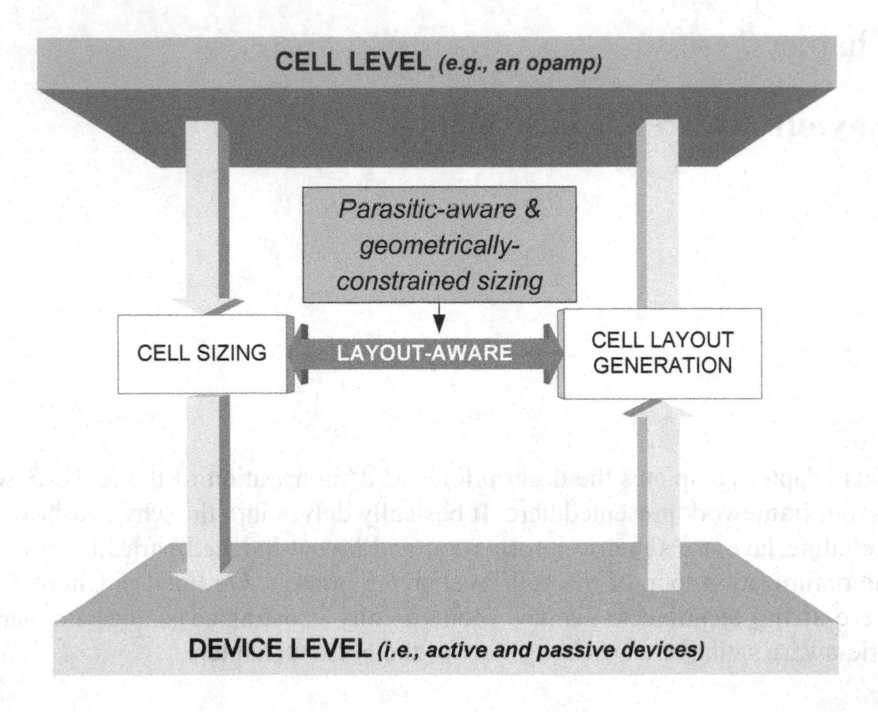

Figure 1. Layout-aware circuit sizing.

2 GEOMETRICALLY CONSTRAINED SIZING

At the layout level, the slicing-based template approach explained in Chapter 5 facilitates the creation of fully reusable analog and mixed-signal (AMS) circuit blocks. This further improves fast and efficient reuse of the circuit block and allows incorporation of layout designer's expertise within the database representing the circuit block as well.

The quality of the eventually produced layout, however, may be suboptimal since the relative placement of the template building blocks on the two-dimensional layout plane (i.e., the floorplan) remains unchanged during the reuse process (either a technology migration or a performance retargeting). To illustrate this point, consider the examples shown in Fig. 2 (for the sake of simplicity, routing wires are not shown). Although the default sizing of the original circuit layout in Fig. 2(a) is rectangular-shaped and the relative area loss[1] is very low, when this circuit is reused resulting, for instance, the layout

1. Area loss is defined as the portion of the total circuit layout area that is unused, that is, it is not occupied by building block layouts nor routing wires and the process-mandatory spacing layout rules.

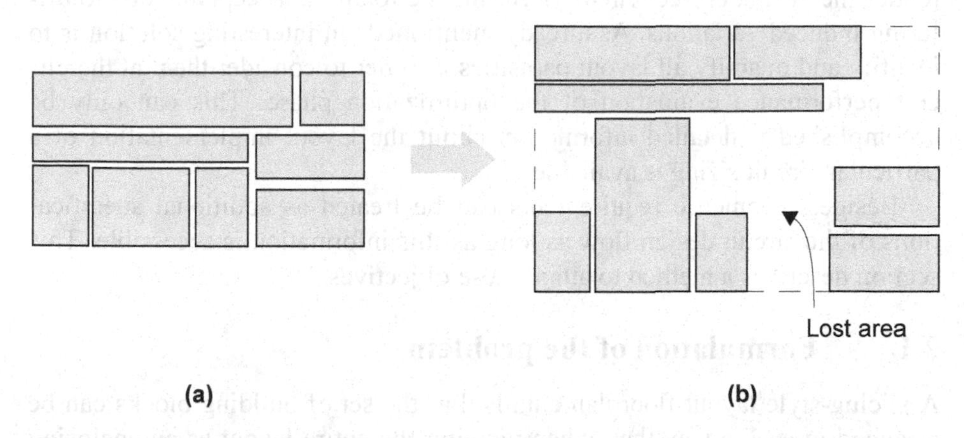

Figure 2. Circuit sizing including geometric layout constraints.

of Fig. 2(b), the total area may be too large (especially when compared with the area occupied by the building blocks) and the aspect ratio, though secondary compared with area concerns, may become very different from user's initial expectations.

To overcome this drawback it is necessary to extend the parameterization of the layout template in order to account for every single combination of block sizes and adjust their relative positions in accordance. Doing so, however, would increase the complexity of the parameterization task exponentially. Moreover, there are combinations of block sizes that prevent achievement of user's geometric requirements, such as minimal area, minimal area loss, or a desired layout aspect ratio. Another option to improve the quality of the resulting layout consists in finding out the exact shape of each building block (e.g., by folding transistors and resistors and changing the size of capacitors while maintaining the intended capacitance) such that the layout template instance meets the geometric requirements. Such an approach cannot be efficiently undertaken manually (e.g., by trying a block shape, then adjusting the shape of the remaining blocks to meet certain aspect ratio while minimizing the area) due to the large number of building block shape combinations and geometric variables for a relatively complex analog circuit.

It is also worth noting that different geometric variable values entail different layout-induced effects (e.g., the diffusion capacitance of the drain and source of a transistor change with the number of folds[2]). Therefore, any eventual modification in the corresponding layout to improve its geometric characteristics (by following either of the two methods explained above), may

2. The number of folds of a MOS transistor is also usually referred as the number of fingers, meaning the number of unitary components resulting from the transistor folding.

render the at first correct circuit performance totally unacceptable due to differing induced variations. As already mentioned, an interesting solution is to identify and quantify all layout parasitics in order to consider them at the circuit performance evaluation of the optimization phase. This can only be accomplished if detailed information about the layout implementation of a particular circuit sizing is available.

Besides, geometric requirements can be treated as additional specifications of the circuit design flow as long as this information is accessible. This section describes a method to attain these objectives.

2.1 Formulation of the problem

A slicing-style layout floorplan entails that the set of building blocks can be arranged in such a way that, when viewing the entire layout as an enclosing rectangle subdivided by horizontal and vertical lines (from the abutting building blocks, as shown in Fig. 3(a)), the two following conditions hold true [Otten82] [Wong86]:

- There are no overlapping rectangles.
- Either each basic rectangle is a building block or there is a line segment (a slicing cut) dividing it into two pieces such that each piece fulfils these two conditions.

A slice is, thus, a combination of two or more components, either building blocks or further slices. A useful way to describe the slicing floorplan is by representing the hierarchy structure of the slices in an oriented rooted binary[3] tree called a *slicing tree*. Fig. 3 shows an example of a slicing tree. In any slicing tree, there are m non-leaf nodes labeled either h or v, specifying whether the slice is horizontal or vertical (from S1 to S7 in Fig. 3(b)), and $n = m + 1$ leaf nodes, each corresponding to one of the eight basic building blocks in Fig. 3(b).

Each basic building block (i.e., each leaf node of the slicing tree) has several possible shapes, i.e., different values of the pair (width, height). Given an electrical sizing of the built-in devices of the building block and a fabrication process, the collection of different shapes can be calculated by varying one o more geometric parameters (GP) defining the block shape (e.g., the number of transistor folds). This collection of (width, height) pairs forms the **shape function** of the building block, by which it is possible to obtain the height value corresponding to a particular width value, and vice versa. An

[3] Requiring trees to be binary simplifies the description of the problem without loss of generality.

example of shape function is shown in Fig. 4, for a polysilicon folded resistor building block whose resistance is equal to 200Ω. The geometric parameter here is the number of folds or resistor strips, from one to eight (the four-strip resistor solution is explicitly shown in Fig. 4). As shown in the plot legend, the width of each strip as well as the strip separation remain constant, being the length of each strip the only parameter strictly dependent on the number of resistor strips.

Then, for a specific electrical sizing of the building blocks and for a given fabrication process, the problem is to find the exact shape (i.e., the building

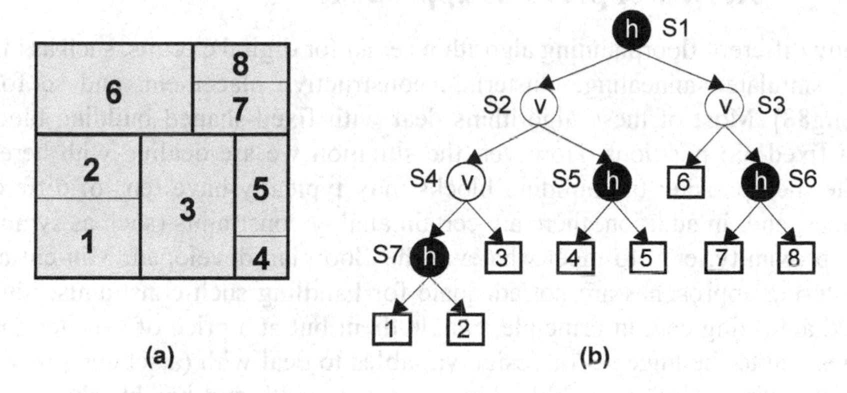

Figure 3. Example of a binary slicing tree representation of a layout template.

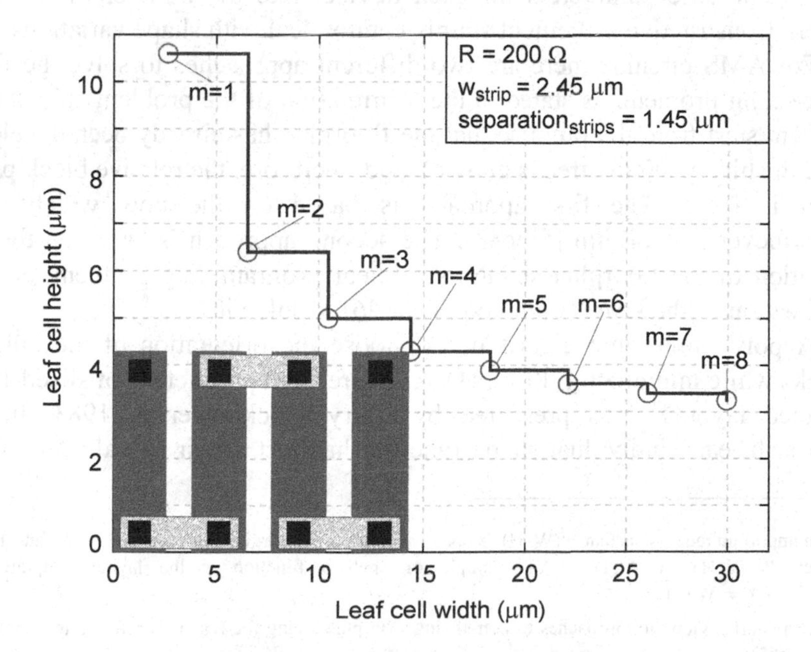

Figure 4. A folded resistor shape function.

block width and height, and, correspondingly, the value of the geometric parameters) of each building block such that some objective function $\Psi(W, H)$ [4] of the circuit layout width, W, and height, H, is minimized. This problem is known as the **floorplan-sizing problem**. At the same time, it should be very appealing that other geometric aspects such as the layout aspect ratio (W/H) or the layout width and/or height would be set to user-specified values. Then, a typical problem should be to minimize the area and set the aspect ratio close to 1.

2.2 Review of previous approaches

Many different floorplanning algorithms exist for digital circuits, such as min-cut, simulated annealing, clustering, constructive placement, and so forth [Leng88]. Most of these algorithms deal with fixed-shaped building blocks and fixed pin positions. However, the situation we are dealing with here is quite the opposite: the building blocks may typically have tens of different shapes, and, in addition, there are certain analog constraints (such as symmetry, proximity, etc.) to meet whatever the floorplan developed. Min-cut and clustering approaches are not adequate for handling such constraints. Simulated annealing can, in principle, handle them but at a price of very long run times due to the huge set of design variables to deal with (a set composed of every device's design variable, like transistor width and height, plus one or more geometric parameters for each device, like the number of transistor folds). Constructive placement simply cannot deal with shape variations.

For AMS circuits, there are two different approaches to solve the floorplan-sizing problem, as stated in the formulation of the problem above. Both of them start from the premise that the floorplan has already been developed (i.e., the binary slicing tree is known), and, therefore, the relative block placement is fixed. The first approach is based on the now widely cited Stockmeyer's algorithm [Stoc83]. The second approach is based on the formulation of the floorplan sizing as a linear programming problem, and the application of the *simplex* method [Tokh96] to solve it.

A polynomial-time algorithm to choose the orientation of the building blocks while minimizing $\Psi(W, H)$ (e.g., area and perimeter), of sliced-floorplanned layouts[5] was presented by Larry Stockmeyer in 1983. In this approach, each individual shape function has just a pair of values or shape

4. An important requisite is that $\Psi(W, H)$ is nondecreasing in both arguments (i.e., if $W \leq W'$ and $H \leq H'$ then $\Psi(W, H) \leq \Psi(W', H')$. An example of such a function is the layout template area, $\Psi(W, H) = W \cdot H$.

5. For a short review of approaches to non-slicing floorplan sizing the reader is referred to [Shi96] and [Leng88].

combinations, that is, if the cell height and width are h and w respectively, and $h \neq w$, the shape function is formed by the pair of values $\{(h, w), (w, h)\}$ (the non-rotated and rotated combinations). The proposed algorithm works in two phases. In the first phase, the tree is processed bottom-up, beginning by associating to each leaf node of the tree a list that represents its respective pair of values. It is very important that each list be sorted according to the rules:

$$\begin{aligned} h_i &> h_{i+1} \\ w_i &< w_{i+1} \end{aligned} \qquad (1)$$

If the corresponding component is square-shaped, then there is only one possible orientation and, thus, only one pair. It then proceeds by associating to each non-leaf node v (a similar procedure is used for horizontal non-leaf nodes) a list of s pairs, $\{(h_1, w_1), \ldots, (h_s, w_s)\}$, satisfying the following properties:

1. $s \leq |L(v)| + 1$, where $L(v)$ stands for the set of leaf nodes of the subtree rooted at v.
2. $h_i > h_{i+1}$ and $w_i < w_{i+1}$.
3. For each of these pairs there is an orientation ρ of $L(v)$ in terms of the defined slicing cuts.
4. For each orientation ρ of $L(v)$ there is a pair (h_i, w_i) such that $h_i \leq h(\rho)$ and $w_i \leq w(\rho)$, which means that $(h(\rho), w(\rho))$ is kept in the list unless there is another orientation ρ' that is strictly better than ρ in the h or w dimension (or both) and is not worse than ρ in either dimension.

Therefore, if $\{(h_1, w_1), \ldots, (h_k, w_k)\}$ and $\{(h_1', w_1'), \ldots, (h_m', w_m')\}$ are the sorted lists (according to property 2) of the two children of the non-leaf node v, the list associated to this vertical cut is constructed by using the Stockmeyer's algorithm. For the sake of clarity, the pseudocode of this algorithm is shown in Fig. 5 for the case of a vertical slicing cut[6].

It is important to notice that there is no need to consider all $k \cdot m$ combinations since many pairs are suboptimal in regard with the nondecreasing function $\Psi(W, H)$. For instance, if $h_i > h_j'$, there is no reason to join (h_i, w_i) with (h_z', w_z') for any $z > j$ since, recalling Eq. (1),

$$\max(h_i, h_z') = \max(h_i, h_j') = h_i, \qquad (2)$$

and

6. The algorithm can be easily modified to a horizontal non-leaf node.

```
1--  i ← 1, j ← 1
2--  while i ≤ k and j ≤ m
3--      Add (max{h_i , h_j'}, w_i + w_j') to the list with pointers to
         (h_i , w_i) and (h_j', w_j')
4--      if h_i > h_j' then i ← i+1
5--      elseif h_i < h_j' then j ← j+1
6--      else i ← i+1 and j ← j+1
```

Figure 5. Stockmeyer's algorithm for a vertical non-leaf node.

$$w_i + w_z' > w_i + w_j' \tag{3}$$

The running time of this algorithm is $O(k + m)$[7] and the length of the list produced is, at most, $k + m - 1$. By recursively applying this algorithm, the first phase ends with the associated list or shape function of the overall slicing tree T.

The second phase starts by choosing the minimum of the function $\Psi(W, H)$ (e.g., the area minimum) by simple inspection of the generated list T and finding out the combination that minimizes the Ψ function. Afterwards, the tree is traversed top-down for selecting the corresponding shape (width and height) of each leaf node, by using the pointers associated (see line 3 of Stockmeyer's algorithm).

The running time and storage requirements of the complete Stockmeyer's approach is $O(nd)$, where n is the number of leaf nodes (i.e., the number of the layout building blocks) and d is the depth of the binary slicing tree. As the maximum depth is $d = n - 1$ (every non-leaf node has exactly one leaf child), the worst case running time is $O(n^2)$. For balanced trees (every non-leaf node has exactly two leaf nodes), the running time upper bound is $O(n \log_2 n)$. Stockmeyer's algorithm can be easily extended to allow an arbitrary number of realizations for each building block. The implementation described below in Section 2.3 serves as example.

Reported applications in the open literature of the Stockmeyer's algorithm on analog and/or mixed-signal design is, to the knowledge of the author, limited to those in [Koh90] and [Conw92]. None of these approaches, however,

---

7. O-notation is used to describe an asymptotic upper bound, to within a constant factor, of the running time of an algorithm. For a given function $g(n)$, we denote by $O(g(n))$ the set of functions

$O(g(n)) = \{f(n)$: there exist positive constants c and n_0 such that $0 \le f(n) \le cg(n)$ for all $n \ge n_0$ $\}$.

That is, $O(k + m)$ means that the worst-case running time is defined by sum of the lengths of the leaf node lists, k and m.

consider the floorplan-sizing problem within a circuit sizing optimization loop.

The second approach to solve the floorplan-sizing problem tries to formulate it as a linear programming problem[8] and, makes use of the above explained shape function of the building blocks too. Consider the layout template depicted in Fig. 6. If the shape functions of each of the three blocks are known, and the maximum layout height, H, is given, then the problem can be formulated as follows.

$$\text{Given}\begin{cases} H \\ w_i = f_i(h_i) \qquad \text{for } i = 1, 2, 3 \end{cases} \tag{4}$$

$$\text{minimize} \qquad W = \max(\overline{w}) \tag{5}$$

$$\text{subject to}\begin{cases} \sum_i h_i \le H \\ h_i^{min} \le h_i \le h_i^{max} \end{cases} \tag{6}$$

where h_i and w_i are, respectively, the height and width of each building block, f_i is its shape function, W is the width of the layout template and $\overline{w}$ is the set of building block widths.

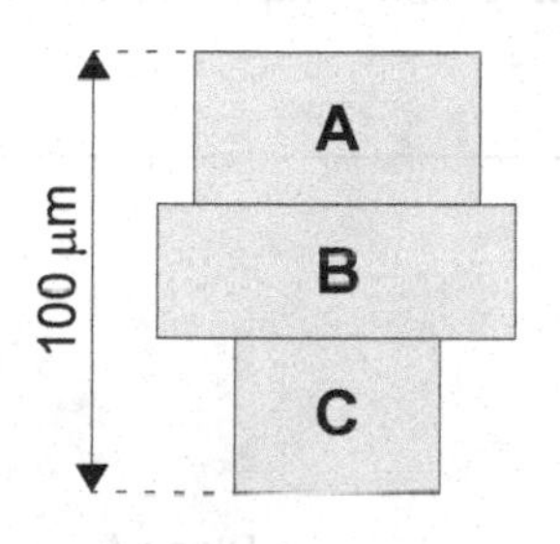

Figure 6. Example of floorplan.

This linear programming technique is used to solve the floorplan-sizing problem in [Onod92] by means of the simplex method, and in [Dess01a], by using a similar solving algorithm. In the latter, the floorplan area is optimized at the slicing level, i.e., the algorithm is applied only to a vertically (or, correspondingly, horizontally) stacked set of horizontally (vertically) arranged devices (what above was called a building block), called groups. The shape function of each group is computed from the individual shape functions of each device, although no method to carry out this composition is detailed. The execution of the algorithm has two phases. Consider once more the template in Fig. 6, and the area

---

8. A linear programming problem is a special case of a mathematical programming problem. From an analytical perspective, a mathematical program tries to identify an extreme (i.e., minimum or maximum) point of a function $f(x_1, x_2, \ldots, x_n)$, which furthermore satisfies a set of constraints, e.g., $g(x_1, x_2, \ldots, x_n) \ge b$. Linear programming is the specialization of mathematical programming to the case where both, function f –to be called the objective function– and the problem constraints are linear. A review of the linear programming concepts and solving techniques such as the simplex and the interior-point methods can be found in [Vand96].

optimization problem. In the first phase, the estimation phase, an initial estimate of the group heights is done such that the slice height H (which, as pointed out in the figure, must fulfill that $H \leq 100$ μm) is divided between the groups proportionally to the estimated area (e.g., the product $\mathrm{width} \times \mathrm{length}$ of a transistor). Then, by using the shape function of each group, an initial solution (e.g., obtaining the multiplicity of a folded transistor) is deduced from each height estimate. The width of the widest group determines the slice width. In the ensuing optimization phase, the algorithm tries to reduce that maximum width.

Unfortunately, neither the approach in [Dess01a] nor that in [Onod92] can fully ensure area minimization, since they actually do no tackle minimization of the function $\Psi(W, H)$ in a comprehensive way (that is, global minimization of $\Psi(W, H)$ is not carried out during the sizing phase) and strongly depend on the initial estimates done. As a minor downside, they may result in suboptimal solutions in regard with the area loss. To bear out this statement, consider the example in Fig. 6, where the blocks A, B, and C implement three folded transistors M_A, M_B, and M_C, respectively. The transistors aspects $S_i = \mathrm{width}_i/\mathrm{length}_i$ [9] (for $i = A, B, C$) and the block shape functions[10] are shown in Fig. 7.

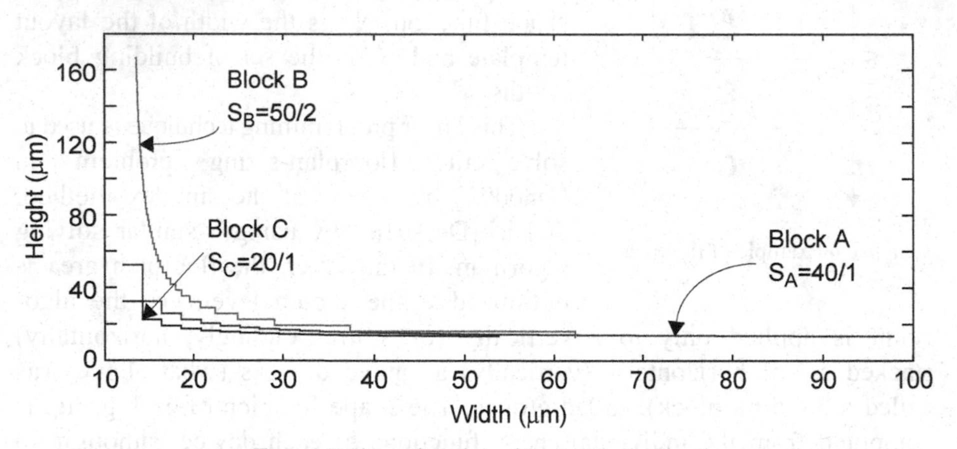

Figure 7. Shape function of the blocks in Fig. 6.

Given that $H \leq 100$ μm, an initial estimate yields that $h_A \approx 25$ μm, $h_B \approx 62.5$ μm and $h_C \approx 12.5$ μm, which, by simple inspection of each shape function, returns the following initial solution:

---

[9] Note that, whereas width_i and length_i in the aspect expression refer to the transistor geometry, the block width w_i and height h_i refer to the geometry of the layout block implementing the folded transistor. Likewise, while width_i and length_i are fixed, the block w_i and h_i depend upon the number of folds.

[10] Block B is rotated 90° with respect to blocks A and C.

$$
\begin{aligned}
w_A &= 18.6\mu m & h_A &= 22.4\mu m & m_A &= 4 \\
w_B &= 15.55\mu m & h_B &= 61.0\mu m & m_B &= 16 \\
w_C &= 53.8\mu m & h_C &= 13.4\mu m & m_C &= 20
\end{aligned} \tag{7}
$$

The template width, determined by w_C, is optimized according to the constraints expressed in Eq. (6). Fig. 8(a) shows the final achieved solution. Although the template width has been minimized while keeping $H < 100$ μm, area minimization has not been taking into consideration all along. Furthermore, the final solution largely depends on the initial estimate.

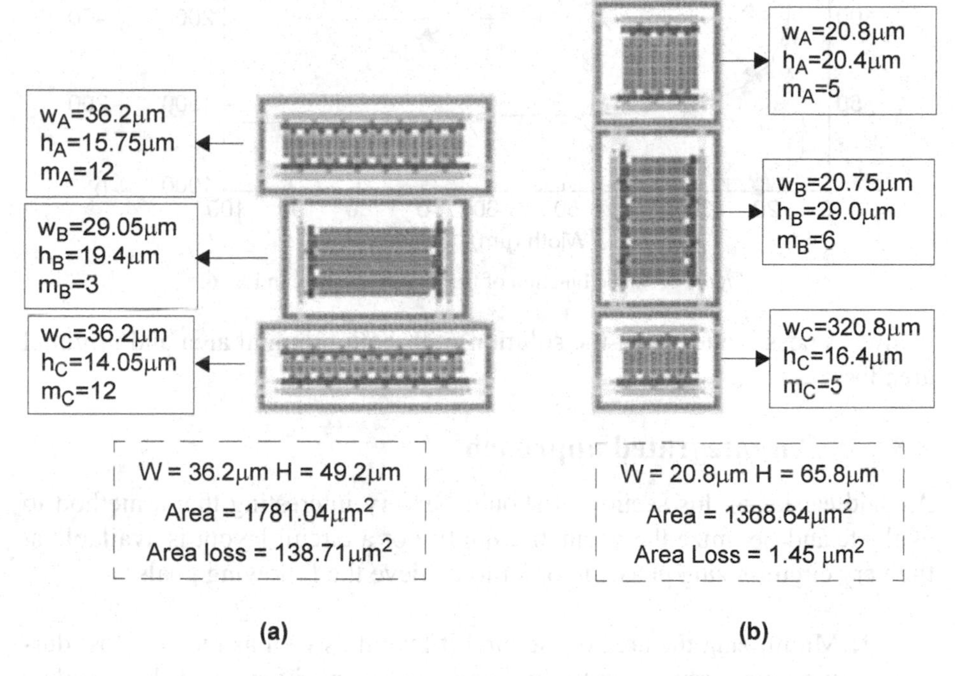

Figure 8. Solution of the example in Fig. 6: (a) obtained solution with the algorithm reported in [Dess01a]; (b) obtained solution with the approach based on Stockmeyer's algorithm.

If floorplan-sizing problem of this same example is tackled by using the Stockmeyer's approach, it results in effective area minimization while fulfilling the constraint $H \leq 100$. Fig. 9 illustrates this process. The shape function of the full template, obtained with the algorithm shown in Fig. 5, as well as the area figure computed for each optimal combination, allows selecting the solution minimizing area occupation, as explicitly pointed in the plot of Fig. 9. Fig. 8 compares this solution with the result obtained by using the linear programming approach. For illustration purposes, Fig. 9 also shows the area loss figure for each value of the shape function, which suggests that, if desired, it

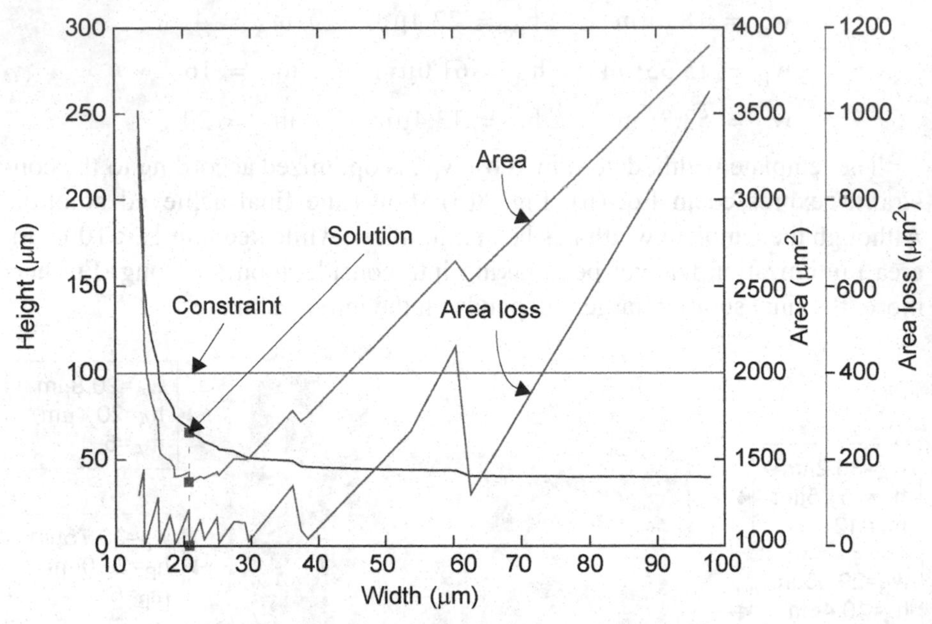

Figure 9. Shape function of the layout template in Fig. 6.

is always possible to select the solution with both minimal area and minimal area loss[11].

2.3 An integrated approach

As said earlier in this section, it should be very interesting that a method to evaluate and optimize the geometric quality of a circuit layout is available at the very circuit-sizing phase, in order to achieve the following goals:

1. Minimizing the area of the circuit layout as well as the area lost during layout reuse due to the fixed relative positioning of the template building blocks.
2. Addressing user's geometric constraints imposed on the aspect ratio, width, and height of the circuit layout.
3. Obtaining specific information on the circuit layout geometry (such as the number of fingers/strips, routing lines, and so on) in order to correctly evaluate the circuit parasitics and add them to the circuit optimization phase.

---

[11] Although area loss figure is not formulated as an objective of the design process (inasmuch as the main concern is area minimization), it can serve as a good indicative of the regularity and area usage of the layout template instance.

The first goal is reached effectively, in terms of storage and running time, by using floorplan-sizing procedures based on shape functions and **binary slicing trees** and area minimization algorithms based on or inspired by Stockemeyer's algorithm [Stoc83]. Solutions to achieve the second goal have been reported in [Koh90] and [Dess01a]. Regarding the third and last goal, none of the reviewed approaches includes the floorplan-sizing problem within the synthesis process, thus being impossible to include the layout-induced effects correctly. The only exception is the methodology reported in [Dess01a], but, as demonstrated above, the algorithm described fails to attain minimal area.

This section describes a methodology to include such geometric concerns to systematically improve, on the one hand, the quality of the obtained layout and, on the other hand, the robustness against circuit parasitics of the designed analog circuit. This methodology is based on Stockmeyer's algorithm and it includes techniques to integrate the floorplan-sizing procedure within a simulation-based optimization process.

Figure 10 shows the flow diagram of the optimization procedure for AMS circuits followed to implement the design reuse methodology. Its core is an optimization engine based on performance evaluation, which guarantees, providing that an accurate performance evaluator and accurate models are used, the accuracy of the predicted performances during the optimization process. As shown in Fig. 10, this kind of optimization process is an iterative procedure, design parameters being updated at each iteration, until an equilibrium point is reached [Mede94] [Mede99]. The degree of compliance of the specifications at each iteration is quantified through a cost function, which, in this case, is defined in the minimax sense. Different weights give different priority levels to the fulfillment of some specifications over others.

In general, the design space of any circuit is a multidimensional space defined by all the design parameters. For example, the physical parameters of transistors, resistors, and capacitors, whose interconnection defines the circuit netlist, define the design space in a simple device-level optimization. The procedure in Fig. 10 features the means to add designers' expertise to the iterative optimization process. This is done to bind the exploration of the circuit design space to only those regions yielding more suitable solutions, thereby improving the efficiency of the procedure. By making use of powerful tools like embeddable C++ programs[12], it is possible to incorporate valuable design expertise in the form of constraint satisfaction equations. This capability has just allowed carrying out the floorplan-sizing task at each iteration of the optimization process. The C++ program labeled **Geometric Constraints module**

[12] It is important to notice that such design constraints, written in C++ language, and associated to each circuit in a library database, are compiled in run-time, thus being fully independent of the optimization core. Therefore, they can be easily expanded, deleted or modified, making the methodology fully open.

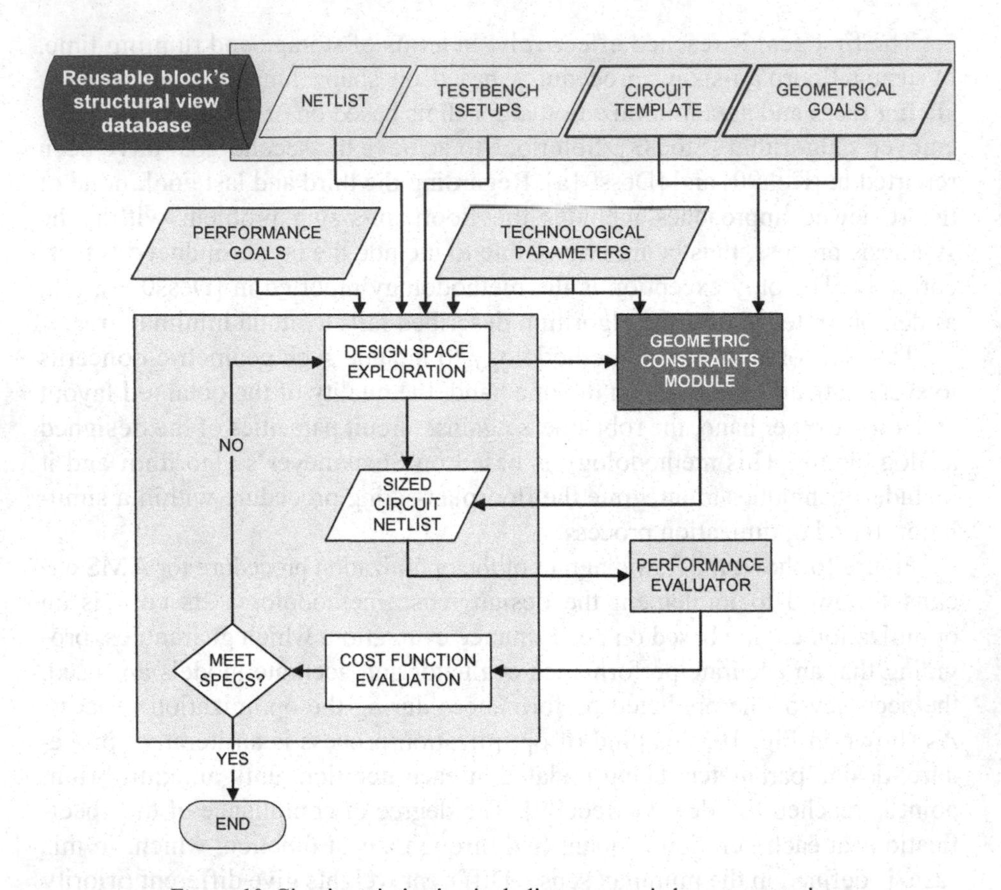

Figure 10. Circuit optimization including geometric layout constraints.

(hereafter referred as the GC module) in Fig. 10, has been written to perform the floorplan-sizing task following the technique (based on shape functions, binary slicing trees, and the Stockemeyer's algorithm) explained above.

The interaction of the GC module and the optimization engine is as follows: prior to the circuit performance evaluation[13], and right after selecting a new point of the design space, the GC module is executed. The module returns parameter values such as the number of fingers or the physical dimension of passive devices, all defining geometric characteristics of the circuit devices. Circuit optimization is subsequently accomplished with the confidence that for the specific visited point of the design space the resulting layout has minimal area. The GC module attains other geometric goals as well. Area can be also formulated as a design objective, so that the exploration of the

[13] This point reveals an important difference with [Dess01a], as this approach solves the floorplan-sizing problem after performance evaluation, with which it is not possible to accurately determine the value of induced parasitics and, therefore, accurately evaluate the circuit performance.

design space is driven towards the minimization of this geometric figure. Note that this minimization refers to an optimization between the visited points of the design space rather than between the possible shape combinations at a specific point of the design space. That is, after the optimization engine has addressed all performance specifications (e.g., $GBW \geq GBW_0$), it performs a fine-tuning of the design variables in order to minimize the area.

Fig. 10 also specifies the collection of input data the optimization engine requires. Particular information needed by the GC module is:

1. **Circuit template description**: in order to generate the shape function of the entire circuit layout, and, thus, being able to select the optimal shape, it is necessary to input the template structure. The GC module works the binary slicing tree representation of the parameterized layout template.
2. **Geometric goals**: these are the user-specified objectives concerning the geometric characteristics of the eventually generated circuit layout. Once the user has defined the goals, the GC module deals with all the possible shape combinations of the circuit layout template and selects, following a searching algorithm, which combination (or combinations) better attains the geometric goals.
3. **Fabrication process data**: for identical built-in devices in a building block, the generated shape functions may be different due to differences of the fabrication processes. Therefore, to generate the shape function of each layout building block it is necessary to particularize the value of technology-specific parameters, such as the poly-resistor sheet resistance. This is a very important concern since design reuse may imply a technological migration.

The following geometric goals are considered:

a. **Aspect ratio**. The user can specify the ratio between circuit layout width and height, $AR = W/H$. Since it is almost impossible to achieve an exact aspect ratio, an acceptable deviation, Δ_{AR}, from the aspect ratio can also be defined. In this way, as illustrated in Fig. 11(b), suitable shape combinations considered are those with height to width ratio fulfilling the following inequality:

$$AR - \Delta_{AR} \leq W/H \leq AR + \Delta_{AR} \tag{8}$$

b. **Maximum and minimum width values**. There are some cases when the circuit layout has to fit into a known area of the chip layout. Accordingly, the user can impose a maximum allowable value of the circuit layout width, W . Only those shape combinations rendering a layout whose width is below the defined maximum width value W_M are accepted. In like manner, the circuit layout width can be constrained to be above a minimum width value W_m. The maximum and minimum limits can also be simultaneously defined, so the layout width is bounded to lie in the following range:

$$W_m \le W \le W_M \tag{9}$$

Figure 11(c) illustrates the three cases.

c. **Maximum and minimum height values**. Similar goals can be defined with respect to the circuit layout height. Then, it is possible to set a maximum height value, H_M, a minimum height value, H_m, or both a maximum and minimum limits such that:

$$H_m \le H \le H_M, \tag{10}$$

These geometric goals are shown in Fig. 11(d).

d. **Simultaneous width and height limit values**. The four previous goals (i.e., maximum width or height and minimum width or height) can be defined in pairs; e.g., maximum width and maximum height values. As illustrated in Fig. 11(e), they can be also defined simultaneously such that the circuit layout shape should satisfy both Eq. (9) and Eq. (10). This is the most constrained scenario and there may be points of the design space where finding a solution is not possible.

Note that, due to the approach used to obtain both the shape function of the circuit layout and the right[14] shape(s) of each individual building block, area is always minimized (i.e., for any circuit sizing, any shape combination of the building blocks has minimal area when compared with any other combination not considered in the shape function).

Figure 12 shows the operation flow of the GC module. The execution of this module at each iteration of the optimization process comprises two phases. The flow diagram at the right-hand side of Fig. 12 illustrates the bottom-up shape annotation phase. As explained at the beginning of this section, the shape function of each building block in the circuit layout must be properly combined to generate first the corresponding shape function of each slice and, eventually, the shape function of the circuit as a whole (Fig. 5 shows the basic underlying algorithm to carry out these combinations). The database shape

14. Here "right" means that the building block shape makes the full circuit layout to achieve the intended geometric goal.

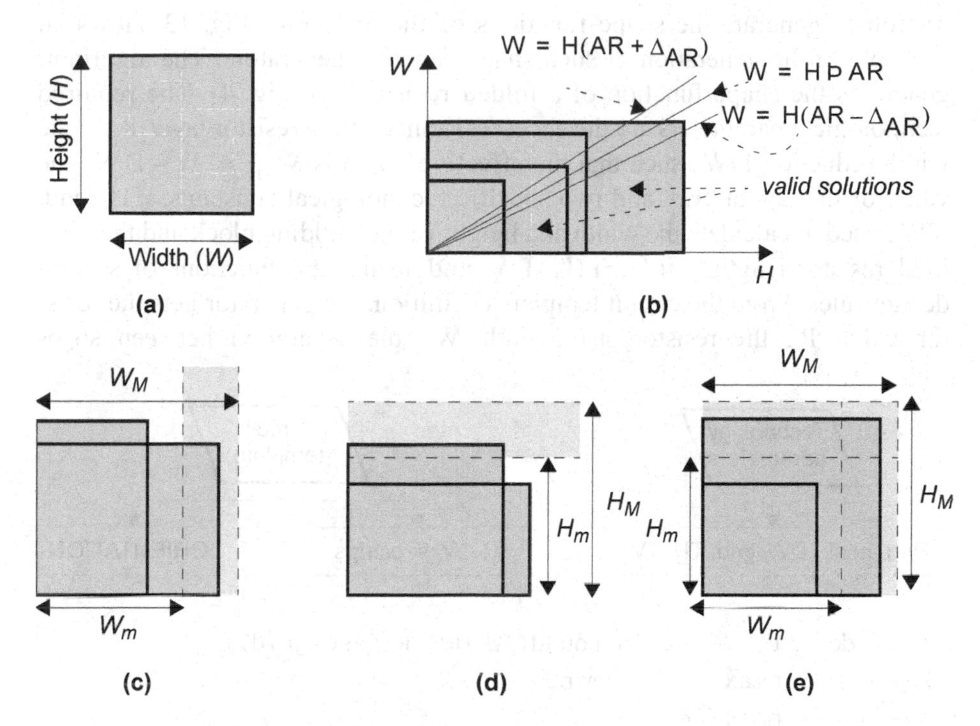

Figure 11. Illustration of the different geometric goals: (a) layout template; (b) aspect ratio; (c) width maximum and minimum values; (d) height maximum and minimum values; (e) simultaneous width and height limits.

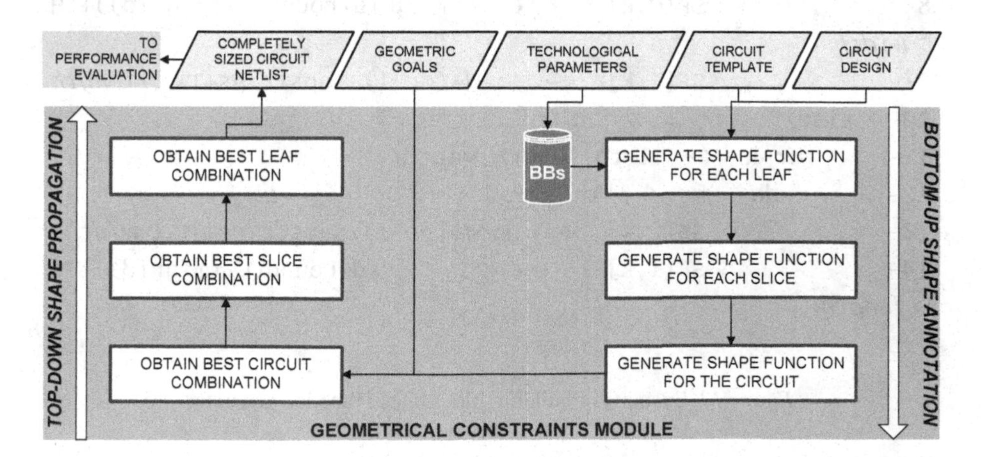

Figure 12. Operation flow of the Geometric Constraint module.

labeled BBs in Fig. 12 represents the database containing the required information to generate the shape functions of the leaf cells. Fig. 13 shows an example of the generation of such shape function generators. The algorithm generates the shape function of a folded resistor (see Fig. 4). The required technological parameters are the sheet resistance of the resistor layer $R_{\square}$, the width reduction DW, such that the effective width is $W_{eff} = W - DW$, the value of the layout grid, and two specific technological constants, TH, and, TW, used to calculate the width and height of the building block and the minimal resistor length minL. TH, TW and minL are functions of several design rules. From the circuit template definition, the generator gets the resistor value R, the resistor strip width W, the separation between strips

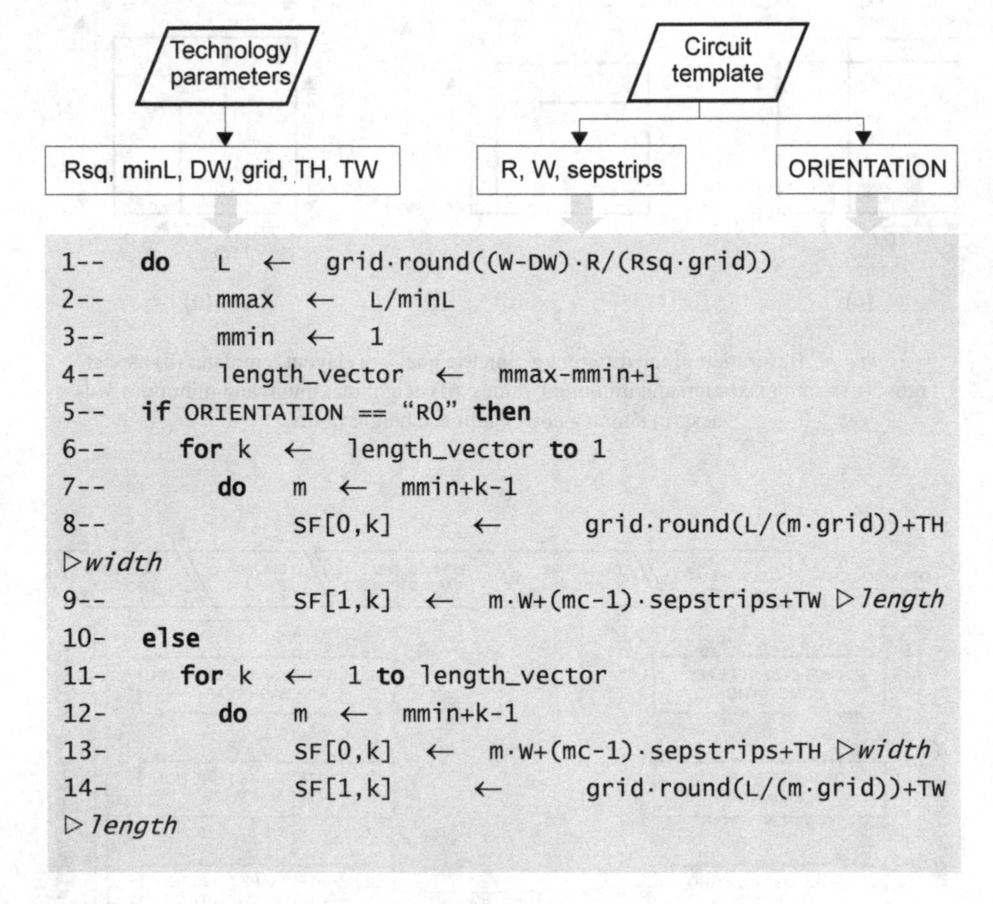

Figure 13. Example of building block shape function generator.

sepstrips, and the orientation of the block (either the resistors are placed vertically –orientation is R90 – or horizontally –orientation is R0)[15]. Recalling again Fig. 4, it is straightforward to understand the generator pseudocode in Fig. 13. After defining the maximum length of the single strip resistor (line 1), the maximum number of strips can be computed by just noting that the minimal length of the strip is given when the number of strips is maximum (line 2). The shape function, SF, is stored as a 2-column array with as many rows as the number of resistor strips, the length of this array defined in line 4. From line 6 to line 9, the algorithm calculates the corresponding width and height of each point of the shape function, for a horizontally oriented resistor, and, likewise, from line 11 to line 14, for a vertically oriented resistor. The round() function rounds x to its nearest integer; it is used to adjust every physical dimension into a realizable, grid-correct layout geometry.

Examples of shape function generators for basic building blocks used are the folded transistor (FT), the MOS differential pair (MDP), the MOS cascode group (MCG), the current mirror group (MCMG), the unit capacitor (UC), and the already explained folded resistor (FR). Each block features a set of electrical and geometric parameters that provides them with a high degree of flexibility to improve device matching, shielding, and reliability. The reader is referred to Section 2.4 for an illustration of these generators and how they are used to describe a layout template in a practical design example.

Following the generation of the shape function for each building block in the circuit layout, the GC module proceeds to generate the shape function of each slicing cut of the binary slicing tree, i.e., the shape function of each slice. To do so, the circuit template information includes which elements compose each slice, which is the physical separation between them, and the type of slicing cut (either vertical or horizontal). Depending on this type, the implemented version of the Stockmeyer's algorithm processes vertical or horizontal cuts, respectively. Fig. 14 plots several examples of generated shape functions. Consider the slicing structure at the top of this figure, with block #1 being a folded transistor block (whose width = 10.0 μm, and length = 1.5 μm and of horizontal orientation), block #2 a folded resistor (with R = 500 Ω, W = 1.15 μm, strip separation equal 1.0 μm and vertical orientation), and block #3 a 0.1-pF capacitor. Note that the shape function of block #3 is smoother than the other functions since no multiplicity is involved but simply grid steps.

---

[15] Unlike Stockmeyer's algorithm, orientation is now a fixed building block parameter. Although nothing in the described approach keeps orientation from being another geometric parameter (such as the number of transistor folds), the parameterization of the layout routing required to accommodate free orientations, would be very much complex. Then, for the sake of simplicity, orientation is fixed in the layout template description and its value is left to the layout expert designer (thereby improving device matching).

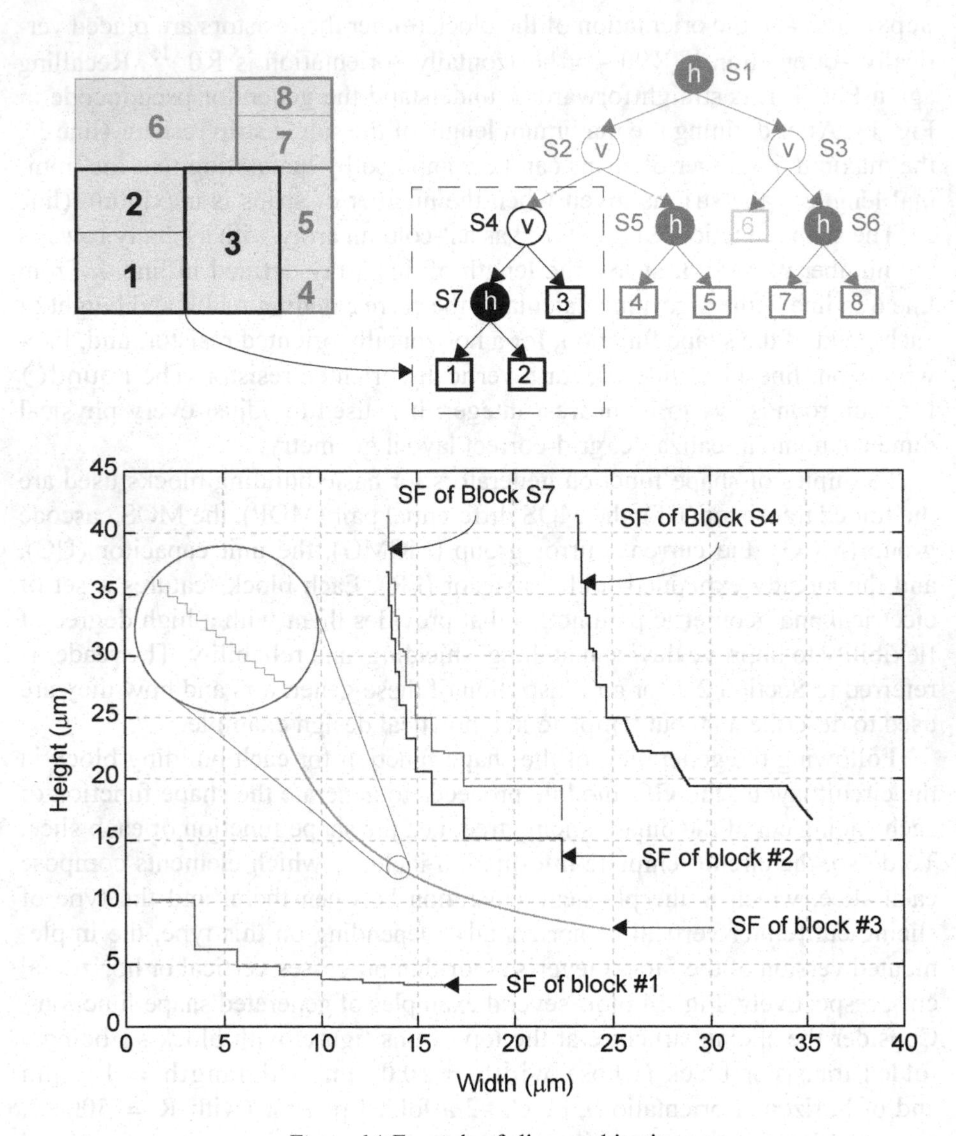

Figure 14. Example of slice combination.

This process is repeated until the top-level slicing cut (slice S1 in example of Fig. 14) is reached, which means that the shape function of the entire circuit has been generated, hence completing the bottom-up shape annotation process. The output of the GC module to this process is a **floorplan-sizing matrix** whose first two columns are the width W (in increasing order of magnitude) and height H (in decreasing order of magnitude) of each point of the shape function of the circuit layout. This matrix has as many rows as points in the shape function. The rest of columns are pointers to the properties of each of those points. The last three columns collect the following information:

- Area (defined as $W \times H$).
- Aspect ratio (defined as W/H).
- Area loss.

The area loss (measured in percentage of the area occupied by the building blocks, without taking into account area for routing, as illustrated in Fig. 15) is computed as the summation of the area loss resulting from any combination of two nodes (leaf and non-leaf) of the slicing tree.

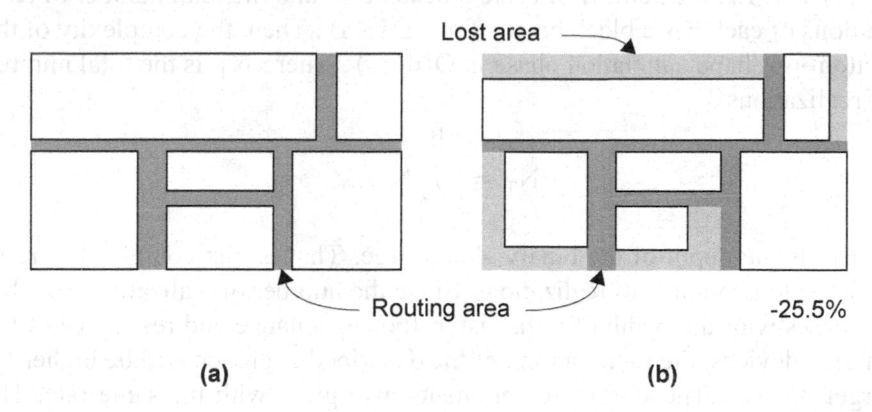

Figure 15. Example of slice combination.

The rest of columns are pointers to specific values of multiplicity (in case of transistors and resistors) or physical dimensions (for capacitors) yielding circuit layout width and height in the first two columns. This matrix structure allows very fast top-down shape propagation, the second phase in the GC module execution.

Prior to this, however, it is necessary to select the set of combinations $\bar{S}$,for which the user-defined geometric goals are met. By using basic searching algorithms [Corm01], working on columns one and two as well as on the last three columns, it is possible to obtain the point of the circuit shape function with which the geometric goals are achieved from all the combinations in $\bar{S}$. Fig. 16 illustrates an example of this selection process. The shape function in Fig. 16(a) corresponds to the layout template in Fig. 3, and the geometric goals are the following:

$$54.0 \leq W \leq 66.0 \tag{11}$$

and

$$72.0 \leq H \leq 88.0 \tag{12}$$

(i.e., $W \in [60 + 10\%, 60 - 10\%]$ and $H \in [80 + 10\%, 80 - 10\%]$). The shaded zones in the plot of Fig. 16(a) represent the points of the shape function fulfilling both Eq. (11) and Eq. (12). Plotted vs. circuit layout width, the area and area loss are also shown in Fig. 16(a). It is then possible to select the solution achieving less area. Fig. 16(b) illustrates the same reasoning by using the shape function matrix and the pointers to the geometric parameter (GP), of each building block, the area, the percentage of area loss and the aspect ratio. Fig. 16(b) highlights the final solution as well.

As mentioned previously when the Stockmeyer's algorithm was roughly described, the running time of the algorithm, for a combination of two slices (or building blocks) each with k and m shape combinations respectively, is $O(k + m)$. Let the number of basic blocks be B and the total number of realizations of each basic block be N_i, for $1 \le i \le B$. Then, the complexity of the bottom-up shape annotation phase is $O(dN_T)$, where N_T is the total number of realizations

$$N_T = \sum_{i=1}^{B} N_i \tag{13}$$

and d is the depth of the binary slicing tree. That is, the complexity grows with the total number of realizations. Since the number of realizations roughly increases with the width of a transistor, the capacitance and resistance of the passive devices, the running time of the described approach will be higher for larger devices. The storage requirements also grow with the same rate. The eventually created matrix has at most $N_T - B + 1$ rows and $B + 4$ columns (2 to store the width and height of the circuit layout, B for pointing out to the geometric parameter of each building block –assuming that each block has only one geometric parameter–, and 2 for the area and aspect ratio characteristics), being thus necessary to store up to $(N_T - B + 1)(B + 4)$ elements[16]. For the second phase of the GC module execution, the top-down shape propagation, its complexity is also bounded as $O(N_T)$, since it is just a basic searching process over $N_T - B + 1$ elements.

By way of example, $B = 16$, $N_T = 3207$ for the design experiments described in the ensuing section. The number of rows of the shape function matrix is 916 and the number of columns is 24 (there are more than one geometric parameter for some building blocks). The average execution time[17] of the GC module was 0.2 seconds of CPU time. Note that the GC module operates during the optimization process, so 0.2 seconds of CPU time is added to the simulation time at each iteration of the optimization process.

---

[16] This number is $(N_T - B + 1)(B + 5)$ providing that area loss is considered relevant.

[17] The example was performed on a PentiumIV@1.3GHz Pc with a 512MB RAM.

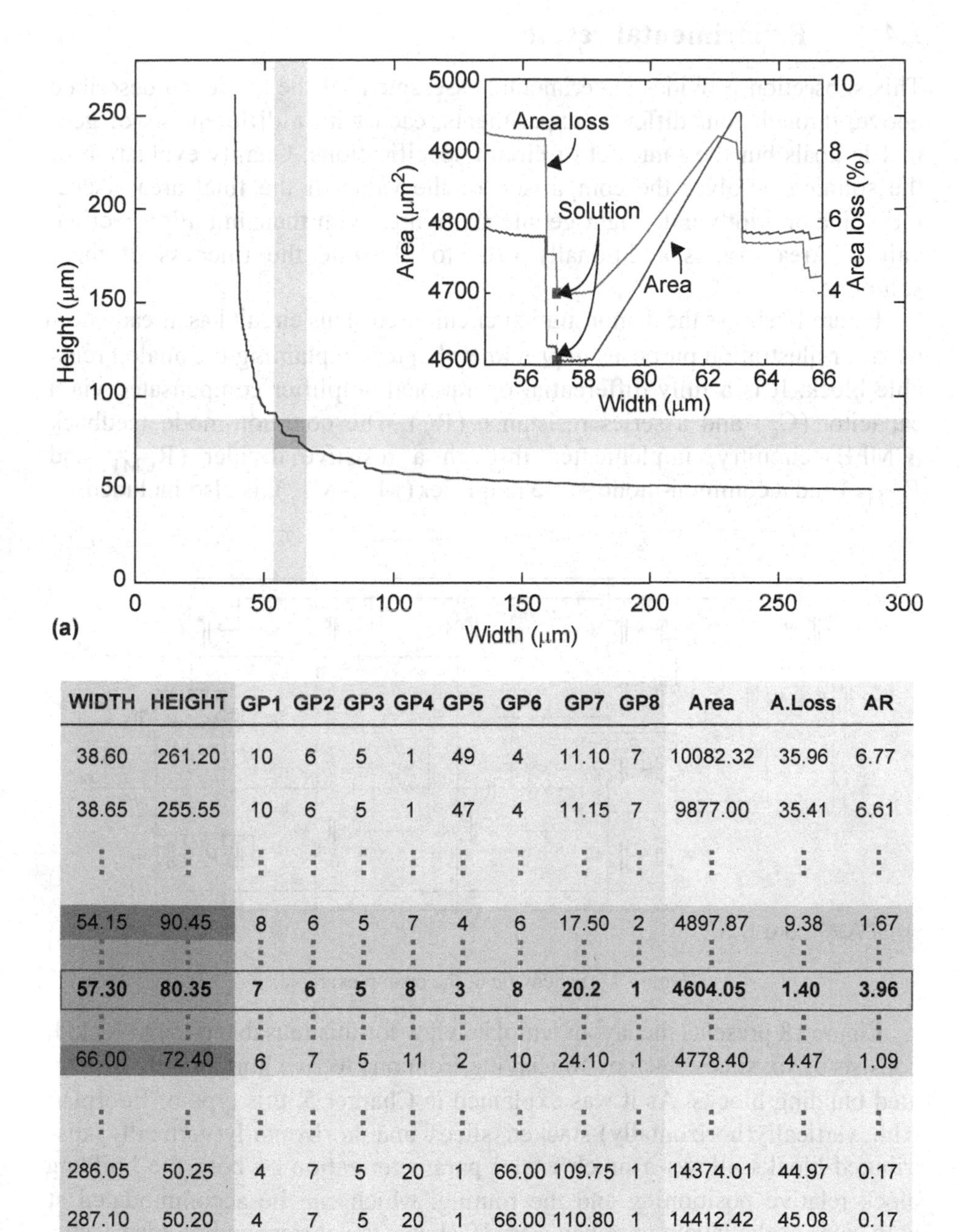

WIDTH	HEIGHT	GP1	GP2	GP3	GP4	GP5	GP6	GP7	GP8	Area	A.Loss	AR
38.60	261.20	10	6	5	1	49	4	11.10	7	10082.32	35.96	6.77
38.65	255.55	10	6	5	1	47	4	11.15	7	9877.00	35.41	6.61
⋮	⋮	⋮	⋮	⋮	⋮	⋮	⋮	⋮	⋮	⋮	⋮	⋮
54.15	90.45	8	6	5	7	4	6	17.50	2	4897.87	9.38	1.67
⋮	⋮	⋮	⋮	⋮	⋮	⋮	⋮	⋮	⋮	⋮	⋮	⋮
57.30	**80.35**	**7**	**6**	**5**	**8**	**3**	**8**	**20.2**	**1**	**4604.05**	**1.40**	**3.96**
⋮	⋮	⋮	⋮	⋮	⋮	⋮	⋮	⋮	⋮	⋮	⋮	⋮
66.00	72.40	6	7	5	11	2	10	24.10	1	4778.40	4.47	1.09
⋮	⋮	⋮	⋮	⋮	⋮	⋮	⋮	⋮	⋮	⋮	⋮	⋮
286.05	50.25	4	7	5	20	1	66.00	109.75	1	14374.01	44.97	0.17
287.10	50.20	4	7	5	20	1	66.00	110.80	1	14412.42	45.08	0.17

(b)

Figure 16. Geometric goals achievement: (a) graphical illustration, (b) the floorplan-sizing matrix and its pointers.

2.4 Experimental results

This subsection provides experimental assessment of the approach described above, through four different experiments, each with a different set of geometric goals but the same set of circuit specifications. Quality evaluation of the solution involves the comparison of the value of the final area, aspect ratio, and/or width and height geometric figures with their initially specified values. Area loss is additionally used to illustrate the fineness of these solutions.

Figure 17 shows the demonstration circuit used. This circuit has already been used for illustration purposes in previous chapters explaining the analog reusable block. It is a fully differential operational amplifier compensated via a capacitor (C_c) and a series resistance (R_z). The common-mode feedback (CMFB) circuitry, implemented through a resistive divider (R_{CM1} and R_{CM2}) and a common-mode sense amplifier (M_{1c}-M_{5c}), is also included.

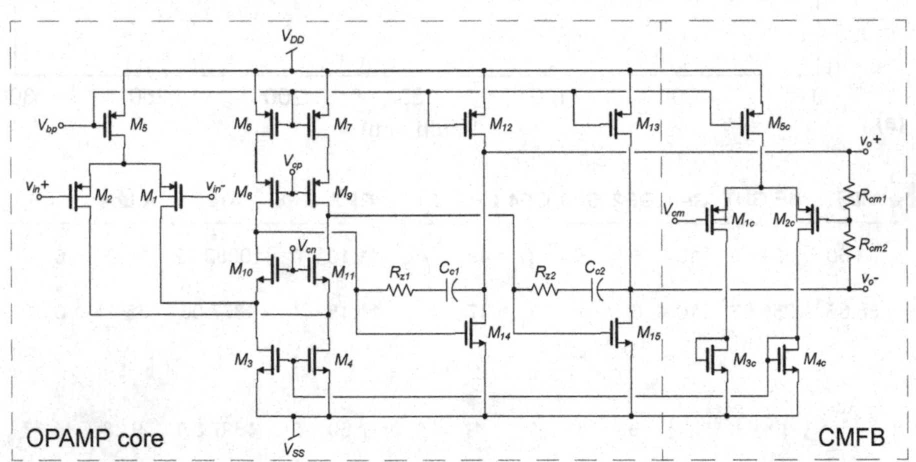

Figure 17. Schematic of the opamp example.

Figure 18 presents the layout template view for this reusable circuit block. It consists of six slices, each of one having from one to five horizontally-distributed building blocks. As it was explained in Chapter 5, this type of floorplan with vertically(horizontally)-stacked slices and horizontally(vertically)-distributed blocks allows a much easier parameterization of both the building block relative positioning and the routing, which can be accommodated at both sides of the layout template. Fig. 19 shows the corresponding binary slicing tree[18].

---

[18] Selecting a binary slicing tree whose depth is minimal allows a faster execution of the GC module. It is left to the reusable block designer to define the minimal-depth slicing tree, that, for circuits of this size, is almost straightforward.

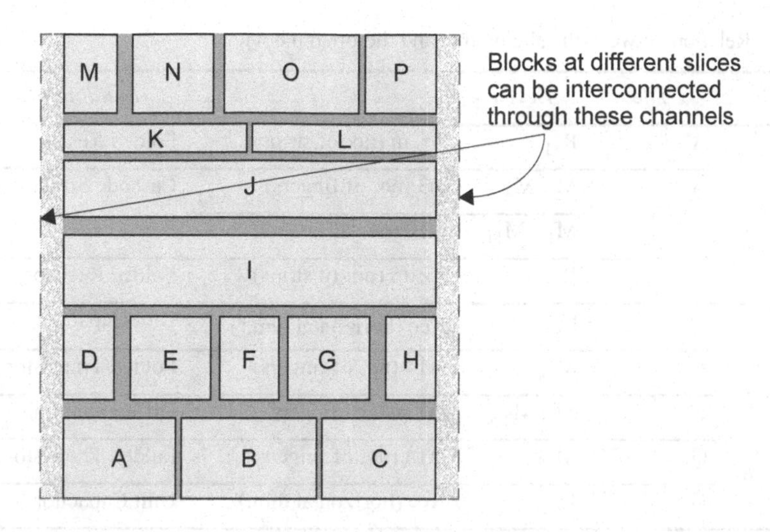

Figure 18. Opamp layout template.

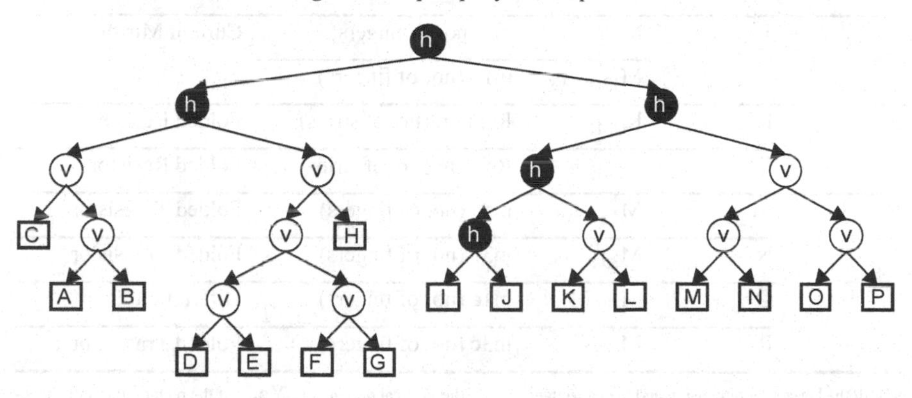

Figure 19. Binary slicing tree representation of the opamp in Fig.18.

The relation between the leaf cells of the slicing tree and the devices (or group of devices) in the schematic of Fig. 17 is explained in Table 1.

In this table, the GP (the variable of each shape function) is also detailed. The devices located in the same table cell are physically identical (e.g., transistors M1 and M2 of the differential input pair share the same width, length, and number of fingers). However, the group may implement several devices, as it happens in some cascode and current mirror structures. A closer inspection of Table 1 and the schematic of the opamp in Fig. 17 allows realizing that analog layout features such as symmetries in the signal path[19] or minimal-

[19] The input differential pair is placed in the vertical symmetry axis while the devices of both differential paths are at both sides of the input pair.

Table 1. Relation between the slicing tree and the opamp devices.

Slice	*Tile*	*Device*	*GP*	*Building Block*
1	A	R_{z1}	Rz_m (no. of strips)	Folded Resistor
	B	M_3, M_4	m3 (no. of fingers)	Cascode Structure
		M_{10}, M_{11}	m10 (no. of fingers)	
	C	R_{z2}	Rz_m (no. of strips)	Folded Resistor
2	D	C_{c1}	Xcc (horizontal dim.[a])	Unit Capacitor
	E	M_{14}	m14 (no. of fingers)	Folded Transistor
	F	M_1, M_2	m1 (no. of fingers)	Differential Pair
	G	M_{15}	m14 (no. of fingers)	Folded Transistor
	H	C_{c2}	Xcc (horizontal dim.)	Unit Capacitor
3	I	M_6-M_9	m6 (no. of fingers)	Cascode Structure
4	J	M_5	m5 (no. of fingers)	Current Mirror
		M_{12}, M_{13}	m12 (no. of fingers)	
5	K	R_{CM1}	Rcm_m (no. of strips)	Folded Resistor
	L	R_{CM2}	Rcm_m (no. of strips)	Folded Resistor
6	M	M_{5c}	m5c (no. of fingers)	Folded Transistor
	N	M_{3c}	m3c (no. of fingers)	Folded Transistor
	O	M_{1c}, M_{2c}	m1c (no. of fingers)	Differential Pair
	P	M_{4c}	m3c (no. of fingers)	Folded Transistor

a. With the horizontal dimension and the capacitance value, the vertical dimension, Ycc, of the rectangular unit capacitor can be easily derived with the help of the capacitance equation $C = C_A X_{cc} Y_{cc} + 2C_P(X_{cc} + Ycc)$, where C is the capacitor value and C_A and C_P are the area and perimeter capacitance, respectively, of the material used for building capacitors.

length routing paths have been considered at the planning phase of the layout template (see Section 7.3 in Chapter 5 on page 182). Thus, layout expertise is stored such that, at any point of the optimization process, both the parameterized nature of the template and the use of the GC module guarantee the preservation of the analog layout features.

Table 2 shows the optimization variables for the set of experiments. The particular variation range of each one of these variables defines the design space to be explored. As explained in Chapter 4, constraints between the circuit's design variables are used to reduce the explorable design space. These constraints, expressed with equations representing opamp-specific design knowledge and stored in the reusable block database, relate the optimization

Table 2. Optimization variables of the example in Fig. 17.

Design variable	*Related devices*	*Description*
W1	M_1, M_2	Transistor width
W3	M_3, M_4	Transistor width
W5	M_5, M_6, M_7	Transistor width
W8	M_8, M_9	Transistor width
W10	M_{10}, M_{11}	Transistor length
W12	M_{12}, M_{13}	Transistor width
L1	M_1, M_2	Transistor length
L5	$M_3, M_4, M_5, M_6, M_7, M_{12}, M_{13}, M_{5c}$	Transistor length
L8	M_8, M_9	Transistor length
L10	M_{10}, M_{11}	Transistor length
L14	M_{14}, M_{15}	Transistor length
Cc	C_{c1}, C_{c2}	Capacitor value
Rz	R_{z1}, R_{z2}	Resistor value
Rz_w	R_{z1}, R_{z2}	Resistor strip width
Rz_sep	R_{z1}, R_{z2}	Resistor strip separation
Rcm	R_{cm1}, R_{cm2}	Resistor value
Rcm_w	R_{cm1}, R_{cm2}	Resistor strip width
Rcm_sep	R_{cm1}, R_{cm2}	Resistor strip separation
W1c	M_{1c}, M_{2c}	Transistor width
L3c	M_{3c}, M_{4c}	Transistor length
L1c	M_{1c}, M_{2c}	Transistor length
IB	--	Biasing current

variables in Table 2 with the rest of the opamp's design variables. The reader is referred to Chapter 4 for further details on its structural database.

As shown before in Fig. 18 and Fig. 19, the layout template structure is another important input of the embedded GC module. As shown in the example of Fig. 20, this information is organized in three groups. First, each one of the leaf nodes of the slicing tree in Fig. 19 is described by setting its building block type, its shape function parameters, and its orientation. Second, the non-leaf nodes are listed (from the bottom to the top level) and described by defining the two blocks or slices composing the slice, the type of slicing cut it

```
// TEMPLATE DESCRIPTION
// -- blocks
A = (FR, Rz, Rz_w, Rz_sep, R90);
B = (MCG, w10, l10, w3, l3, R90);
C = (FR, Rz, Rz_w, Rz_sep, R90);
D = (UC, Cc, R90);
E = (FT, w14, l14, R90);
F = (MDP, w1, l1, R90);
G = (FT, w15, l15, R90);
H = (UC, Cc, R90);
I = (MCG, w6, l6, w6, l6 R90);
J = (MCMG, w12, l12, w5, l5, w13, l13, R90);
K = (FR, Rcm, Rcm_w, Rcm_sep, R0);
L = (FR, Rcm, Rcm_w, Rcm_sep, R0);
M = (FT, w5c, l5c, R90);
N = (FT, w3c, l3c, R90);
O = (MDP, w1c, l1c, R90);
P = (FT, w4c, l4c, R90);
// -- slices
S11 = D E v sep_DE;// level 1
S12 = F G v sep_FG;
S21 = A B v sep_AB;// level 2
S22 = I J h sep_IJ;
S23 = K L v sep_KL;
S24 = M N v sep_MN;
S25 = O P v sep_OP;
S26 = S11 S12 v sep_S11S12;
S31 = S21 C v sep_S21C;// level 3
S32 = S26 H v sep_S26H;
S33 = S22 S23 h sep_S22S23;
S34 = S24 S25 v sep_S24S25;
S41 = S31 S32 h sep_S31S32;// level 4
S42 = S33 S34 h sep_S33S34;
S5 = S41 S42 h sep_S41S42;// level 5 (whole opamp)
// GOALS
AR = 1.0;
DeltaAR = 0.1;
Wm = 0;
WM = 0;
Hm = 0;
HM = 0;
```

Figure 20. Input description of the opamp layout template.

represents (horizontal, h, or vertical, v) and the physical separation of the slice components; finally, the user-defined geometric goals are defined [Eq. (8)-Eq. (10)].

Four different experiments illustrate the different user-available geometric goals. Table 3 shows the specification list for each of the experiments.

Table 3. Description of experiments.

Specification	*Description*	*Exp.#1*	*Exp.#2*	*Exp.#3*	*Exp.#4*
add_db ≥	DC gain	85 dB	85 dB	85 dB	85 dB
ft ≥	Unity-gain frequency	50 MHz	50 MHz	50 MHz	50 MHz
pm ≥	Phase margin	50°	50°	50°	50°
add_db_cm ≥	DC gain of the CMFB circuit	85 dB	85 dB	85 dB	85 dB
ft_cm ≥	Unity-gain frequency of the CMFB circuit	25 MHz	25 MHz	25 MHz	25 MHz
pm_cm ≥	Phase margin of the CMFB circuit	50°	50°	50°	50°
os ≥	Output swing	5.5 V	5.5 V	5.5 V	5.5 V
sr ≥	Slew rate	55 V/μs	55 V/μs	55 V/μs	55 V/μs
Rload	Resistive load	50 kΩ	50 kΩ	50 kΩ	50 kΩ
Cload	Capacitive load	5 pF	5 pF	5 pF	5 pF
AR	Aspect ratio	1.0	2.0		
Δ_{AR}		0.1	0.05		
W_M	Maximum width values			150	300
W_m	Minimum with value				
H_M	Maximum height values				200
H_m	Minimum height value				

Three groups of specifications have been defined. A first group, from add to Cload, is the electrical performance requirements, e.g., ft ≥ 50MHz, as well as setup conditions, e.g., Cload = 50pF. The second group, from aspect ratio (AR) to minimum height value (H_m), is the set of geometric restrictions. The last group (not shown in Table 3) is composed of the design objectives to undertake once the electrical and geometric specifications are met, namely minimization of area[20] and minimization of power consumption.

20. Do not confuse this objective with the minimization that is automatically done at every point of the design space (defined by the optimization variables) by the GC module. The mentioned minimization of area refers to the optimization of this geometric figure from point to point of the explored design space, which is done by the optimization engine.

As it can be noticed from Table 3, whereas electrical specifications are the same in the four experiments, the first and second experiments try to optimize the opamp circuit to obtain layout aspect ratio within $[0.9, 1.1]$ and $[1.95, 2.05]$ respectively. The geometric goal in the third experiment is to obtain a opamp layout of width below 150 μm while its height is unconstrained. In the fourth experiment, the width and height of the resulting layout have to be below 300 and 200 μm, respectively.

In all the experiments, the fabrication process selected has been a three-metal, double-poly, 0.35-μm CMOS process. The performance evaluator used (see Fig. 10) is HSPICE® (in principle, any electrical simulator could be used as well). The MOS transistor models used are BSIM3v3 level-49 models, while, for passive devices, technology parameters are $R_{\square} = 50\Omega/\square$, $DW = 0.35$ μm (for resistors) and $C_A = 0.86$ fF/μm and $C_P = 0.092$ fF/μm^2 (for capacitors)[21].

To obtain the electrical performance at each iteration of the optimization process, the testbench setup associated to the opamp reusable block described in Section 2.2.2 of Chapter 4 (page 109) and Section 3.1 of Chapter 6 (page 230), is used. The experimental results are shown in Table 4, Table 5, Table 6, and Table 7. The first of these tables, Table 4, summarizes the obtained electrical performance at each experiment. Notice that every optimization experiment has reached the required electrical performance specs.

Table 5 shows the obtained values of the geometric goals for the resulting opamp layout of each experiment, and Table 6 lists the attained design objectives. All the geometric goals have been successfully addressed whereas minimization of layout area and power consumption has been carried out. For illustration purposes, area loss is also listed. From all the possible sizes of the opamp layout at the final solution meeting the geometric goals, the described approach can ensure minimization of the area and area loss.

To illustrate this point, Fig. 21 plots the shape function for the final solution of the first experiment. In the enlarged view of those points of the shape function fulfilling that $0.9 \le AR \le 1.1$, both area and area loss become minimal at the solution of the experiment, shown in Table 5.

The physical implementation of the solution of each experiment is shown in Fig. 22 (first experiment), Fig. 23 (second experiment), Fig. 24 (third experiment), and Fig. 25 (fourth experiment). Table 7 lists the corresponding obtained device sizes as well as all the geometric parameters.

---

21. $R_{\square}$ is the poly sheet resistance and DW is the is the width reduction (such that the effective width is $W_{eff} = W - DW$ with W being the drawn width of the resistor strip); C_A and C_P are the area and perimeter capacitances of PIP capacitors.

Table 4. Achieved electrical performance.

Specification	*Goal*	*Exp.#1*	*Exp.#2*	*Exp.#3*	*Exp.#4*
add_db ≥	85 dB	110.37 dB	89.85 dB	77.91 dB	100.82 dB
ft ≥	50 MHz	65.79 MHz	50.64 MHz	70.23 MHz	50.37 MHz
pm ≥	50°	57.26^{o}	71.78^{o}	75.4^{o}	68.94^{o}
add_db_cm ≥	85 dB	114.07 dB	91.29 dB	90.0 dB	111.04 dB
ft_cm ≥	25 MHz	27.64 MHz	25.15 MHz	68.8 MHz	27.52 MHz
pm_cm ≥	50°	50.13^{o}	50.06^{o}	52.37^{o}	52.25^{o}
os ≥	5.5 V	5.86 V	5.85 V	5.66 V	5.76 V
sr ≥	55 V/μs	59.15 V/μs	57.25 V/μs	50.93 V/μs	59.96 V/μs
Rload	50 kΩ				
Cload	5 pF				

Table 5. Obtained values of the geometric characteristics.

Experiment	Aspect Ratio		Width (μm)		Height (μm)	
	Goal	*Achieved*	*Goal*	*Achieved*	*Goal*	*Achieved*
Exp.#1	[0.9,1.1]	1.0		165.7		165.85
Exp.#2	[1.95,2.05]	1.96		229.0		116.7
Exp.#3			<150	143.2		171.1
Exp.#4			<300	186.9	<200	138.1

Table 6. Attained value of the design objectives and achieved area loss.

Feature	*Goal*	*Exp.#1*	*Exp.#2*	*Exp.#3*	*Exp.#4*
Power (mW)	Minimize	3.18	3.21	4.93	2.98
Area (μm^2)	Minimize	27481	26724	24502	25811
Area Loss (% of Area)	Minimize	1.94	3.06	0.93	0.57

Table 7. Obtained sizes and geometric parameters of the opamp devices.

Slice	*Tiles*	*Devices*	*Parameter*	*Exp.#1*	*Exp.#2*	*Exp.#3*	*Exp.#4*
1	A,C	R_{z1}, R_{z2}	Rz	873.91 Ω	887.5 Ω	2455.88 Ω	3706.9 Ω
			Rz 1 (strip)	13.4 μm	10.65 μm	16.7 μm	21.5 μm
			Rz_w	2.65 μm	2.15 μm	2.05 μm	1.8 μm
			Rz_sep	0.9 μm	1.3 μm	2.9 μm	2.4 μm
			Rz_m	3	3	5	5
	B	M_3, M_4	W3/L3	78.0/1.5	28.75/1.5	35.25/1.5	15.8/1.5
			m3	13	25	5	2
		M_{10}, M_{11}	W10/L10	90.0/0.55	16.5/0.55	98/0.55	389.4/0.55
			m10	15	15	14	33
2	D, H	C_{c1}, C_{c2}	Cc	1.21 pF	0.85 pF	1.08 pF	0.92 pF
			Xcc	24.9 μm	42.7 μm	37.35 μm	44.25 μm
			Ycc	55.9 μm	22.75 μm	33.75 μm	23.75 μm
	E, G	M_{14}, M_{15}	w14/l14	370.4/0.35	263/0.35	47.7/0.35	98.35/0.35
			m14	8	20	2	7
	F	M_1, M_2	w1/l1	308/1.2	43.2/1.2	36.2/1.2	40.5/1.2
			m1	16	12	4	10
3	I	M_6-M_9	w6/l6	13.2/1.5	25/1.5	107.4/1.5	28/1.5
			m6	12	20	12	16
4	J	M_5	w5/l5	22.2/1.5	42.4/1.5	179.2/1.5	46.8/1.5
			m5	4	8	8	4
		M_{12}, M_{13}	w12/l12	124.3/1.5	177.6/1.5	416.8/1.5	332.8/1.5
			m12	22	32	16	26
5	K, L	R_{CM1}, R_{CM2}	Rcm	4150 Ω	8591.38 Ω	3618.75 Ω	3578 Ω
			Rcm_l (strip)	78.85 μm	83.05 μm	67.55 μm	89.45 μm
			Rcm_w	3.2 μm	1.8 μm	3.15 μm	2.85 μm
			Rcm_sep	3.0 μm	0.6 μm	0.6 μm	3 μm
			Rcm_m	3	3	3	2
6	M	M_{5c}	w5c/l5c	29.4/1.5	56.4/1.5	238.5/1.5	62.4/1.5
			m5c	3	8	18	8
	N, P	M_{3c}, M_{4c}	w3c/l3c	80.4/2.55	56.4/4.9	36.3/1.85	31.6/4.95
			m3c	8	8	2	4
	O	M_{1c}, M_{2c}	w1c/l1c	43.0/1.2	21/1.2	46.2/1.2	28/1.2
			m1c	20	20	14	28

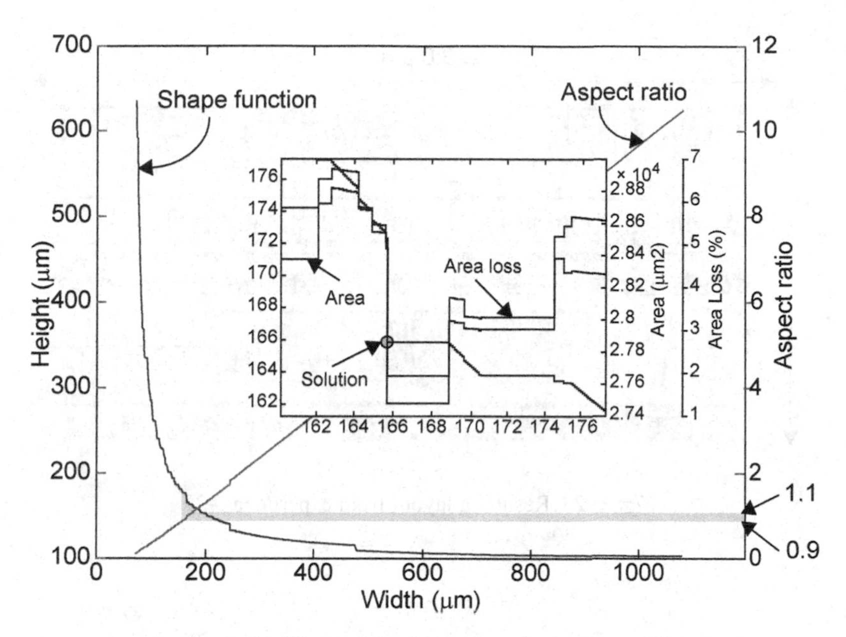

Figure 21. Illustration of solution of experiment #1.

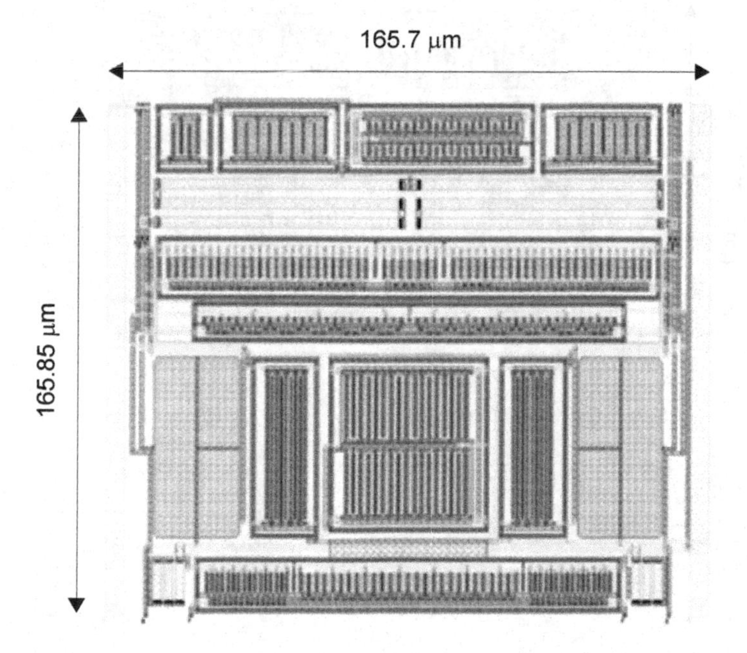

Figure 22. Resulting layout from experiment #1.

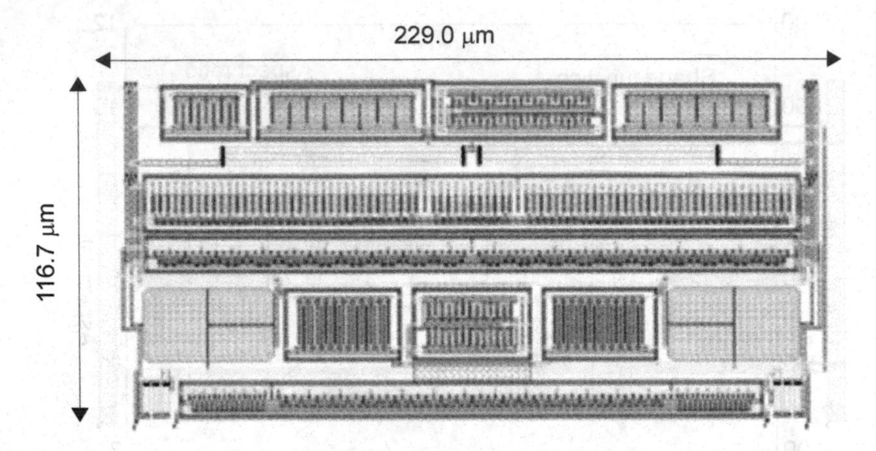

Figure 23. Resulting layout from experiment #2.

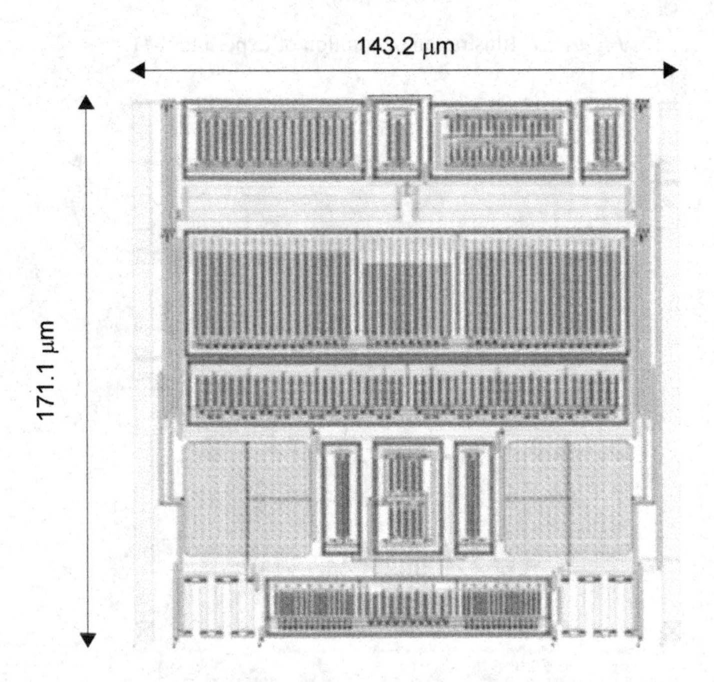

Figure 24. Resulting layout from experiment #3.

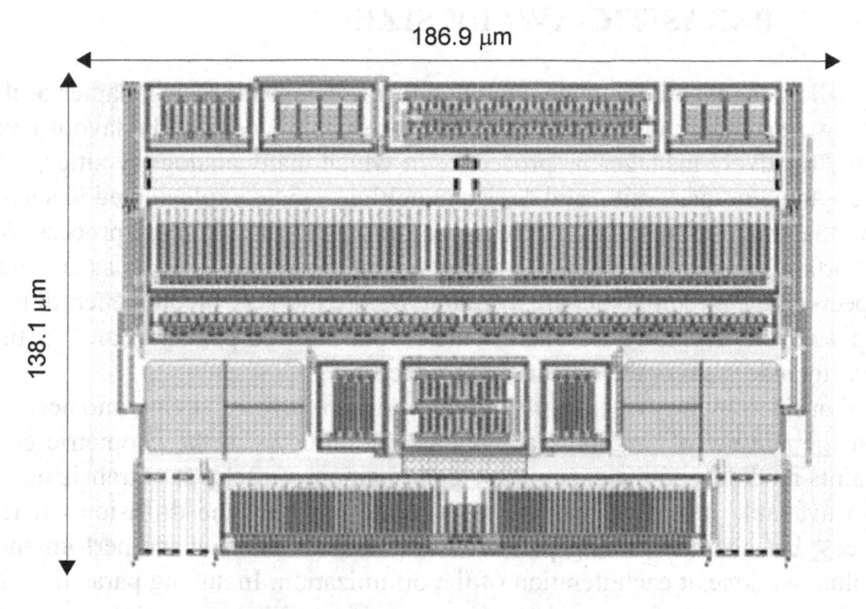

Figure 25. Resulting layout from experiment #4.

Finally, Table 8 displays the elapsed CPU time[22] and the number of iterations of each experiment. As these results demonstrate, taking into account geometric features of the circuit layout earlier in the optimization process, allows to obtain optimum results in terms of area and area loss and, also, with regard to certain user-defined goals, such as specific aspect ratio and width or height specified values.

Table 8. Number of iterations and elapsed CPU time.

Figure[a]	*Exp.#1*	*Exp.#2*	*Exp.#3*	*Exp.#4*
No. of iterations	1708	5778	6462	4824
CPU time (s)	409.92	1380.9	1563.8	1157.76

a. These figures includes both the opamp core and the CMFB circuitry sizing processes.

22. The experiments were performed on a PentiumIV@1.3GHz PC with 512Mb RAM.

3 PARASITIC-AWARE SIZING

To follow the layout template approach turns, as explained in Chapter 5, the otherwise complex task of creating a fully reusable blocks at the layout level into a relatively manageable procedure in which many analog layout aspects like symmetry, proximity, and device matching, can be implemented in such a way that they are successfully preserved during any layout reuse process. An important derived benefit from using layout templates is that, as detailed aspects of the circuit layout are available before the very layout generation, it is possible to accurately estimate[23] the layout-induced parasitics and, in this way, include them in the circuit sizing process.

Consider, by way of example, a two-transistor layout. Taking into account that the number of fingers is already known –thanks to the geometric constraints module–, so the exact shape and position of each MOS transistor are also available it is then feasible to estimate not only the diffusion capacitances, but also the routing parasitics and include them in the performance evaluation done at each iteration of the optimization. Including parasitic estimates early in the sizing process, known as **parasitic-aware sizing**, reveals very useful at avoiding unwieldy iterations between layout generation and sizing design steps, which incur repeated translations and other interfacing costs, eventually leading, as said in Chapter 2, into product-to-market failure[24].

This section describes the problem of parasitic-aware sizing of analog circuits. First, it is discussed why parasitics are important and how are they estimated and included in the analog design flow. Then, a solution is proposed and demonstrated that integrates the parasitic estimation directly in the sizing process of the design reuse flow presented in this book.

3.1 Layout parasitics

All integrated circuits contain passive electrical elements not required for their normal operation. These **parasitic** components, usually jeopardizing the correct performance of the integrated circuit, are consequence of the "imperfect" nature of the materials used in the physical implementation of the circuit.

---

23. In reference to parasitics, the words *estimation* and *extraction* are indistinctly used in this book.

24. As it is explained later, this technique benefits from the fact that design variables can provide information on the eventually instanced layout template at a certain hierarchical level and at each iteration of the optimization-based sizing process. For instance, such information can be directly retrieved from the design variables (e.g., device sizes) at the cell hierarchical level.

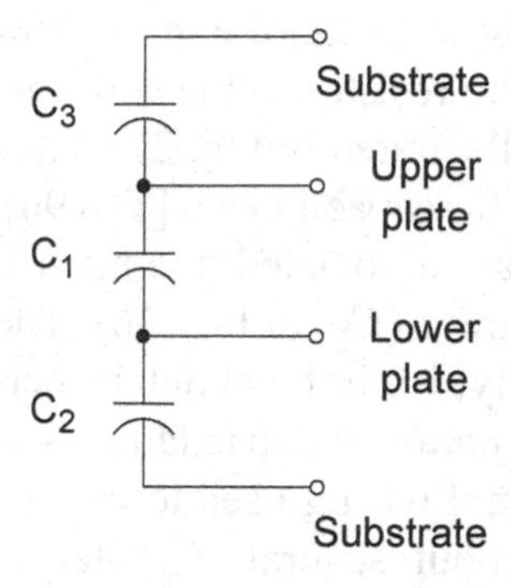

Figure 26. PIP capacitor.

For instance, consider the PIP (poly-insulator-poly) capacitor model in Fig. 26. The intended capacitance C_1 results from the electrostatic interaction of the lower and upper plates of the capacitor. This same interaction occurs between both plates and the rest of the circuit, producing unwanted capacitive parasitics, C_2 and C_3. The parasitic elements are usually capacitive, as in the foregoing example, and resistive, and, at sufficiently high frequencies, inductive parasitics arise as well. These unwanted components can be classified as follows [Prie01]:

a. **Device parasitic components.** Since no practical circuit device is completely isolated from the surrounding environment, a number of interactions unavoidably appear between the environment and the device itself. These interactions can be modeled by replacing each device with a sub-circuit with ideal elements. A group of these elements represents the ideal operation of the device, while the rest are parasitic elements, as illustrated in Fig. 26. The ultimate effect of the parasitic elements is the degradation of the device ideal operation. For example, the oxide surrounding a POLY resistors, whose cross-section is shown in Fig. 27, acts as a

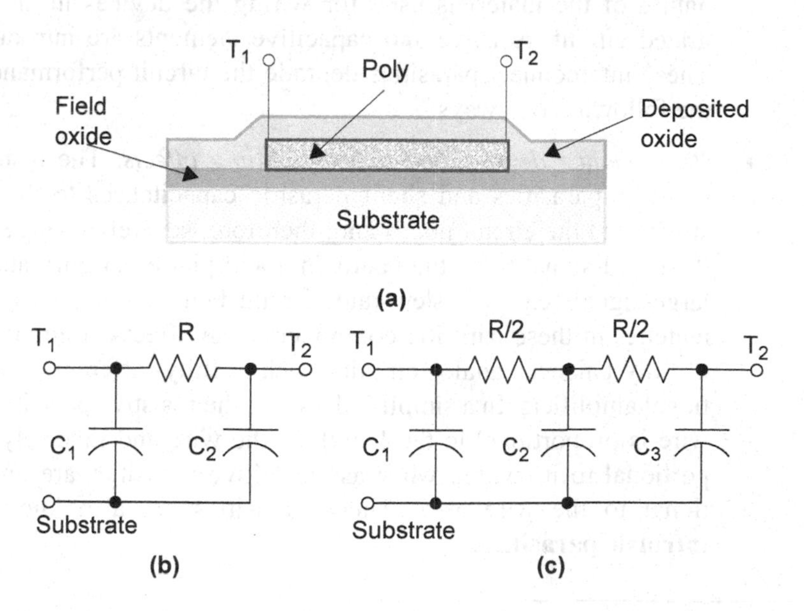

Figure 27. (a) Cross-section of a poly resistor; (b), (c) parasitic-inclusive models.

capacitive dielectric coupling the resistor to adjoining components. The resulting capacitance from the resistor to the substrate is modeled as distributed elements, as illustrated in Fig. 27(b) and Fig. 27(c), where C_1-C_3 are the parasitic capacitances [Tsiv96]. For capacitors, two parasitic elements are associated: the main is between the lower plate and the substrate (C_2 in Fig. 26). The value of this parasitic capacitance can typically be about 10 percent of the main capacitance. Another parasitic capacitance is at the upper plate, consequence of the metal wiring used to contact this plate, and its value is usually about several femtofarads [Tsiv96]. These parasitics are dependent upon how the capacitor is constructed. For MOS transistors, the situation is similar: as shown in Fig. 28 (where the NMOS transistor can be considered ideal): there are undesirable parasitic capacitances at each node of the device. Some of these parasitic elements (such as the gate capacitances) are function of the gate area and, hence, cannot be minimized by using special layout techniques. On the other hand, there are other parasitic capacitances, such as the junction capacitances C_{DB} and C_{SB}, whose effect can be alleviated by following special layout techniques since their values depend on the area and perimeter of the pn junction [Tsiv96].

b. **Interconnect[25] parasitic components**. Due to the non-ideal nature of the materials used for wiring the devices in an integrated circuit, resistive and capacitive elements are introduced. These interconnect parasitics degrade the circuit performance in the following two ways:

- *Performance degradation due to loading effects.* The resulting resistive parasitics and shunt parasitic capacitances to the substrate load the circuit nodes and, therefore, severely change both the small-signal (e.g., the bandwidth and phase margin), and the large-signal (e.g., the slew-rate) circuit behavior. Besides, mismatches in these parasitic components can affect certain aspects of fully-balanced analog circuits, such as fully-differential operational amplifiers. In a simplified model, the resistive parasitic of a wire is proportional to the length of the wire and inversely proportional to its width, whereas capacitive parasitics are proportional to the wire area. These parasitics are also known as **intrinsic parasitics**.

25. Also known as routing parasitics.

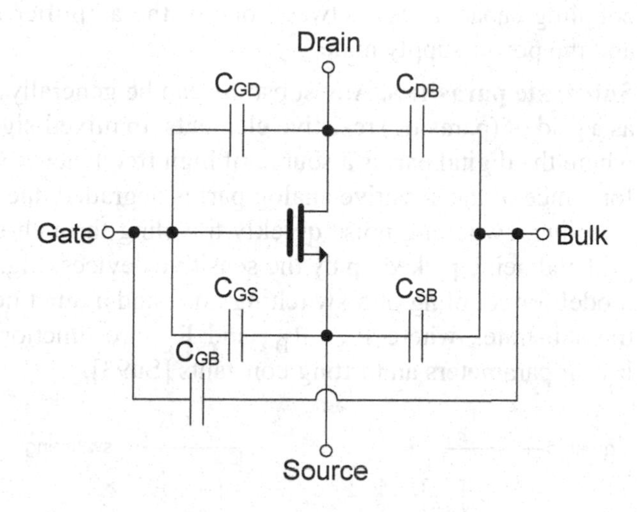

Figure 28. Parasitic model of a NMOS transistor.

- *Performance degradation due to coupling effects between routing wires.* As shown in Fig. 29, for nearby wires running in parallel or crossing in different layout layers, and implementing different nodes, the resulting capacitive coupling may severely degrade the circuit intended performance. Roughly speaking, such coupling is inversely proportional to the separation of the wires (if they are in the same layer) and proportional to the crossover area (for wires in different layers). This capacitive coupling may introduce unexpected noise on certain nodes (effect known as *crosstalk*), which can be very important in mixed-signal circuits. They can even degrade circuit stability due to unforeseen feedback. An example of a performance characteristic that can be dangerously degraded is the amplifier's power supply rejection ratio (PSRR), due to

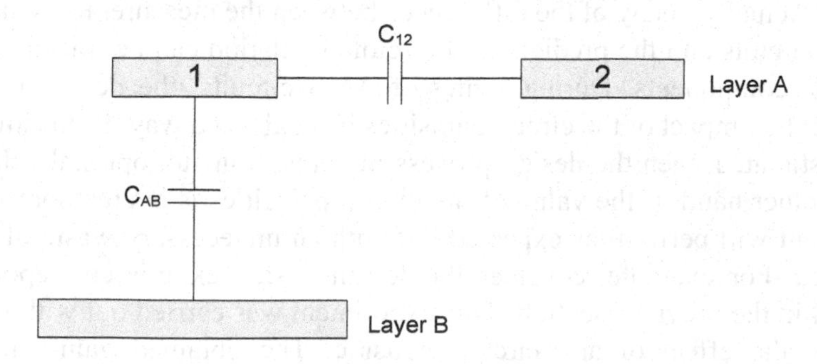

Figure 29. Illustration of interconnect capacitive coupling.

coupling capacitances between one of the amplifier input nodes and the power supply node.

c. **Substrate parasitics**. Any substrate can be generally represented as a grid of (parasitic) resistive elements. In mixed-signal circuits, where the digital part is a source of high frequency noise, the performance of the sensitive analog part is degraded due to substrate coupling, switching noise quickly traveling through the resistive grid and being picked up by the sensitive devices. Fig. 30 shows a model for coupling of a switching node and a quiet node through the substrate., where R_A, R_B, and R_C are function of various layout parameters and fitting constants [Su93].

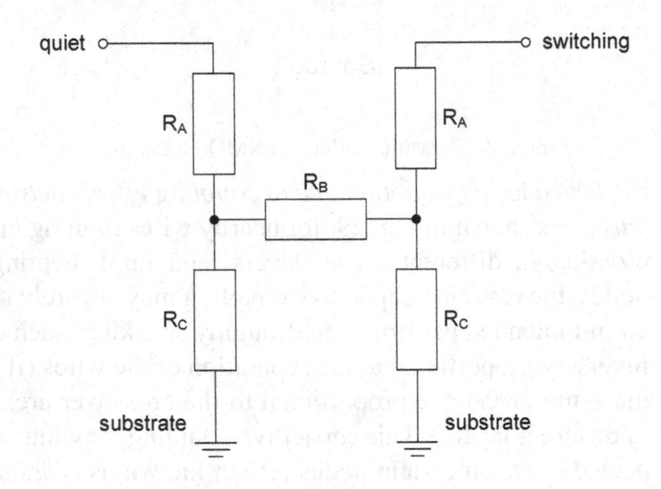

Figure 30. Circuit model for substrate coupling.

Layout-induced parasitic components can be, as it can be concluded from the above relation of unwanted effects, of crucial importance for AMS circuit design. Actually, many of the differences between the measurements made on silicon circuits and the prediction of circuit simulation can be traced to these parasitic components. During sizing of AMS circuits, the designer should consider the impact of the circuit parasitics in a balanced way: if this impact is underestimated, then the design process may result in sub-optimal solutions; on the other hand, if the value of the circuit parasitics is overestimated, then the circuit will perform as expected but with an unnecessary waste of power and area. For example, consider the fourth design experiment reported in Table 4 in the previous section. This experiment was carried out without considering the effect of any circuit parasitic. The obtained gain-bandwidth product (50.37 MHz) is very close to the specification value (50.0 MHz), the "safety margin" being thus very small. Therefore, it is likely that, due to the impact of several capacitive parasitics present in sensitive nodes of the opamp circuit, measurements on silicon yield a gain-bandwidth value under the intended minimum value.

3.2 Extraction methods

The parasitic extraction process aims at, first, detecting all the (or, at least, the most important) capacitive and resistive[26] parasitics and, second, estimating their values. There exists a trade-off between accuracy of the parasitic values and the computation time the extraction process takes. The ultimate goal of parasitic extraction is to obtain a device-level description or netlist of the circuit layout that includes the circuit parasitics in order to perform a simulation of the circuit behavior and its deviation from nominal performances.

Estimation of most layout-induced parasitics requires detailed knowledge on the circuit layout[27], either routing, placement or device realization. An example of this kind of parasitics is the MOS junction capacitances C_{DB} and C_{SB}, which, require exact evaluation of drain/source area and perimeter as follows:

$$C_{XB} = \frac{A_X C'_{jo}}{\left[1 + \left(\frac{V_{XB}}{\phi_1}\right)\right]^{\eta_1}} + \frac{P_X C^*_{jo}}{\left[1 + \left(\frac{V_{XB}}{\phi_2}\right)\right]^{\eta_2}} \tag{14}$$

C'_{jo} and C^*_{jo} are the bottom-wall junction zero-bias capacitance per unit of bottom-wall area and the sidewall junction zero-bias capacitance per unit of sidewall perimeter, respectively. ϕ_1 and ϕ_2 are the bottom-wall and sidewall junction built-in potential, η_1 and η_2 are the bottom-wall and sidewall junction characteristic exponents, and A_X, P_X are the diffusion node (drain or source) bottom-wall area and sidewall perimeter, respectively. To accurately evaluate drain area requires knowing the number of fingers realizing the MOS transistor. Another parasitic component of the MOS transistor whose calculation requires the number of fingers is the extrinsic gate-bulk capacitance, C_{GBe}, expressed as follows:

$$C_{GBe} = LC^*_1 \tag{15}$$

---

26. The inductive parasitics, crucial to RF design, are beyond the scope of this book.

27. On the other hand, there are circuit parasitics whose evaluation does not require specific knowledge on the circuit layout. Examples of this type of parasitics are the MOS transistor extrinsic gate capacitances, C_{GSe}, C_{GDe}, directly proportional to the transistor width W, as well as the intrinsic capacitances, C_{GSi}, C_{GDi}, C_{GBi}, C_{SBi}, and C_{DBi}, which are function of the $W \cdot L$ product. Thus, calculation of these parasitics can be easily carried out during the circuit sizing process without knowing the number of transistor fingers.

for C_l^* being the gate-substrate overlap capacitance per unit of channel length. This parasitic capacitance requires knowing the number of fingers as well. Other examples are: the resistor and capacitor parasitic components, interconnect resistive and capacitive parasitics, interconnect coupling capacitances and substrate parasitics. Typically, data required for their evaluation are not available at the circuit-sizing phase[28].

Parasitic associated to MOS transistors[29], are estimated by computing the area and perimeter of drain and source nodes, as expressed in Eq. (14). These figures depend on the number of fingers and some technology-specific constants. Extraction of these parasitics can be done in the following two ways:

1. **Geometric method**. This method consists on directly measuring the area and perimeter of the mask layers whose interaction defines each drain or source. The area and perimeter of those masks from a same node are summed up. To carry out this type of extraction, the description of the circuit layout, e.g., in GDSII format, must be accessible. Consequently, the circuit layout has to be previously generated. This method is used in many commercial layout tools and extractors [Diva05] [Mayo90].
2. **Analytical method**. This method uses analytical expressions for the MOS transistor drain(source) area and perimeter as a function of the number of fingers, the exact implementation style, and fabrication process data. Appendix A provides expressions to carry out the analytical extraction of drain/source parasitic capacitances for folded transistors as well as for arrays of stacked transistors. Note that, as long as these expressions are available and the number of fingers, the implementation style, and the fabrication process characteristics are known, there is no need to actually perform the circuit layout generation.

Regarding the interconnect wires, they can be viewed, as illustrated in Fig. 31, as a dense system of conductive layers in a stratified dielectric medium over a silicon substrate. As indicated above, interconnect capacitive parasitics can be divided into three components: ground, fringing, and cou-

---

28. Some MOS transistor models already incorporate the diffusion capacitances, which are computed from the diffusion area and perimeter (parameters of the transistor model). Some simulators calculate the diffusion area and perimeter of a single transistor as a function of its width and length. Unfortunately, most circuit simulators (if not all) provide but a very limited facility to describe the layout implementation of each device in the circuit netlist.

29. Parasitics associated to resistive and capacitive devices, such as `POLY` or `NWELL` resistors and PIP or MIM capacitors, can be treated as interconnect layers, discussed below.

pling (wire-to-wire) capacitances. The ground capacitance depends on the wire's width and the distance from the wire to the substrate. The coupling and fringing capacitances are complex function of various layout parameters, such as the wire thickness, the distant between adjacent wires, and the exact geometry of the wires on the stratified medium [Onoz95].

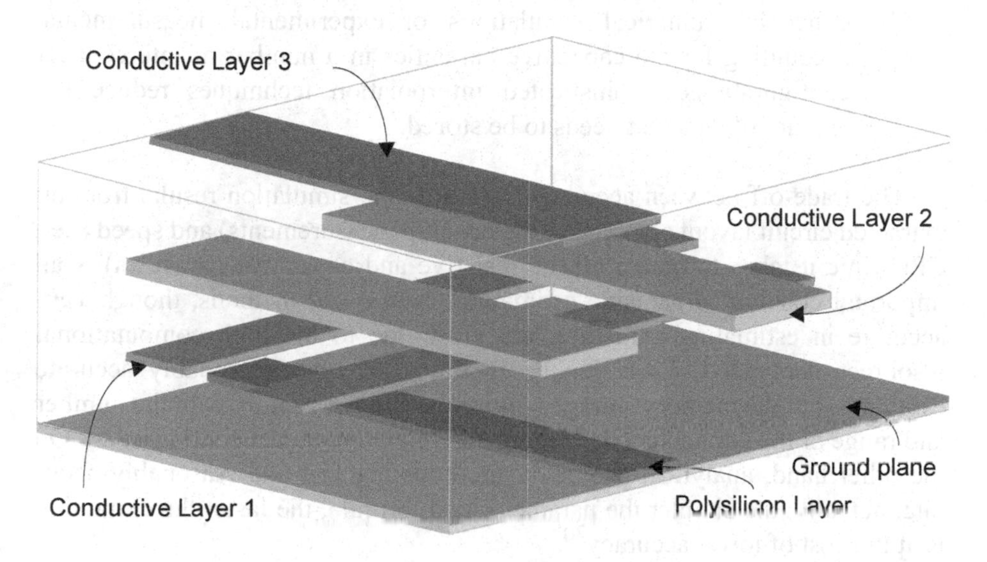

Figure 31. 3-D view of an interconnect structure.

So far, extraction of interconnect parasitics has been tackled by means of three different methods. These are the following:

1. **Numerical methods**. These methods attempt to solve the Laplace equation over the system of stratified layers to determine the exact relationship between the voltage and charge on the conductor surfaces, for capacitive parasitics (by either solving the differential equations directly [Dengi97] [Cost87] or formulating the problem with an integral-equation approach [Dengi98] [Nabo92] [Wei84]) or between the wire potential and the wire current, for resistive parasitics [Harb86] [Hall87].
2. **Analytical-geometric methods**. These methods creates analytical parameterized formulas or models of the capacitive parasitics for a number of commonly encountered interconnect configurations, stored in a database. The formulas fit to the data generated by numerical simulation or experimental measurements. Geometric methods are subsequently used to extract the layout parameters directly from the layout geometries to evaluate each analytical function. These extrac-

tion methods have advanced from 1-D, 2-D, 2.5-D to 3-D effects to meet the required accuracy [Kao01]. On the other hand, resistive parasitics of an interconnection are based on a decomposition of the wire into rectangles along equipotential lines [Horo83].

3. **Table lookup methods**. Look-up tables store the data generated either by numerical simulations or experimental measurements, accounting for the capacitive parasitics in a number of interconnect configurations. Sophisticated interpolation techniques reduce the amount of data that needs to be stored.

The trade-off between accuracy (i.e., how the simulation results from an extracted circuit layout compare to the on-chip measurements) and speed (i.e., CPU time it takes to obtain all the resistive and capacitive parasitics) is an important concern in parasitic estimation. Numerical methods, though very accurate in estimating parasitics, are slow due to the high computational resources demanded. Table lookup methods provides reasonably accurate estimates, but the memory storage requirements grow rapidly with the number and range of the parameters describing a given interconnect configuration. On the other hand, analytical-geometric methods are fast and reasonably accurate; actually, the simpler the parameterized formula, the faster the extraction is at the cost of lower accuracy[30].

3.3 Extraction of parasitics in the design process

Figure 32 illustrates the traditional design flow at the cell level, where layout generation follows the sizing process.

The layout is optimized without real quantification of the cell performance degradation due to, among others, unavoidable capacitive and resistive parasitic effects. Afterwards, the design flow continues with formal verification (DRC, ERC, and LVS, not shown in Fig. 32) and parasitic extraction, followed by performance checking of the extracted layout. If the difference between intended and extracted layout performances is not permissible, one or more redesign loops are needed. This trial-and-error approach reveals several drawbacks. First, no systematic method exists that guides the correction process of the circuit sizing, its layout, or both: the problem may not be necessarily due to a single parasitic, but a combined effect of a group of them. Second, the list of extracted parasitics may be so large that such correction process can be extremely laborious. Third, the number of iterations in the redesign loop may

[30] Most modern extraction tools base their operation on analytical-geometric methods; many of them are complemented by embedded capacitance field-solvers (i.e., numerical methods) for critical parasitic extraction requiring higher accuracy.

be rather large for commonly designed analog circuits [Chou93]. This all means that efficiency of the redesign process relies, largely, on the designer's expertise and the heuristics used, and, consequently, redesign iterations can be too time-consuming.

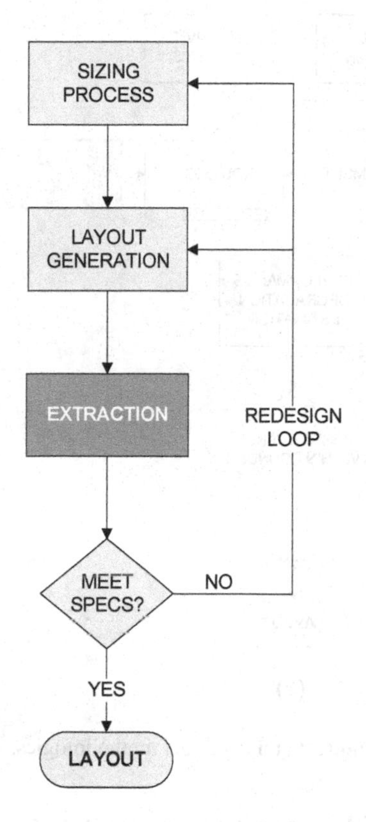

Figure 32. Traditional flow.

Layout parasitics cannot be evaluated at earlier stages of manual design, since no information on layout is available. Therefore, it is considered impractical to undertake both sizing and layout at the same time. For this reason, the splitting up of the analog design flow of Fig. 32 has long been a de facto standard.

With the first analog placement and routing tools, the extraction process has been, however, gradually included in the layout generation phase. The ILAC tool [Rijm89], for instance, routes nets following a given priority order (power nets, sensitive nets, and, last, noisy nets) with distance and crosstalk penalties. KOAN/ ANAGRAM II tool [Cohn91] performs placement and routing based on weighted parasitics minimization[31]. Both tools use a very simple analytical-geometric extraction method to estimate parasitics within their layout optimization loops. Later approaches try to reduce the number of redesign loops of Fig. 32 by including parasitic extraction directly in the layout generation phase and constraining both placement and routing with quantitative knowledge on parasitics and the relation between these and the performance degradation induced. Two alternatives exist here: indirect and direct methodologies. Fig. 33 outlines both design flows.

Indirect methods [Fig. 33(a)] compute sensitivity of performance specifications with respect to all possible layout parasitics before the layout is generated [Chou93]. In accordance with this sensitivity analysis [Chou90c], the method generates a set of parasitic constraints[32] which are then enforced by the placement and routing tools [Chang97] [Mala96]. Once an intermedi-

[31] No clear strategy to define parasitic weights is, however, reported.

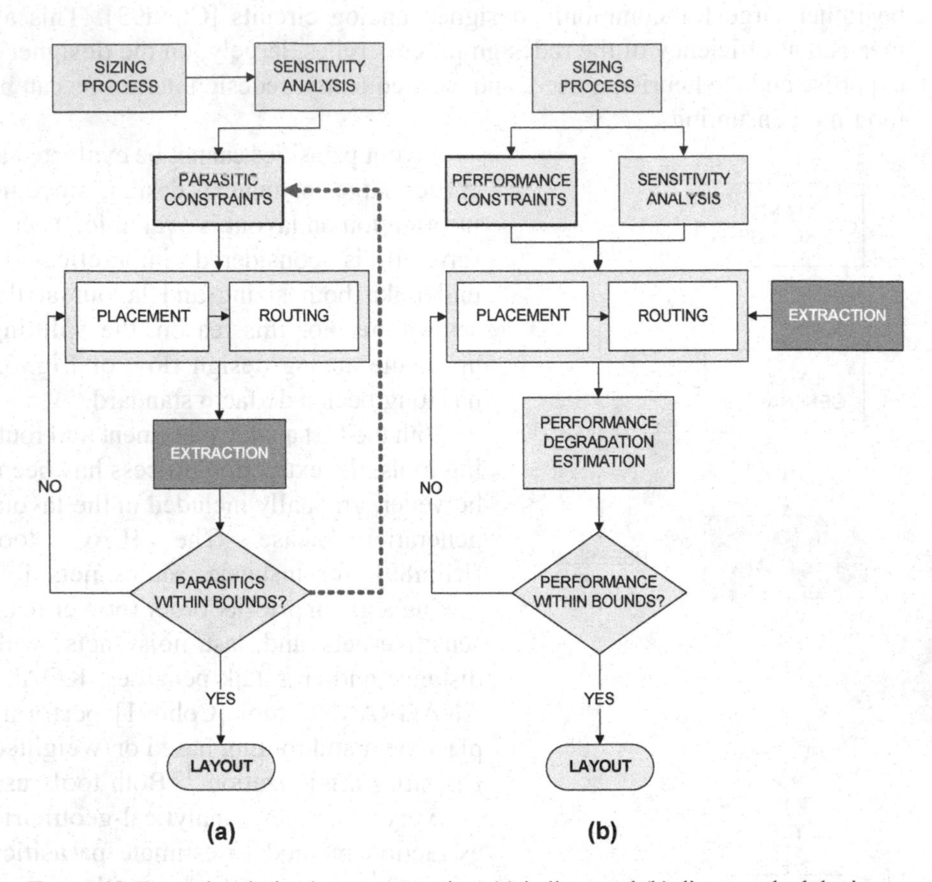

Figure 33. Extraction during layout generation: (a) indirect and (b) direct methodologies.

ate placement or routing solution is available, parasitics are extracted via analytical-geometric methods and compared against the set of constraints. Failing to meet any of these constraints means that one or more performance specifications are not fulfilled. Indirect constraint-based methods, however, overly constrain the layout tools since not all the parasitic constraints can be easily realized. Besides, as only one set of parasitic constraints is generated, if the method fails to comply with them, another set of constraints, with the resulting CPU time overhead, must be generated. Direct methods overcome this drawback by eliminating the intermediate generation of parasitic constraints and directly driving the layout tools with performance constraints[33] [Lamp99] [Prie01]. As illustrated in Fig. 33(b), the relevant geometric infor-

[32] Based on the sensitivity analysis an estimate of the maximum and minimum values $(p_{j,min}, p_{j,max})$ for each parasitic p_j is obtained.

mation is extracted from an intermediate solution of the placement or routing tools, and the parasitics are calculated by using the analytical-geometric approach. Using the sensitivity analysis, the effects that layout parasitics have on the circuit's performance features are evaluated. If all the performance specifications fall within its allowed range, the layout is finally generated. Otherwise, the method executes one more iteration.

Whereas indirect methods may lead to a number of iterations –between generation of constraints and layout, as pointed out in Fig. 33(a)– with different constraint values, direct methods yield a *correct* layout or flag the specifications as being impossible to meet without any of these iterations.

Note also that the above described methods use an already sized design. Even if the designer uses direct methods, which should return correct-by-constructions layouts, circuit sizing needs, however, to leave sufficient performance margin to accommodate the eventual layout-induced degradation on performance, which could lead to unnecessary waste in power and area [Giel01]. This suggests that layout parasitics should be somehow included in the sizing process, as the parasitic-aware sizing flow illustrated in Fig. 34.

Parasitic-aware circuit sizing aims at improving the automatic design of AMS circuits by (1) avoiding unsystematic and time-consuming iterations between layout extraction and circuit redesign, and (2) arriving at better optima of the circuit design space., which is possible because parasitics are

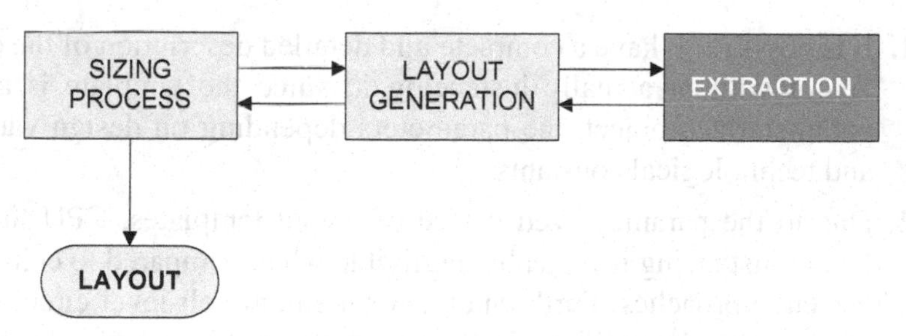

Figure 34. Extraction during circuit sizing.

33. For a performance specification P, layout-induced effects result in an additional performance variation ΔP_{lay}. For the circuit to be functional, the value of P after layout extraction has to be in the interval $[P_{max}, P_{min}]$. Then, it results the following constraints for ΔP_{lay}: $\Delta P_{lay} \in [P_{min} - P_{proc}^{min}, P_{max} - P_{proc}^{max}]$, with P_{proc}^{min} and P_{proc}^{max} being the minimum and maximum deviations of P due to process variations. If the circuit has N_p parasitics p_i, $(i = 1 \ldots N_p)$, the layout induced performance degradation for P can be modeled by the linear approximation:

$$\Delta P_{lay} = \sum_{i=1}^{N_p} S_{p_i} p \text{ where } S_{p_i} \text{ is the sensitivity of performance characteristic } P \text{ to parasitic layout } p_i:$$

$$S_{p_i} = \frac{\delta P}{\delta p_i}$$

included early in the sizing process rather than merely consider them in subsequent redesign loops.

As explained above, accurate extraction of parasitics requires knowing the circuit layout in full detail, which involves obtaining information on the implementation style of each device, the interconnect structure, and their relative positioning. Furthermore, this layout knowledge must be generated or retrieved at each iteration of the sizing process. Therefore, whichever the method used to obtain this knowledge, it must be rapid enough to prevent circuit sizing from being prohibitively long. A way to obtain such knowledge is to actually generate the layout at each iteration of the sizing process, thereby being critical to reduce the CPU time of such layout generation. For this reason, constraint-driven approaches for layout generation are currently too slow to be called in the inner loop of an automated parasitic-aware sizing process. For instance, the tools reported in [Lamp99] yield CPU times from 550 to 800 seconds for opamp-like circuits. Considering that a typical circuit sizing involves a few thousand iterations, and neglecting the CPU time for the rest of processes (simulation, extraction, and so on), it would take several days to complete the parasitic-aware circuit sizing, which can be comparable (if not worst) to manual design.

Procedural layout based on templates is, on the contrary, a more suitable solution for **parasitic-aware circuit sizing** [Lamp99] [Dess01a] [Vanc01] [Tang02][34] for the following two worthwhile properties:

1. It is possible to have a complete and detailed description of the circuit layout without actually instancing it, since the template is a fully parameterized object, the parameters depending on design variables and technological constants.
2. Due to the parameterized nature of layout templates, CPU time for layout instancing is typically negligible when compared to constraint-driven approaches. Furthermore, for medium cell-level circuits such as operational amplifiers, instancing the layout template is faster than simulating the circuit or extracting the layout[35]. For instance, to instance a typical opamp layout template with the Cadence® Virtu-

34. Minimization and control of parasitics can be also done by imposing constraints directly on the layout template, as described in Chapter 5. Some guidelines are: (i) to control layers, (ii) to reduce the number of vias, (iii) to control geometry of the interconnect wires, (iv) to route carefully symmetrical paths of fully-differential circuits, (v) to avoid parallel-running wires –especially those from noisy sources–, and (vi) to shield sensitive devices with guard-rings.

35. In this context, to instance a layout template means to generate either the layout from the parameterized description of the template –i.e., assign each parameter a numerical value– or a list with the updated size and position of every polygon in the layout.

oso® layout editor [Virt00a] [Virt00b] takes 0.01 seconds, while extraction and simulation takes around 300 times more.

3.4 Demonstration of the parasitic-aware design flow

Figure 35 outlines the general design flow of parasitic-aware circuit sizing in more detail. The sizing engine, knowledge or optimization-based, returns first relevant electrical information on the visited point of the design space being explored: MOS transistor sizes, passive device values, and biasing currents and voltages. This information is required by the procedural layout generation to allow estimation of parasitics. A specific process determines the value of geometric parameters (e.g., as the number of MOS transistor folds, the exact shape of passive devices, or the optimum width of the routing wires) prior to parasitic extraction. These geometric parameters may be constrained by some kind of user's defined restrictions, like the aspect ratio of the circuit layout. Such information is then processed to estimate the layout-induced parasitics. This estimation process can be carried out following several methods, explained above in Section 3.2. Note that some of these methods may require the instancing of the layout template at each iteration of the sizing process. Afterwards, the electrical description of the circuit is completed with the parasitic estimates and the overall performance is checked against intended performance specifications, which requires evaluation of the circuit performance. A correct layout (that is, one whose degradation on the circuit performance is acceptable) is eventually generated.

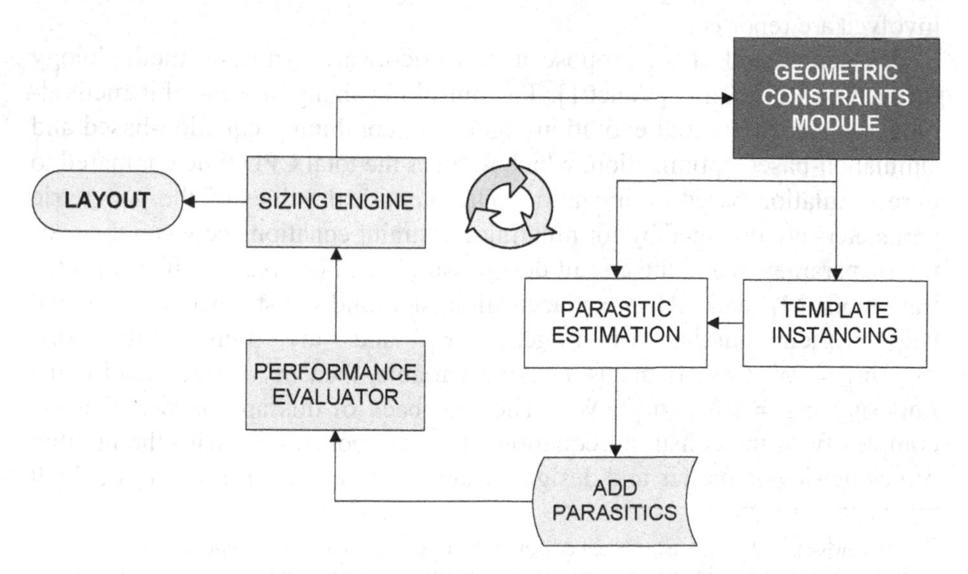

Figure 35. Parasitic-aware sizing flow.

Parasitic-aware circuit sizing was first proposed in [Onod90]. The operational amplifier compiler presented there uses a procedural layout generation method and a set of functions supporting the estimation of parasitics. The sizing engine is composed of two phases: global design, for architecture/topology selection and, detailed design, for performance optimization. Whereas the former phase uses a knowledge-based procedure, the latter uses a simulation-based optimization procedure. Parasitics are considered only in this second phase. Nevertheless, neither a description of the extraction method is provided, nor details of the method to decide on geometric parameter are given. Examples reported show that including parasitics is efficient, but at the cost of increasing a 20% the total design time.

The methodology reported in [Dess01a] uses a knowledge-based[36] approach with specific design plans for the circuit to size. The performance evaluation is done by using developed design equations. Geometric parameters are decided upon following a linear programming technique, explained in Section 2. Procedural layout is the approach used to generate the layout and estimate the circuit parasitics, done by using simple analytical equations. The extraction of these parasitics is carried out without actually generating the layout: the parameterized, codified dimensions of all layers in the layout template are the variables of the parasitic analytical equations. Apart from MOS transistor capacitances, only the routing parasitics to substrate are considered. The sizing process is done by using a knowledge-based design plan that is repeated until parasitics convergence is reached, i.e., the calculated parasitics remain unchanged. No CPU time figures of any of the processes involved are reported.

Vanconrenland et al. propose a parasitic-aware synthesis methodology focused on RF circuits [Vanc01]. To control the sizing process, this methodology uses a differential evolution algorithm, combining equation-based and simulation-based optimization, which reduces the total CPU time compared to pure simulation-based optimization. The numerical values of the geometric parameters are obtained by formulating constraint equations between the geometric parameters and the circuit design variables. For instance, if two folded transistors M_1 and M_2 are placed alongside and constrained to be equal height, their number of fingers, m_1 and m_2 must fulfill that $W_1/m_1 = W_2/m_2$. If m_1 is a design variable, then m_2 is obtained by just imposing $m_2 = (W_2 \cdot m_1)/W_1$. The drawback of this approach is that the complexity of the constraint equations grows exponentially with the number of geometric parameters and design variables, thereby resulting very difficult

36. The drawback of the knowledge-based design methods is, in contrast to simulation-based optimization methods, the higher effort required to develop both the design equations and the design plan for each reusable block.

to formulate and solve. After procedural-based layout generation, parasitics on critical nets are extracted using previously parameterized 2.5-D formulas[37]. With this method, evaluation of parasitics takes around 0.3 seconds of CPU time on a standard workstation[38]. The methodology is suitable for typically small but hard-to-design circuits, as the RF mixer circuit used for demonstration. Design expertise is required to reduce the overall design time, both in the optimization and the parasitic extraction phases. For instance, to speed up parasitic extraction, it only comes down to consider those at the most critical nodes of the mixer circuit example, for which design expertise on the circuit behavior is certainly necessary.

The approach in [Tang02], although not based on procedural layout generation but on layout constraint graphs[39], employs a technique based on analytical-geometric methods to estimate (intrinsic and coupling) routing and substrate parasitics. The models used are those reported in [Lamp99] and [Stan94]. These parasitic estimates are compared with extracted values from a modern, off-the-self extractor. The error was within 20% from the extracted values.

The salient characteristics of the above reviewed approaches are listed in Table 9. The table collects, for each reference, the sizing method, the performance evaluator, and the extraction technique used. It specifies whether layout generation is needed or not as well as the process employed (if so) to decide on the geometric parameter values.

In the work reported here, two approaches are described, both using a simulation-based optimization process with a numerical simulator in the loop (to perform circuit sizing) and the geometrically constrained sizing technique presented in Section 2 (to obtain suitable values of the geometric parameters). The first solution relies, on the one hand, on accurate calculation of the MOS transistor diffusion areas and perimeters by using the appropriate analytic equations (available in Appendix A); on the other hand, estimates of the routing parasitics are obtained by sampling the layout template [Cast02b] [Fern03]. Prior to circuit sizing, a number of different instances of the circuit layout template are extracted. To do so, each one of the n design variables is uniformly sampled with m data points. The instances are simulated (including parasitics), the critical parasitics are identified, and their values stored in a look-up table. Then, the value of the parasitic is retrieved from such table depending upon the value of the design variables. The problem with this tech-

[37] Extraction is, thus, tailored to the circuit to size. This implies a considerable, thorough effort in creating the circuit database (e.g., layout template, parasitic models, optimization constraints).

[38] Technical specifications are not reported.

[39] This method, similar to the one explained in Chapter 5, consists of constrained placement (with constraints such as symmetry) and channel routing. It is used for cell-level layout only.

Table 9. Parasitic-aware approaches.

Reference	*Sizing method*	*Performance evaluation*	*Estimation*	*Layout generation?*	*Geometric parameters?*
[Onod90]	Knowledge & simulation-based opt.[a]	Numerical simulation (SPICE)	A-G[b] models	No	No
[Dess01a]	Plan-based	Equations	A-G[c] models	No	Linear programming
[Vanc01]	Differential evolution genetic algorithms	Equations/ numerical simulation	A-G (2.5-D)[d] models	Yes	Equations
[Tang02]	Simulation-based opt.	Numerical simulation	A-G[e] models	Yes	No
This book [Cast02b] [Fern03]	Simulation-based opt.	Numerical simulation (HSPICE®)	Layout sampling & fixed estimates[f] + accurate diffusion calculations	No	Slicing tree algorithm
			Accurate off-the-shelf extractor	Yes	
[Agar04]	Simulation-based opt.	Numerical simulation (NG-SPICE)	Layout sampling	No	No
[Ranj04]	Simulation-based opt.	Symbolic analysis	Accurate off-the-shelf extractor	Yes	No

a. Parasitic extraction is only carried out during simulation-based optimization.

b. A-G stands for analytical-geometric, both for device and routing parasitics.

c. Neither interconnect coupling capacitive nor resistive parasitics are considered.

d. Only the overlap and fringing routing parasitics of critical nets are extracted.

e. Only interconnect and substrate parasitics are considered.

f. The estimates are obtained through detailed extraction, done with off-the-shelf tools available with the Cadence® *DFII* suite.

nique is that, since complex analog circuits typically have a "large" number of design parameters, both generation time and storage requirements for the look-up table may be quite large[40]. For that reason, only $N < m^n$ instances from all the m^n possible combinations are generated and extracted, and only critical parasitics (e.g., those in the signal path) are considered. Identifying which are the most critical parasitics and the most significant layout instances require a big deal of design expertise to be called upon.

To further improve accuracy of the parasitic-aware sizing technique illustrated above, a dedicated, high-accuracy extraction process can be included in the optimization loop. Such a solution, whose functional flow is shown in Fig. 36, has been implemented as well. After a selection of a new point of the design space, the geometric parameters are decided by using the module explained in Section 2. This module returns a file with parameter values for the instancing of the layout template. Afterwards, extraction is done by using an accurate off-the-shelf extractor, and, last, the design is simulated including the extracted layout-induced parasitics. As it can be seen in the figure, from layout instancing to performance evaluation is done with the resources available in the *DFII* environment from Cadence® (such as OCEAN scripts [Ocean05] to perform batch simulation), while selection of the geometric

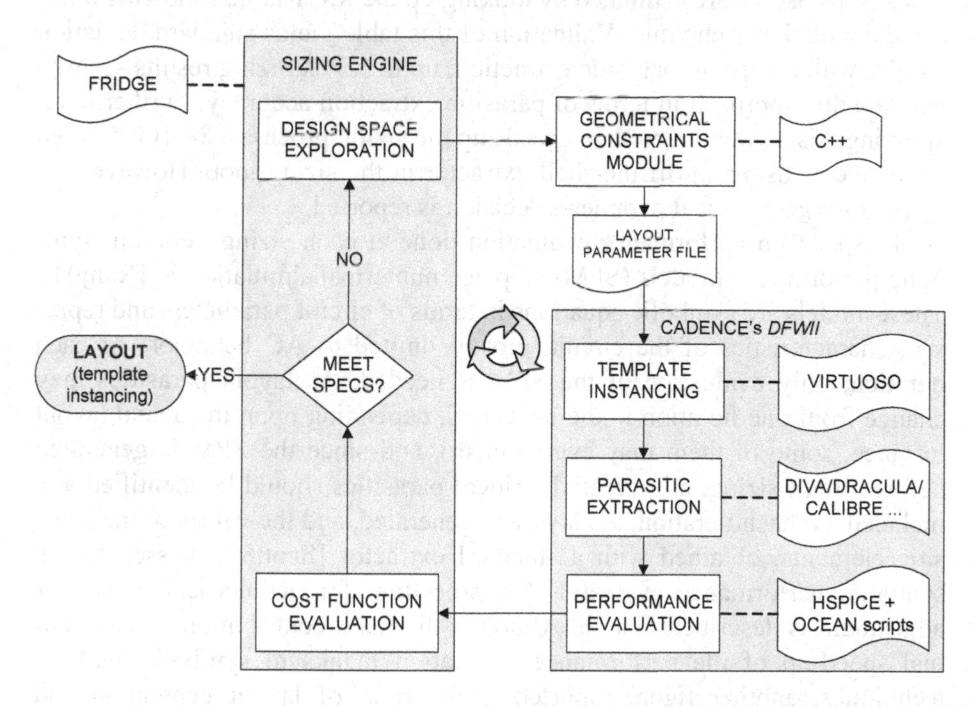

Figure 36. Parasitic-aware sizing flow with in-the-loop accurate extraction.

---

[40] By way of example, the number of instances to generate for a sizing experiment with $n = 20$ and $m = 2$, is $2^{20} = 1{,}048{,}576$, which would take around 290 hours providing that layout generation and extraction takes just 1 second per instance. Considering 20 parasitic elements and 8 bytes per stored parasitic value, 158 Mb are required to store such a look-up table.

parameters and design optimization is done externally. Note that a solution obtained with this parasitic-aware sizing technique will feature the highest accuracy provided the extractor used is the most accurate. That is, the final sized circuit will meet specifications even considering the performance degradation caused by layout parasitics. In contrast, better accuracy implies longer CPU times employed for extraction and, possibly, longer simulation times due to the presence of a larger number of parasitics.

The recent approaches reported in [Agar04] and [Ranj04] present some similarities with the above-described approach. In [Agar04], the solution consists in a simulation-based optimization system with procedural layout generation. As in [Cast02b], and thanks to the layout template approach, based on [Samp03], the circuit layout is also intensively sampled on a uniform grid and the parasitics of each instance are extracted by using an off-the-shelf extractor. With the complete collection of extracted parasitics, a Module Characterization Table (MCT) is constructed. At each iteration of the sizing process, parasitics are estimated by looking up the MCT table following a linear interpolation technique. Validation of this table yields standard deviation values, with respect to parasitic extraction, up to 3.3%. Sizing results seem to validate this approach in terms of parasitic extraction accuracy. Furthermore, applying this extraction method speeds up the sizing process a 88-100% when compared to using an off-the-shelf extractor in the sizing loop. However, no method for geometrical parameter decision is reported.

To speed up performance evaluation done at each sizing iteration, symbolic performance models (SPMs) replace numerical simulation in [Ranj04]. These models are symbolic equations in terms of circuit parameters and represent characteristics of the circuit, mostly limited to AC behavior. At each iteration, only evaluation of the SPM is needed. As layout parasitics may change from one iteration to the next (and, depending upon the actual layout instance, some of them may even vanish), and since the SPM is generated before circuit sizing, all potential, critical parasitics should be identified and included. At each iteration, the layout is generated, and the values of the parasitic elements, obtained with a standard extractor [Scott85], passed to the symbolic performance evaluator. No procedure for geometrical parameter adjustment is described though. Although the main contribution is the eventual speed-up of the performance evaluation thanks to symbolic analysis techniques, another figure reported is the ratio of layout generation and extraction time to performance evaluation time. This ratio provides an idea of the overhead paid for using accurate extraction in the sizing loop. Experiments described show this ratio is around 2 when using a numerical simulator (NG-Spice) and around 5 when using the symbolic approach.

An important factor to account for in the previous review is whether optimization of the layout geometrical parameters has been considered –see last column of Table 9. The impact of automatically optimizing certain aspects of an analog layout (such as area and density) while considering parasitics, is noteworthy. Consider, by way of example, that a circuit is being optimized and no attention is being paid to folding transistors or adjusting the geometry of passive device, in order to improve matching, minimize parasitics, or simply optimize the layout area. Quite the opposite, these concerns are put off until sizing completion. Consider also that the circuit is sizing following a parasitic-aware method and that the obtained solution is optimized for a set of circuit parasitics. Then, any subsequent change in the circuit layout (modify the number of transistor folds, select another geometry for passive devices, or simply change the layout block spacing) may change the value of one or more parasitics in, and, consequently, deteriorate the circuit behavior until malfunction. It is, therefore, of the upmost importance that parasitic-aware circuit sizing is carried out while simultaneously considering optimization layout geometric aspects such as area and density.

Regarding the solutions proposed in this book, Table 10 collects, for the sake of illustration, the results of three design experiments, two of them using the parasitic-aware circuit sizing technique. The circuit to size is the fully-differential two-stage operational amplifier shown above in Fig. 17 (CMFB not included). The three different sizing experiments have been carried out in a 0.35μm CMOS fabrication process with $V_{DD} = 1.5$ V, $V_{SS} = -1.5$ V, $Rload = 100$ kΩ, and $Cload = 8$ pF. These experiments are:

- **Experiment I**. In this experiment, neither layout geometric constraints nor parasitic estimates are considered during the size tuning process. Area minimization is pursued by assuming a direct relationship with device sizes and not using any information from the layout template.
- **Experiment II**. This experiment considers layout geometric constraints and parasitic estimates using the layout sampling technique. The aspect ratio required is $AR \approx 1$.
- **Experiment III**. In this last experiment, a dedicated, high-accuracy extraction tool, in combination with the geometric constraint module, is used to estimate layout parasitics (also pursuing $AR \approx 1$).

Third, fourth and fifth columns of Table 10 show the results of each experiment. For experiments I and II, the numbers between brackets correspond to the values of the performance features obtained after layout template instancing, extraction, and simulation. For experiment III, the numbers already correspond to extracted layout performance.

Template instantiation of the resulting devices sizes are shown in Fig. 37 (notice that all layouts have been captured with the same resolution and, therefore, the relative dimensions are real). It can be seen in Table 10 that the required nominal performance is accomplished in the three experiments. The

Table 10. Experimental results for parasitic-aware sizing.

Feature	*Specified*	*Experiment I*	*Experiment II*	*Experiment III*	*Units*
DC gain	>110	110.0 (110.0)	113.0 (113.0)	111.7	dB
Unity-gain frequency	>90	91.8 (89.8)	105.7 (105.4)	106.6.	MHz
Phase margin	>65	67.6 (63.4)	65.4 (65.0)	66.2	o
Output swing	>5.25	5.3 (5.3)	5.3 (5.3)	5.4	V
Slew rate	>40	46.9 (46.8)	57.2 (57.0)	56.7	V/μs
Power dissipation	minimize	7.3 (7.3)	8.0 (8.0)	8.3	mW
Total area	minimize	195.8 × 358.8	173.8 × 191.25	189.6 × 193.05	μm^2
Aspect ratio	≈1	0.55	0.91	0.98	

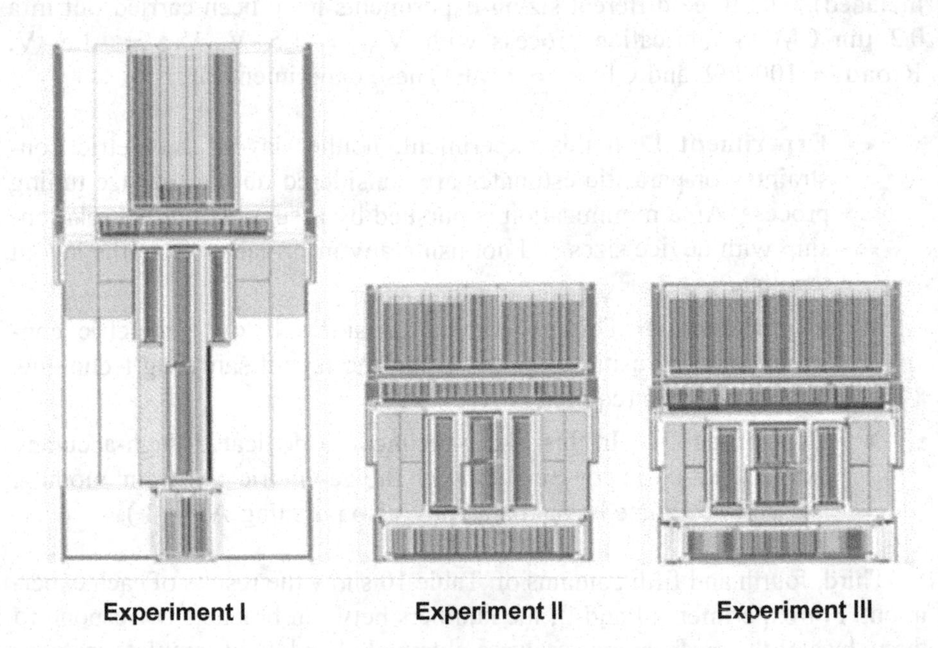

Figure 37. Layout instances of the experiments in Table 10.

exclusion of geometric layout constraints and/or parasitic estimates, however, results in additional re-design iterations (when, due to layout parasitics, the final design does not meet the required specifications –see the non-fulfilled phase margin of experiment I in Table 10) or in hardly compact layouts with large empty areas when the sizing process is geometrically unconstrained.

The number of iterations and the elapsed CPU times required for each experiment are listed in Table 11. Using an off-the-shelf extractor requires around 15 % more CPU time than when using the layout sampling technique (layout instancing and extraction takes 17 % of the total sizing time).

Table 11. Elapsed CPU time for the parasitic-aware sizing experiments.

Figure	*Exp. I*	*Exp. II*	*Exp. III*
No. of iterations	2116	2437	3044
CPU time (s)	507.8	612.31	880.3

4 SUMMARY

In this chapter, a layout-aware sizing methodology has been described. This sizing methodology is composed of two techniques. The first one, called geometrically constrained sizing, is used to solve the floorplan-sizing problem at each iteration of optimization-based sizing process. Such problem consists in finding the value of a set of layout parameters (e.g., number of folds of MOS transistors) such that one or more functions of the circuit layout width and height are optimized. Examples are the layout area or the aspect ratio. The second technique, called parasitic-aware sizing, tries to reduce the number of iterations between the layout and the sizing phases by bringing the layout-induced parasitics in the sizing loop. Two approaches are described that differ in accuracy of the parasitics included and the CPU time required for parasitic estimation. Both the geometrically constrained and the parasitic-aware sizing techniques are illustrated with design examples.

Appendix A

Analog and Mixed-Signal Layout Rules

This appendix collects a number of rules, guidelines, and techniques for layout design of analog and mixed-signal circuits. These rules, regularly followed by the layout designers, can render the analog and mixed-signal layout more robust against unwanted layout-induced effects such as device mismatch, loading and coupling parasitics, reliability loss, and area waste.

Most of the ensuing layout rules can be found at many different papers and textbooks. Rules devised to improve device matching are provided in [Vitt85], [OLea91], [Malo94], and [Tsiv96]. The influence of different layout styles on MOS transistors matching is studied in [Bast96a]. A detailed analysis of the properties of common-centroid arrays can be found in [Hast01]. Resistor matching is treated in [Lane89]. Some specific rules to capacitor matching improvement are provided in [McNu94].

Basic loading issues are addressed in [Tsiv96] and [Malo94]. Stacking algorithms for minimization of diffusion parasitics of MOS transistors are reported in [Basa96], [Mala95], and [Naik99]. Area minimization of MOS transistor stacks is also treated in [Naik99]. Besides, complete expressions for parasitic and area optimization of MOS transistor stacks have been generated and reported in this appendix.

Several rules to enhance reliability of analog layouts are discussed in [Malo94], [Wolf99], [Lamp99], and [Hast01].

The following layout rules are organized in five categories, namely dev*ice matching*, *loading effects*, *coupling effects*, *reliability*, and *area occupation*. Some of these categories are, at the same time, organized with respect the concerned device primitive: MOS transistors, passive resistors, and passive capacitors[1].

[1] Layout rules for inductors are not considered here.

1 DEVICE MATCHING

The following rules employ the terms *minimal*, *moderate*, and *precise* matching to denote the following meanings:

- **Minimal device matching**: for passive resistors and capacitors, it involves approximately $\pm 1\%$ three-sigma mismatch $(3\sigma_\delta)$[2], or 6-7 bits[3] of resolution. Such matched devices are suitable for general-purpose analog circuitry, such as degenerating current mirrors for biasing. For MOS transistors, minimal matching typically corresponds to voltage offsets of $\pm 10\text{mV}$. Minimal matched MOS transistors are therefore used for constructing bias current networks that do not require any particular degree of precision.
- **Moderate device matching**: for passive resistors and capacitors, it involves approximately $\pm 0.1\%$ $3\sigma_\delta$, or 9-10 bits of resolution (suitable for bandgap references, opamps, comparator input stages, and most analog applications). For MOS transistors, moderate matching means typical $3\sigma_\delta$ offset voltages of $\pm 5\text{mV}$ or drain current mismatches of less than $\pm 1\%$. They are useful for input stages of non-critical opamps and comparators.
- **Precise device matching**: for passive resistors and capacitors, it involves approximately $\pm 0.01\%$ $3\sigma_\delta$, or 13-14 bits of resolution (best suited to precision AD and DA conversion and the rest of applications requiring extreme precision). For MOS transistors precise matching means typical $3\sigma_\delta$ offset voltages of $\pm 1\text{mV}$ or drain current mismatch of less than $\pm 0.1\%$.

In short, device matching rules for MOS transistors, resistors and capacitors are: (1) use same device structure; (2) place matched devices on isotherms; (3) use same shape and size; (4) use common-centroid geometries; (5) use same device orientation; (6) place same surroundings. The following part details the device matching rules for each type of device primitive.

---

2. The *three-sigma mismatch* equals the sum of the absolute value of the mismatch mean value m_δ (Eq(2) in Chapter 5) plus three times the mismatch standard deviation σ_δ (Eq(3) in Chapter 5). Three-sigma mismatch provides a confidence interval of 0.9973002 (i.e., less than 1% of all devices should have mismatches greater than the three-sigma value).

3. Since $\log_2 X = 3.32\log_{10} X$, each additional decimal digit is about $3\frac{1}{3}$ bits.

1.1 MOS transistors

A.1.1.1 *Use identical finger geometries.* Most matched transistors require relatively large widths and are usually divided or folded into sections, or *fingers* (see Fig. 1). Each of these fingers should have the same width and length as all others.

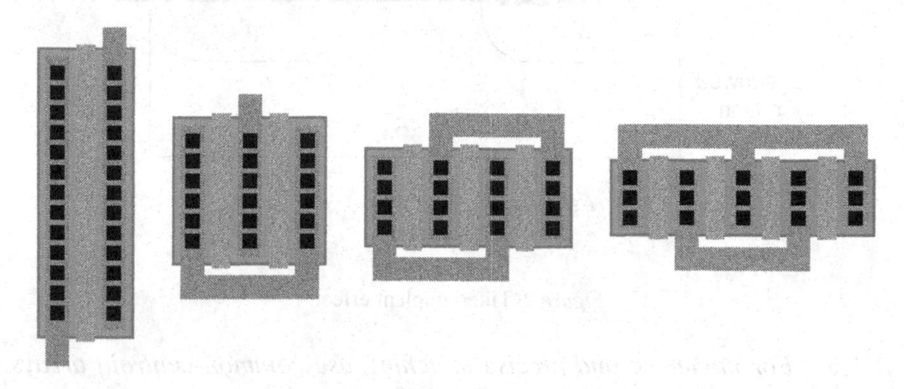

Figure 1. MOS transistor folding.

A.1.1.2 *Use large active areas* $W \times L$. Random fluctuations scale inversely with the square root of the transistor active area. Thus, moderate matching requires active areas of several hundreds square microns, while precise matching requires thousands.

A.1.1.3 *Orient transistors in the same direction.* Due to diagonal shifts in the source/drain implants (done to avoid channeling), the source/drain region on the left side of the MOST gate differs from the source/drain region on the right side, creating, thus, a small asymmetry illustrated in Fig. 2. This is a subtle effect called *gate shadowing* or *tilted implant* effect. The asymmetry caused may lead to a mismatch in the current factor of the transistors due to tilted-induced mobility variations. This mismatch can be reduced by placing transistors parallel to one another. Furthermore, matched transistors should have equal *chirality*[4]. That is, the fraction of right-oriented transistors minus the fraction of left-oriented transistors (chirality) must be equal in all matched transistors.

A.1.1.4 *Place transistors in close proximity.*

A.1.1.5 *Keep the layout of matched transistors as compact as possible.*

---

4. *Chirality* refers to the asymmetry of an object. This term is most commonly used in stereo chemistry.

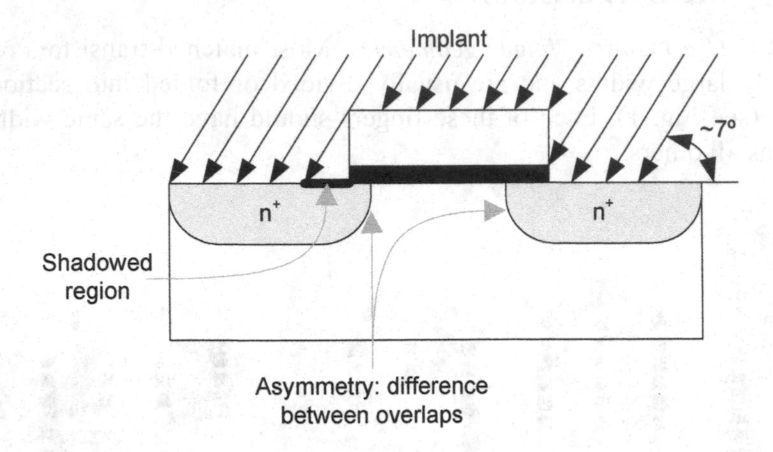

Figure 2. Tilted implant effect.

A.1.1.6 *For moderate and precise matching use common-centroid arrays.* These structures can entirely cancel the effects of long-range variations as long as these are linear functions of the distance[5]. The fingers of the folded transistors in a common-centroid array should fulfill the following rules[6]: *coincidence* (the centroids of the matched devices should exactly coincide), *symmetry* (the array should be symmetric with respect to both the horizontal and vertical axes), *dispersion* (the fingers should be distributed throughout the array as uniform as possible), and *compactness* (ideally, the array should be nearly square). Examples of common-centroid layouts are shown in Fig. 3[7].

A.1.1.7 *Place dummy segments on the ends of arrayed transistors.* Devices with different surroundings can show a considerable mismatch due to polysilicon etching rate variations. Arrayed transistors should include, to reduce such mismatch, dummy gates at either end of the array. These dummies need not to have the same length that the actual gates, but the spacing between actual and dummy gates should be the same as the spacing between actual gates. The dummy gates should be connected to a potential that prevents channel formation underneath, such as the ground potential.

A.1.1.8 *Connect the gate fingers of moderately and precisely matched transistors using metal straps.*

A.1.1.9 *If possible, place matched transistors in a low stress area.*

5. Even for nonlinear variations, they still remain approximately linear over short distances.

6. Plus the orientation rule (Rule A.1.1.3).

7. These examples are also valid for common-centroid layouts of resistors and capacitors.

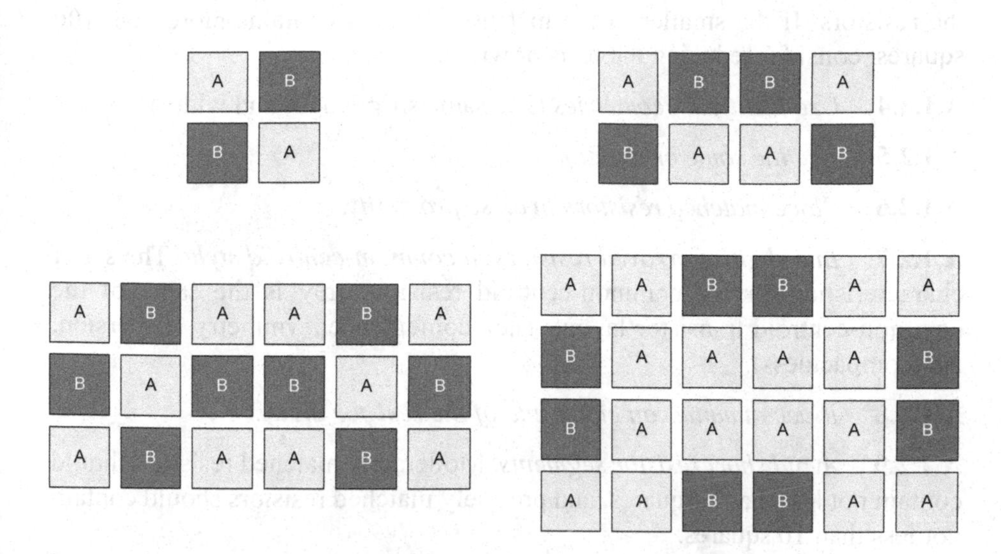

Figure 3. Several examples of common-centroid arrays.

A.1.1.10 *Place matched transistors far away from power devices*[8].

A.1.1.11 *Place precisely matched transistors on the symmetry axes of the die.*

1.2 Passive Resistors

Minimal matching between resistors can be obtained without much difficulty. Moderate matching is attained by using resistor interdigitation (that is, resistors are divided into sections or *strips*, and these are arranged to form a symmetric pattern along one dimension). Due to variations in contact resistance and the presence of thermal and stress gradients, precise matching is very difficult to obtain. The rules for matching improvement of passive resistors are:

A.1.2.1 *Do not construct matched resistors from different materials.*

A.1.2.2 *Make matched resistors the same width.*

A.1.2.3 *Make matched resistors wide enough.* Assume that minimal matching of resistors containing 30 or more squares requires 150% of the minimum allowed width, while moderate matching requires 200% and precise matching requires 400% of the minimum width. If the smallest of the matched resistors contains less than 20 to 30 squares, consider increasing the width of

[8] Any device dissipating more than 50mW should be considered a power device.

the resistors. If the smaller of the matched resistors contains more than 100 squares, consider reducing the resistor width.

A.1.2.4 *Use identical geometries* (i.e., same strip length and width).

A.1.2.5 *Use the same orientation.*

A.1.2.6 *Place matched resistors in close proximity.*

A.1.2.7 *Interdigitize arrayed resistors in common-centroid style*. The set of characteristics for the common-centroid resistor array is the same of the common-centroid transistor layouts (i.e., coincidence, symmetry, dispersion, and compactness).

A.1.2.8 *Place dummies on either end of the resistor array.*

A.1.2.9 *Avoid short resistor segments*. Moderately matched resistors should contain not less than 5 squares, and precisely matched resistors should contain not less than 10 squares.

A.1.2.10 *Connect matched resistors in order to cancel thermoelectric effects.*

A.1.2.11 *Use poly resistors in preference to diffused resistors.*

A.1.2.12 *Section resistors are superior to serpentine resistors.*

A.1.2.13 *If possible, place matched resistors in a low stress area.*

A.1.2.14 *Place matched resistors far away from power devices.*

A.1.2.15 *Place precisely matched resistors on axes of symmetry of the die.*

1.3 Passive Capacitors

The mismatch between capacitors (or between two devices) is usually expressed as a deviation of the measured capacitor ratio from the intended capacitor ratio. Matched capacitors become insensitive to systematic mismatch when their area-to-periphery ratios equal one another. For capacitors of the same value, this is achieved by using the same geometry for both capacitors. If the capacitors have values that are not in simple ratio, the layout designer should instead resort to arrays of capacitor segments called *unit capacitors*.

A.1.3.1 *Use identical geometries*. If the capacitors are not the same size, consider using a unit capacitor array. The larger capacitor should consist of multiple segments connected in parallel, while the smaller one should have fewer segments connected in parallel. If the intended capacitor ratio is not an integer number, use a non-unitary capacitor inserted into the larger of the matched capacitors. The aspect ratio should not exceed 1.5:1.

A.1.3.2 *Use square geometries for precisely matched capacitors.* The smaller the periphery-to-area ratio, the higher the matching. Therefore, since the square geometry has the lowest periphery-to-area ratio, it also features the highest matching level. Whereas moderately matched capacitors can be made of rectangular capacitors (with aspect ratios of 2:1 or 3:1), precisely matched capacitors always require square (1:1) capacitors.

A.1.3.3 *Make matched capacitors as large as practical* (because increasing the size reduces random mismatch). For several CMOS processes, reported optimum sizes lie between 20 × 20 μm and 50 × 50 μm.

A.1.3.4 *When practical, use common-centroid configurations.* Then, the set of rules of common-centroid arrays, namely coincidence, symmetry, dispersion, and compactness, still apply for capacitors.

A.1.3.5 *Place capacitors adjacent to one another.*

A.1.3.6 *Place dummy capacitors around the outer edge of the array.* The dummy capacitors need not to be the same size that the capacitors of the array as long as Rule A.1.3.7 below is fulfilled. Otherwise, as fringing fields extend 30 and 50μm from the array, so dummies should extend at least this far to ensure precise matching. Moderate matching requires a minimum-width ring of dummy capacitors, while minimal matching does not require dummy capacitors at all. The spacing between dummy and array capacitors should equal the spacing between array capacitors.

A.1.3.7 *Electrostatically shield precisely matched capacitors* (i.e., cover the entire array, including dummies, with a shield of grounded metal). This has four benefits: (1) it contains fringing fields to the capacitor array; (2) it allows routing over the capacitors (Rule A.1.3.9); (3) it prevents coupling; (4) it reduces the effect of packaging stress.

A.1.3.8 *Use two minimum-width wires connecting the top electrode of unit capacitors.* This is needed to consider the capacitance of wire connecting to the precisely and moderately matched capacitors. For non-unitary capacitors, the number of wires should be twice the ratio of its capacitance to the unit capacitance.

A.1.3.9 *Do not run leads over matched capacitors unless they are electrostatically shielded.*

A.1.3.10 *Connect the upper electrode of a matched capacitor to the higher-impedance node.* This is useful because the upper electrode generally exhibits less parasitic capacitance than the lower electrode.

A.1.3.11 *Place capacitors in areas of low stress gradients.*

A.1.3.12 *Place capacitors well away from power devices.*

A.1.3.13 *Place capacitors on axes of symmetry of the die.*

A.1.3.14 *Place matched capacitors over field oxide, far away from well and diffusion regions.*

2 LOADING EFFECTS

2.1 MOS transistors

The two *junction* capacitances of a MOS transistor may deteriorate the circuit operation due to their resulting loading effect. The value of these parasitic capacitances are expressed in the following equation

$$C_{XB} = \frac{A_X C'_{jo}}{\left[1+\left(\frac{V_{XB}}{\phi_1}\right)\right]^{\eta_1}} + \frac{P_X C^*_{jo}}{\left[1+\left(\frac{V_{XB}}{\phi_2}\right)\right]^{\eta_2}} = C^A_{XB} + C^P_{XB}, \tag{1}$$

where X refers to the drain or source node. C'_{jo} and C^*_{jo} are the bottom-wall junction zero-bias capacitance per unit of bottom-wall area and the side-wall junction zero-bias capacitance per unit of sidewall perimeter, respectively. ϕ_1 and ϕ_2 are the bottom-wall and sidewall junction built-in potential. η_1 and η_2 are the bottom-wall and sidewall junction characteristics exponents. A_X and P_X are the diffusion node (drain or source) bottom-wall area and sidewall perimeter, respectively. Then, C^A_{XB} is the fraction of the total junction capacitance due to the total diffusion node area, and C^P_{XB} is the fraction due to the total diffusion node perimeter.

For a single MOS transistor, these parasitics can be reduced by *folding* the transistor, that is, by dividing it into several smaller transistors (called *fingers*) whose diffusion areas overlap, as illustrated in Fig. 1 on page 349.

For a set of MOS transistors sharing one or more drain/source nodes, the *stacking* technique allows reducing the associated diffusion capacitance at the shared nodes. Stacking consists in folding and placing each MOS transistor such that the shared drain/source areas are merged.

For the sake of generality, the stacking technique is first revisited, as folding a MOS transistor is but stacking a set with only one component. The results obtained for the stacking technique can be easily adapted to derive those for the folding technique.

A.2.1.1 *MOS transistor stacking.* There are four basic steps for stacking a given circuit block with connected transistors:

1. Construct a *diffusion graph* $G(V, E)$. This is accomplished by mapping all the m nodes of the circuit block to the vertices, v_i ($i = 1, \ldots, m$), of a graph, and by connecting those vertices using the source-to-drain connection as edges, e_l ($l = 1, \ldots, k$). An example is shown in Fig. 4. It is necessary that all the transistors are of the same type (i.e., N- or P-type). Otherwise, they could not be interdigitized in the same stack. Note that the diffusion graph is also valid for the folding technique: the graph has only two vertices, representing the source and drain nodes, and one edge, representing the transistor gate.

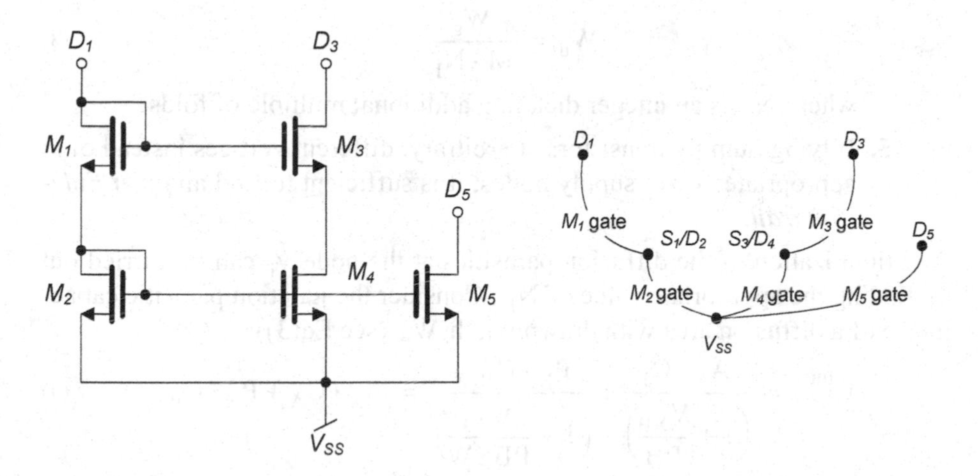

Figure 4. A circuit composed of NMOS transistors and its diffusion graph.

2. Compute the greatest common divisor (GCD), W_g, of all the transistor widths, to obtain the minimum number of folds of each transistor. W_g is the maximum width of the transistor fingers. The number of folds of each transistor, n_i, is obtained by dividing the width by the GCD, thus obtaining the vector of multiplicities $\mathbf{n}$:

$$\mathbf{n} = \frac{1}{W_g}\{W_1, \ldots, W_k\} = \{n_1, \ldots, n_k\} \tag{2}$$

Each of the k edges of the graph G is correspondingly fractured using the integer values in Eq. (2). The degree (odd or even) for each vertex is defined by the degree of the total number of edges connected to it.

3. Find an *Eulerian trail*[9] of the diffusion graph.
4. For common-centroid arrays, it is necessary that the Eulerian trail starts and ends at the same vertex. That is, it is necessary to find a *closed Eulerian trail.* It is possible to find such an Eulerian trail if and only if all elements in $\mathbf{n}$ are of even degree. This means that all vertices in the graph G must be of even degree too. Since $\mathbf{n}$ is obtained by dividing each of the widths by the GCD, the elements in $\mathbf{n}$ are coprime of each other; that is, at least one of the elements in $\mathbf{n}$ is odd. Therefore, it is necessary to fragment again the elements in $\mathbf{n}$ by an integer M such that $\mathbf{n_g} = M \cdot \mathbf{n}$. To avoid excessive fragmentation, M has to be minimum. Thus, it follows that $M = 1$ if each of the vertices in G is of even degree, and $M = 2$ otherwise. The actual width used for drawing the stack is:

$$W_d = \frac{W_g}{M \cdot N_F} \tag{3}$$

   where N_F is an integer dictating additional multiple of folds.
5. If tying dummy transistors at arbitrary, different vertices instead of at appropriate power supply nodes, it is sufficient to find an *open Eulerian trail.*

Minimization of the diffusion parasitics at the node v_i can be carried out by finding the appropriate value of N_F. Consider the junction parasitic capacitance of a diffusion area with drawn width W_d (see Eq(3)):

$$C_{XB}^{unit} = \frac{A_X \cdot C_J}{\left(1 + \frac{V_{XB}}{PB}\right)} + \frac{P_X \cdot C_{JSW}}{\left(1 + \frac{V_{XB}}{PBSW}\right)} = A_X \cdot C_A + P_X \cdot C_P \tag{4}$$

Now, it is necessary to draw a distinction: the diffusion area can be either external or internal to the transistor stack, as illustrated in Fig. 5. K_e and K_i are technological constants determined by the *contact-to-gate spacing*, the *exact contact size*, and the *diffusion enclosure of contact* design rules.

Accordingly, C_{XB}^{unit} can be redefined as

$$C_{XB_i}^{unit} = A_{X_i} \cdot C_A + P_{X_i} \cdot C_P \tag{5}$$

for internal diffusion areas, and as

$$C_{XB_e}^{unit} = A_{X_e} \cdot C_A + P_{X_e} \cdot C_P \tag{6}$$

---

[9] A *trail* in a graph $G(V, E)$ is a finite alternating sequence of vertices and edges, v_o, e_1, v_1, e_2, ..., v_{k-1}, e_k, v_k, with v_{i-1} and v_i being the end vertices of the edge e_i, such that all its edges are distinct. An *Eulerian trail* is a trail containing all the edges of G. When the end vertices are the same, it is called a *closed Eulerian trail*; otherwise, it is an *open Eulerian trail.*

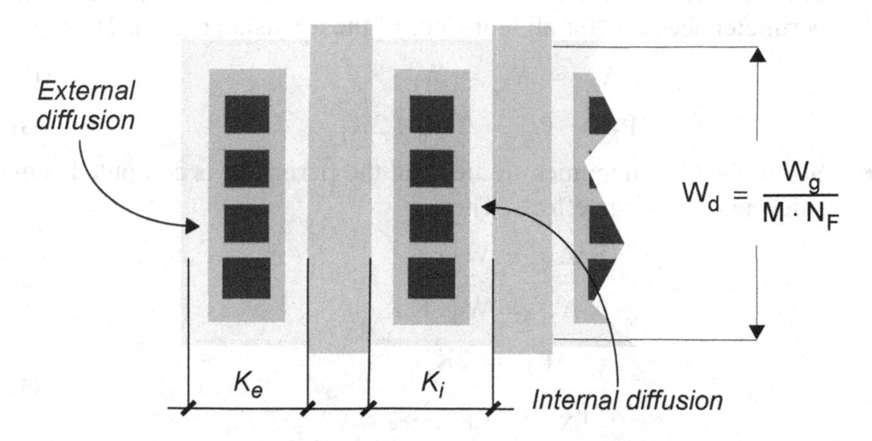

Figure 5. Internal and external diffusions of the MOS transistor.

for external diffusion areas.

The total parasitic capacitance at any vertex v_j can be computed as

$$C_{XB}^{v_j} = \left(\frac{D_{v_j} \cdot M \cdot N_F}{2} - 1\right) C_{XB_i}^{unit} + 2C_{XB_e}^{unit} \tag{7}$$

if v_j is the starting and ending vertex of the Eulerian trail; as

$$C_{XB}^{v_j} = \left(\frac{D_{v_j} \cdot M \cdot N_F}{2}\right) C_{XB_i}^{unit} \tag{8}$$

if v_j is an internal vertex of the Eulerian trail; or as

$$C_{XB}^{v_j} = \left(\frac{D_{v_j} \cdot M \cdot N_F}{2} - \frac{1}{2}\right) C_{XB_i}^{unit} + C_{XB_e}^{unit} \tag{9}$$

if v_j is either starting or ending vertex of the Eulerian trail. Note that for common-centroid arrays, the vertex v_j can only be either internal [Eq. (8)] or the starting and ending vertex [Eq. (7)]. Eq. (9) is needed for non-symmetrical stacks.

Substituting Eq. (5) and Eq. (6) into Eq. (7)-Eq. (9) allows to write $C_{XB}^{v_j}$ as a function of N_F. For the sake of completeness, computation of the area and perimeter of the unit capacitance in Eq. (5) and Eq. (6), is carried out at the four different scenarios:

- **Scenario 1:** there are dummies tied at both sides of the stack and the perimeter is computed considering only fringe effects[10]. Then,

$$A_{X_i} = A_{X_e} = W_d \cdot K_i \tag{10}$$

$$P_{X_i} = P_{X_e} = 2K_i \tag{11}$$

- **Scenario 2:** there are dummies tied at both sides of the stack and the perimeter accounts for all four sides of the diffusion region. Then,

$$A_{X_i} = A_{X_e} = W_d \cdot K_i \tag{12}$$

$$P_{X_i} = P_{X_e} = 2W_d + 2K_i \tag{13}$$

- **Scenario 3:** no dummies are tied and the perimeter is computed considering only fringe effects. Then,

$$\begin{aligned} A_{X_i} &= W_d \cdot K_i \\ A_{X_e} &= W_d \cdot K_e \end{aligned} \tag{14}$$

$$\begin{aligned} P_{X_i} &= 2K_i \\ P_{X_e} &= W_d + 2K_e \end{aligned} \tag{15}$$

- **Scenario 4:** no dummies are tied and the perimeter accounts for all four sides of the diffusion region. Then,

$$\begin{aligned} A_{X_i} &= W_d \cdot K_i \\ A_{X_e} &= W_d \cdot K_e \end{aligned} \tag{16}$$

$$\begin{aligned} P_{X_i} &= 2W_d + 2K_i \\ P_{X_e} &= 2W_d + 2K_e \end{aligned} \tag{17}$$

Table 1 to Table 4 show the resulting parasitic capacitance for the three types of vertex positions and the four scenarios considered. The tables also display the value of N_F for which the minimum value of the parasitic capacitance is obtained.

The circuit block in Fig. 6(a) illustrates an example of minimization of diffusion parasitics. With $W_{M_1} = W_{M_2} = 100\ \mu m$ and $W_{M_3} = W_{M_4} = 200\ \mu m$, suppose that the goal is to minimize the diffusion parasitics at node D_1. The GDC is $W_g = 100\ \mu m$, resulting the multiplicities $n_1 = n_2 = 1$ and $n_3 = n_4 = 2$. The initial diffusion graph is shown in Fig. 6(b), where vertices S, D_3, and D_4 are of even degree (2), and vertices D_1 and D_2 are of odd degree (3). Therefore, if the common-centroid style is used, $M = 2$ to have only vertices of even degree. The resulting diffusion graph is shown in Fig. 6(c). Otherwise, $M = 1$.

For a common-centroid stack, the vertex D_1 can be either internal or the ending and starting vertex of a closed Eulerian trail over the block's diffusion

[10] It is rather common that the diffusion perimeter includes all four sides, i.e., $2W + 2K$, where W is the transistor width and K is one of the technological constants, K_e or K_i, in Fig. 5. This is not strictly correct, as the internal side is not adjacent to the field-oxide area; including all four sides tends to overestimate these parasitics [Tsiv96].

Table 1. Parasitic capacitances for MOS transistor stacks; **scenario 1**: fringe effects and dummies.

Node	*Dependence*	$C_{XB}^{v_j}$	*Minimum*
Internal	$\left(\frac{D_{v_j} \cdot M \cdot N_F}{2}\right) C_{XB_i}^{unit}$	$C_A \cdot \left(\frac{D_{v_j} \cdot W_g \cdot K_i}{2}\right) + C_P \cdot (D_{v_j} \cdot M \cdot N_F \cdot K_i)$	$N_F = 1$
Ending AND starting	$\left(\frac{D_{v_j} \cdot M \cdot N_F}{2} - 1\right) C_{XB_i}^{unit} + 2C_{XB_e}^{unit}$	$C_A \cdot \left(\frac{D_{v_j} \cdot W_g \cdot K_i}{2} + \frac{W_g \cdot K_i}{M \cdot N_F}\right) + C_P \cdot (2K_i + D_{v_j} \cdot M \cdot N_F \cdot K_i)$	$N_F = \frac{1}{M}\sqrt{\frac{C_A \cdot W_g}{C_P \cdot D_{v_j}}}$
Ending OR starting	$\left(\frac{D_{v_j} \cdot M \cdot N_F}{2} - \frac{1}{2}\right) C_{XB_i}^{unit} + C_{XB_e}^{unit}$	$C_A \cdot \left(\frac{D_{v_j} \cdot W_g \cdot K_i}{2} + \frac{W_g \cdot K_i}{2 \cdot M \cdot N_F}\right) + C_P \cdot (K_i + D_{v_j} \cdot M \cdot N_F \cdot K_i)$	$N_F = \frac{1}{M}\sqrt{\frac{C_A \cdot W_g}{2C_P \cdot D_{v_j}}}$

Table 2. Parasitic capacitances for MOS transistor stacks; **scenario 2**: full perimeter and dummies.

Node	*Dependence*	$C_{XB}^{v_j}$	*Minimum*
Internal	$\left(\frac{D_{v_j} \cdot M \cdot N_F}{2}\right) C_{XB_i}^{unit}$	$C_A \cdot \left(\frac{D_{v_j} \cdot W_g \cdot K_i}{2}\right) + C_P \cdot (D_{v_j} \cdot M \cdot N_F \cdot K_i + D_{v_j} \cdot W_g)$	$N_F = 1$
Ending AND starting	$\left(\frac{D_{v_j} \cdot M \cdot N_F}{2} - 1\right) C_{XB_i}^{unit} + 2C_{XB_e}^{unit}$	$C_A \cdot \left(\frac{D_{v_j} \cdot W_g \cdot K_i}{2} + \frac{W_g \cdot K_i}{M \cdot N_F}\right) + C_P \cdot \left(2K_i + D_{v_j} \cdot M \cdot N_F \cdot K_i + D_{v_j} \cdot W_g + \frac{2W_g}{M \cdot N_F}\right)$	$N_F = \frac{1}{M}\sqrt{\frac{2C_P \cdot W_g + C_A \cdot W_g \cdot K_i}{C_P \cdot D_{v_j} \cdot K_i}}$
Ending OR starting	$\left(\frac{D_{v_j} \cdot M \cdot N_F}{2} - \frac{1}{2}\right) C_{XB_i}^{unit} + C_{XB_e}^{unit}$	$C_A \cdot \left(\frac{D_{v_j} \cdot W_g \cdot K_i}{2} + \frac{W_g \cdot K_i}{2 \cdot M \cdot N_F}\right) + C_P \cdot \left(K_i + D_{v_j} \cdot M \cdot N_F \cdot K_i + D_{v_j} \cdot W_g + \frac{W_g}{M \cdot N_F}\right)$	$N_F = \frac{1}{M}\sqrt{\frac{C_P \cdot W_g + (C_A \cdot W_g \cdot K_i)/2}{C_P \cdot D_{v_j} \cdot K_i}}$

Table 3. Parasitic capacitances for MOS transistor stacks; **scenario 3**: fringe effects and no dummies.

Node	*Dependence*	$C_{XB}^{v_j}$	*Minimum*
Internal	$\left(\frac{D_{v_j}\cdot M\cdot N_F}{2}\right)C_{XB_i}^{unit}$	$C_A\cdot\left(\frac{D_{v_j}\cdot W_g\cdot K_i}{2}\right)+C_P\cdot(D_{v_j}\cdot M\cdot N_F\cdot K_i)$	$N_F = 1$
Ending AND starting	$\left(\frac{D_{v_j}\cdot M\cdot N_F}{2}-1\right)C_{XB_i}^{unit}+2C_{XB_e}^{unit}$	$C_A\cdot\left[\frac{D_{v_j}\cdot W_g\cdot K_i}{2}+\frac{W_g}{M\cdot N_F}\cdot(2K_e-K_i)\right]+$ $C_P\cdot\left[D_{v_j}\cdot M\cdot N_F\cdot K_i+\frac{2W_g}{M\cdot N_F}+4K_e-2K_i\right]$	$N_F = \frac{1}{M}\sqrt{\frac{2C_P\cdot W_g+C_A\cdot W_g\cdot(2K_e-K_i)}{C_P\cdot D_{v_j}\cdot K_i}}$
Ending OR starting	$\left(\frac{D_{v_j}\cdot M\cdot N_F}{2}-\frac{1}{2}\right)C_{XB_i}^{unit}+C_{XB_e}^{unit}$	$C_A\cdot\left[\frac{D_{v_j}\cdot W_g\cdot K_i}{2}+\frac{W_g}{2\cdot M\cdot N_F}\cdot(2K_e-K_i)\right]+$ $C_P\cdot\left[D_{v_j}\cdot M\cdot N_F\cdot K_i+\frac{W_g}{M\cdot N_F}+2K_e-K_i\right]$	$N_F = \frac{1}{M}\sqrt{\frac{C_P\cdot W_g+(C_A\cdot W_g\cdot(2K_e-K_i))/2}{C_P\cdot D_{v_j}\cdot K_i}}$

Table 4. Parasitic capacitances for MOS transistor stacks; **scenario 4**: full perimeter and no dummies.

Node	*Dependence*	$C_{XB}^{v_j}$	*Minimum*
Internal	$\left(\frac{D_{v_j}\cdot M\cdot N_F}{2}\right)C_{XB_i}^{unit}$	$C_A\cdot\left(\frac{D_{v_j}\cdot W_g\cdot K_i}{2}\right)+C_P\cdot(D_{v_j}\cdot M\cdot N_F\cdot K_i+D_{v_j}\cdot W_g)$	$N_F = 1$
Ending AND starting	$\left(\frac{D_{v_j}\cdot M\cdot N_F}{2}-1\right)C_{XB_i}^{unit}+2C_{XB_e}^{unit}$	$C_A\cdot\left[\frac{D_{v_j}\cdot W_g\cdot K_i}{2}+\frac{W_g}{M\cdot N_F}\cdot(2K_e-K_i)\right]+$ $C_P\cdot\left[D_{v_j}\cdot M\cdot N_F\cdot K_i+\frac{2W_g}{M\cdot N_F}+4K_e-2K_i+D_{v_j}\cdot W_g\right]$	$N_F = \frac{1}{M}\sqrt{\frac{2C_P\cdot W_g+C_A\cdot W_g\cdot(2K_e-K_i)}{C_P\cdot D_{v_j}\cdot K_i}}$
Ending OR starting	$\left(\frac{D_{v_j}\cdot M\cdot N_F}{2}-\frac{1}{2}\right)C_{XB_i}^{unit}+C_{XB_e}^{unit}$	$C_A\cdot\left[\frac{D_{v_j}\cdot W_g\cdot K_i}{2}+\frac{W_g}{2\cdot M\cdot N_F}\cdot(2K_e-K_i)\right]+$ $C_P\cdot\left[D_{v_j}\cdot M\cdot N_F\cdot K_i+\frac{W_g}{M\cdot N_F}+2K_e-K_i+D_{v_j}\cdot W_g\right]$	$N_F = \frac{1}{M}\sqrt{\frac{C_P\cdot W_g+(C_A\cdot W_g\cdot(2K_e-K_i))/2}{C_P\cdot D_{v_j}\cdot K_i}}$

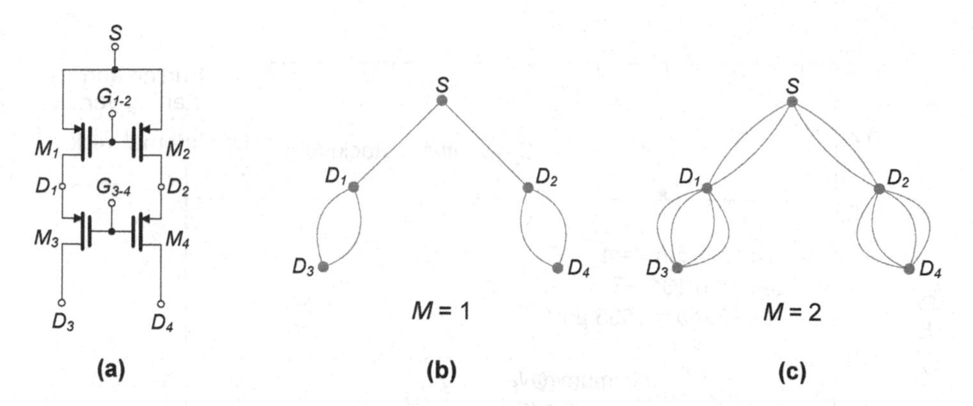

Figure 6. Circuit example illustrating parasitic minimization.

graph G in Fig. 6(c). If a dummy transistor is attached at both ends of the stack, and considering only fringe effects for perimeter calculation, the expressions in Table 1 provide the value of N_F for which a minimum capacitance (i.e., a maximum reduction from the node diffusion capacitance before stacking) is obtained. The results are shown in Fig. 7. A maximum reduction is achieved if the vertex is made internal at the cost of a larger stack area, computed with the following equation:

$$A = K_i \cdot \left(1 + M \cdot N_F \cdot \sum_{l=1}^{k} n_l\right) \cdot W_d + 2 \cdot K_e \cdot W_d + \sum_{l=1}^{k} W_l \cdot L_l + 2 \cdot W_d \cdot L_l \tag{18}$$

with W_l and L_l ($l = 1, \ldots, 4$) being the full width and length of transistors M_1-M_4.

A.2.1.2 *MOS transistor folding*. The previous results can be easily adapted to the case of a folded MOS transistor by just making $M = N_F = 1$, $W_g = W/m$, and $D_{v_j} \equiv m$, where W is the width of the unfolded transistor and m is the number of fingers, called *multiplicity.* The folded transistor is obtained by just splitting the transistor into m unit transistors with width equal to W/m and merging the corresponding drain/source areas. Note that,

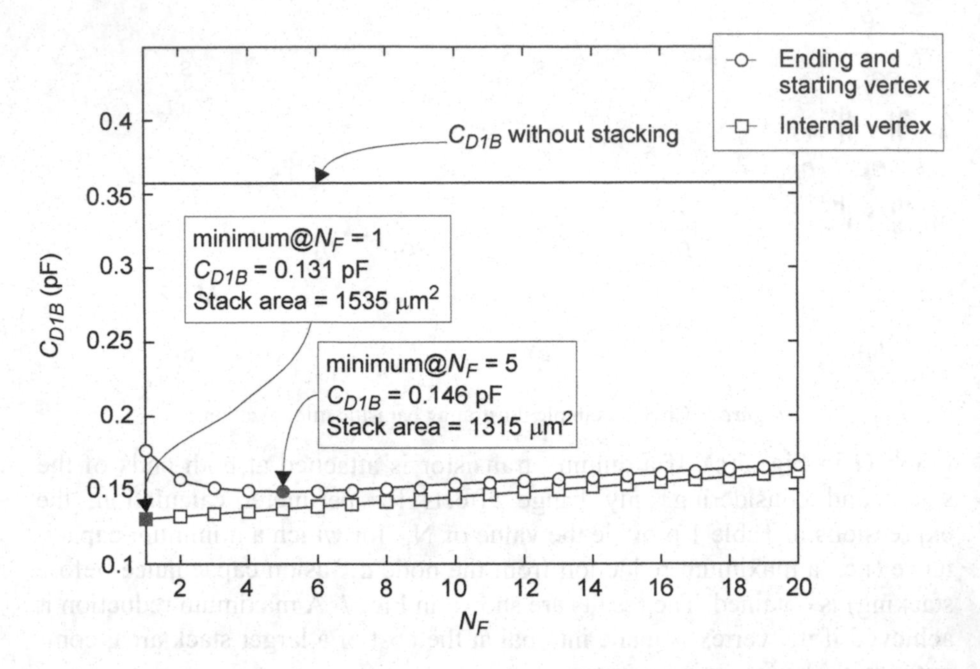

Figure 7. Plots of the diffusion capacitance of node D1 versus the number of folds N_F of the circuit in Fig.6(a), with $W_g = 100\ \mu m$ and M = 2.

as the diffusion graph has only two vertices and one edge, the three types of vertices are now mapped into these three types of diffusion nodes:

- m is of even degree and the node has only internal diffusion areas. Therefore, the exact number of diffusion areas is $m/2$.
- m is of even degree and the node has internal and external diffusion areas. The exact number of diffusion areas is $(m+2)/2$.
- m is of odd degree (consequently the node has always internal and external diffusion areas). The exact number of diffusion areas in this case is $(m+1)/2$.

Since no dummies are attached, only the expressions of the diffusion node capacitance for scenarios 3 and 4 in Table 3 and Table 4 are necessary to obtain equivalent expressions for the folded transistor. In this case, the diffusion node capacitance can be easily derived as a function of C^{unf}, the diffusion capacitance of the unfolded transistor

$$C^{unf} = C_A^{unf} + C_P^{unf} = C_A \cdot W \cdot K_e + C_P \cdot (W + 2K_e) \tag{19}$$

if only fringe effects are considered for computation of the perimeter, or

$$C^{unf} = C_A^{unf} + C_P^{unf} = C_A \cdot W \cdot K_e + C_P \cdot (2W + 2K_e) \tag{20}$$

if all four sides of the diffusion are taken into account. The expressions are shown in Table 5 and Table 6. Table 7 and Table 8 display the same expressions for $K_e = K_i = K$.

A.2.1.3 The width of each finger must be chosen such that the resistance of the finger is less than the inverse transconductance g_m associated with the finger. In low-noise applications, the gate resistance must be one-fifth to one-tenth of $1/g_m$.

3 COUPLING EFFECTS

3.1 Substrate coupling

Modern CMOS technologies uses heavily-doped p^+ substrates[11] to minimize latch-up susceptibility. Unfortunately, its low resistivity produces unwanted paths between various devices in the circuit, corrupting the sensitive signals. This effect is known as *substrate coupling.* Although less severe, this effect is also present at lightly-doped substrates. At the layout level, there are two ways of minimizing the effect of noise injection at lightly-doped substrates[12]

A.3.1.1 *Increase the physical separation between the noise injectors and the sensitive receivers.* If lightly doped, the susbtrate operate as a high-impedance resistance plane. The isolation between noisy and sensitive circuits improves as the physical separation is increased. The only inconvenience is that, most times, the analog and digital functions are so heavily blended that it is unfeasible or very difficult to separate their corresponding circuits. Heavily-doped substrates operate, on the contrary, as a low-impedance resistive plane, distributing a relatively uniform potential across the chip regardless of the position of the noise generators.

A.3.1.2 *In lightly doped substrates, use guard-rings to isolate the noise injectors from the sensitive receivers.* A guard-ring may be simply a continuous ring made of substrate connections that surrounds the sensitive circuits providing a low-impedance path to ground for the charge carriers produced in the substrate. If the guard-ring is also biased using dedicated

---

11. These substrates have a resistivity of the order of 0.1 Ω·cm.

12. There are also several design methods to minimize substrate coupling in heavily-doped substrates (for instance, see [Raza01] page 662)

Table 5. Parasitic capacitances for MOS transistor folding; **scenario 1**.

Node	C_{XB}^{folded}	*Minimum*
Even & internal	$C_A^{unf}\cdot\frac{K_i}{2K_e}+C_P^{unf}\cdot m\cdot\frac{K_i}{W+2K_e}$	$m = 2$
Even & external	$C_A^{unf}\cdot\left(\frac{K_i}{2K_e}+\frac{2K_e-K_i}{m\cdot K_e}\right)+C_P^{unf}\cdot\left(\frac{m\cdot K_i+\frac{2W}{m}+4K_e-2K_i}{W+2K_e}\right)$	$m = \sqrt{\frac{2C_P^{unf}\cdot W\cdot K_e+C_A^{unf}\cdot(W+2K_e)\cdot(2K_e-K_i)}{C_P^{unf}\cdot K_i\cdot K_e}}$
Odd	$C_{XB}^{v_i} = C_A^{unf}\cdot\left(\frac{K_i}{2K_e}+\frac{2K_e-K_i}{2m\cdot K_e}\right)+C_P^{unf}\cdot\left(\frac{m\cdot K_i+\frac{W}{m}+2K_e-K_i}{W+2K_e}\right)$	$m = \sqrt{\frac{2C_P^{unf}\cdot W\cdot K_e+C_A^{unf}\cdot(W+2K_e)\cdot(2K_e-K_i)}{2C_P^{unf}\cdot K_i\cdot K_e}}$

Table 6. Parasitic capacitances for MOS transistor foldings; **scenario 2**.

Node	C_{XB}^{folded}	*Minimum*
Even & internal	$C_A^{unf}\cdot\frac{K_i}{2K_e}+C_P^{unf}\cdot\frac{W+m\cdot K_i}{2W+2K_e}$	$m = 2$
Even & external	$C_A^{unf}\cdot\left(\frac{K_i}{2K_e}+\frac{2K_e-K_i}{m\cdot K_e}\right)+C_P^{unf}\cdot\left(\frac{m\cdot K_i+\frac{2W}{m}+W+4K_e-2K_i}{2W+2K_e}\right)$	$m = \sqrt{\frac{2C_P^{unf}\cdot W\cdot K_e+C_A^{unf}\cdot(2W+2K_e)\cdot(2K_e-K_i)}{C_P^{unf}\cdot K_i\cdot K_e}}$
Odd	$C_A^{unf}\cdot\left(\frac{K_i}{2K_e}+\frac{2K_e-K_i}{2m\cdot K_e}\right)+C_P^{unf}\cdot\left(\frac{m\cdot K_i+\frac{W}{m}+W+2K_e-K_i}{2W+2K_e}\right)$	$m = \sqrt{\frac{2C_P^{unf}\cdot W\cdot K_e+C_A^{unf}\cdot(2W+2K_e)\cdot(2K_e-K_i)}{2C_P^{unf}\cdot K_i\cdot K_e}}$

Table 7. Parasitic capacitances for MOS transistor folding; **scenario 1** with and $K_e=K_i=K$.

Node	C_{XB}^{folded}	*Minimum*
Even & internal	$C_A^{unf}\cdot\frac{1}{2}+C_P^{unf}\cdot m\cdot\frac{K}{W+2K}$	$m = 2$
Even & external	$C_A^{unf}\cdot\left(\frac{m+2}{2m}\right)+C_P^{unf}\cdot\left(\frac{m\cdot K+\frac{2W}{m}+2K}{W+2K}\right)$	$m = \sqrt{\frac{2C_P^{unf}\cdot W+C_A^{unf}\cdot(W+2K)}{C_P^{unf}\cdot K}}$
Odd	$C_A^{unf}\cdot\left(\frac{m+1}{2m}\right)+C_P^{unf}\cdot\left(\frac{m\cdot K+\frac{W}{m}+K}{W+2K}\right)$	$m = \sqrt{\frac{2C_P^{unf}\cdot W+C_A^{unf}\cdot(2W+2K)}{2C_P^{unf}\cdot K}}$

Table 8. Full perimeter (scenario 2) and $K_e=K_i=K$.

Node	C_{XB}^{folded}	*Minimum*
Even & internal	$C_A^{unf}\cdot\frac{1}{2}+C_P^{unf}\cdot\frac{W+m\cdot K}{2W+2K}$	$m = 2$
Even & external	$C_A^{unf}\cdot\left(\frac{m+2}{2m}\right)+C_P^{unf}\cdot\left(\frac{m\cdot K+\frac{2W}{m}+W+2K}{2W+2K}\right)$	$m = \sqrt{\frac{2C_P^{unf}\cdot W+C_A^{unf}\cdot(2W+2K)}{C_P^{unf}\cdot K}}$
Odd	$C_A^{unf}\cdot\left(\frac{m+1}{2m}\right)+C_P^{unf}\cdot\left(\frac{m\cdot K+\frac{W}{m}+W+K}{2W+2K}\right)$	$m = \sqrt{\frac{2C_P^{unf}\cdot W+C_A^{unf}\cdot(2W+2K)}{2C_P^{unf}\cdot K}}$

package pins, it has the effect of creating a zero potential around the sensitive circuit, hence isolating it from the noisy sources.

3.2 Coupling between routing wires

A.3.2.1 *Shielding*. Crosstalk between a noisy and a sensitive interconnect wire can be reduced through shielding techniques. One first approach [Fig. 8(a)] consists in placing grounded wires on the two sides of a sensitive wire, forcing most of the electrical field lines emanating from the "noisy" wires to terminate on ground rather than on the sensitive signal. This method is more effective than simply allowing more space between the noisy and the sensitive wires [Fig. 8(b)]. The shielding, however, is obtained at the cost of more complex routing and greater coupling capacitance between the signals and ground. Another shielding technique, illustrated in Fig. 9, consists in surrounding the sensitive wire by grounded higher and lower metal layers. However, the signal experiences higher capacitance to ground and the routing results, also, more complicated.

4 RELIABILITY

Circuit reliability is based on the mean time to failure of a sample of integrated circuits under a specified set of worst-case environment conditions.

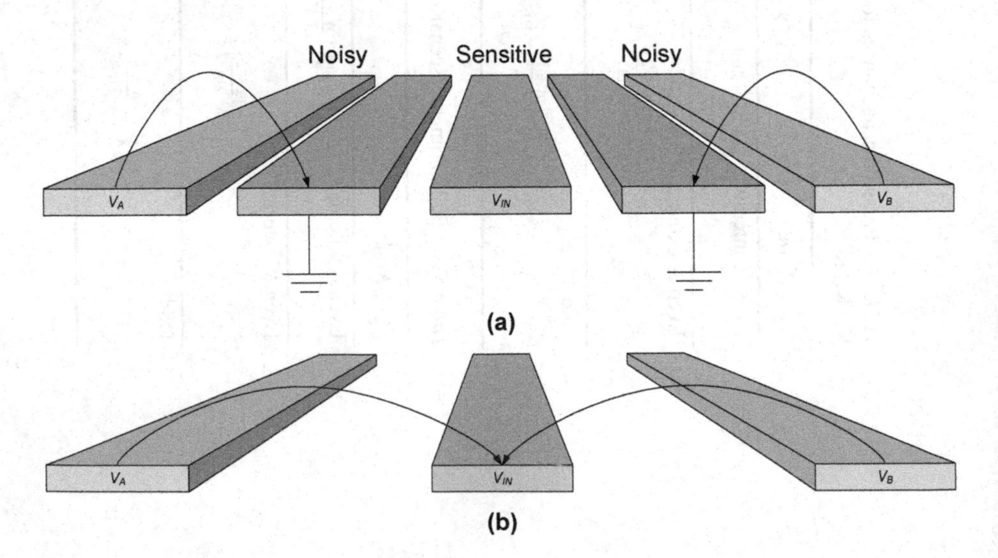

Figure 8. Shielding by additional grounded wires.

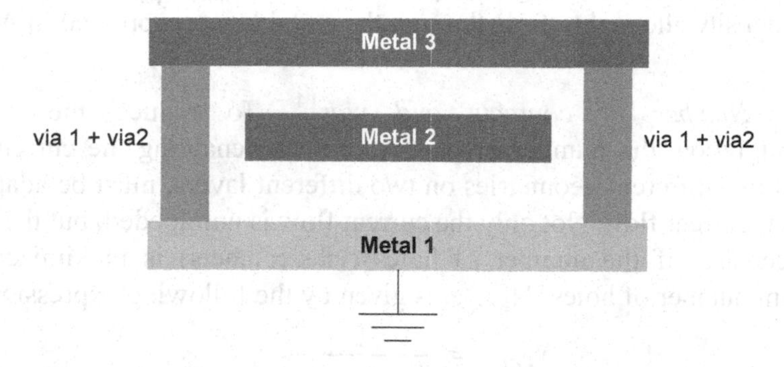

Figure 9. Shielding by lower and upper ground planes.

The following rules are intended to increase the reliability of any integrated circuit.

4.1 MOS transistors

A.4.1.1 *Number of diffusion-metal contacts*. In some layout styles, only one contact is placed over the entire source or drain area to contact this diffusion with metal. Instead, placing several contacts at the minimum spacing allowed by the process design rules is usually preferred because it leads to a reduced curvature of the metal surface, thus reducing the risk of micro-fractures (potential sources of failure) in the body of the metal connections.

4.2 Passive Resistors

A.4.2.1 *Number of termination contacts*. For the reason explained in Rule A.4.1.1, it is very important to carefully lay out the ends of passive resistor strips, placing as many contacts as possible. If the current is disturbed from its laminar flow, a localized resistance at the end points can result, which can be as high as one square of material.

4.3 Routing

A.4.3.1 *Wire width.* To increase circuit's reliability, the width of the metal wire must be adapted to accommodate the current flowing through the wire, thus preventing electromigration[13]. Thus, the right width must be:

$$W = \frac{I}{I_{max,layer}} \tag{21}$$

[13] Electromigration is caused by the impact of moving carriers with stationary metal atoms, leading to a gradual displacement of the metal, until the wire is ultimately severed.

where I is the current flowing through the wire and $I_{max,layer}$ is the maximum current density allowed to flow through the metal layer, expressed in Ampere per meter.

A.4.3.2 *Number of contacts and vias*[14]. To reduce the risk of electromigration, the number of vias or contacts, enabling the current flow between two different geometrics on two different layers, must be adapted to the actual current flow. Not only the current flow is unimpeded, but the ohmic drop decreases if the number of holes (vias/contacts) is maximized. The minimum number of holes, N_{holes}, is given by the following expression:

$$N_{holes} = \frac{I}{I_{max,hole\ type}} \tag{22}$$

with I being the current flowing though the hole array and $I_{max,hole\ type}$ being the maximum allowed current in a hole connecting the two layers, usually expressed as Ampere per hole.

5 AREA OCCUPATION

5.1 MOS transistors

A.5.1.1 *Folding*. For transistors with large width, the total occupied area can be minimized by folding it into m fingers and merging the diffusion areas. If the area of the unfolded transistor is

$$A = W \cdot L + 2W \cdot K_e \tag{23}$$

the resulting area of the folded transistor is

$$A_{folded} = W \cdot L + \frac{2W \cdot K_e}{m} + \frac{W \cdot K_i \cdot (m-1)}{m}, \tag{24}$$

where W and L are the transistor width and length, respectively, and K_e, K_i have been explained in Section 2. Eq. (24) equals Eq. (23) for m =1 and it monotonically tends towards $W \cdot L + W \cdot K_i$, which, as long as $K_i < 2K_e$, is smaller than the area of the unfolded transistor.

A.5.1.2 *Stacking*. Area occupation can also be optimized through adequate selection of the number of the additional folding parameter N_F [15]. In this way, the area occupied by a set of stacked transistors is

---

14. A contact hole is usually referred as the mask layer connecting the diffusion and metal mask layers; a via hole is used to connect different levels of metal layers.

15. See Rule A.2.1.1.

$$A = K_i \cdot \left(1 + M \cdot N_F \cdot \sum_{l=1}^{k} n_l\right) \cdot \frac{W_g}{M \cdot N_F} + 2 \cdot K_e \cdot \frac{W_g}{M \cdot N_F} + \sum_{l=1}^{k} W_l \cdot L_l + 2 \cdot \frac{W_g}{M \cdot N_F} \cdot L_l, \tag{25}$$

if dummy transistors are placed at both ends of the stack, or

$$A = K_i \cdot \left(M \cdot N_F \cdot \sum_{l=1}^{k} n_l - 1\right) \cdot \frac{W_g}{M \cdot N_F} + 2 \cdot K_e \cdot \frac{W_g}{M \cdot N_F} + \sum_{l=1}^{k} W_l \cdot L_l \tag{26}$$

otherwise.

5.2 Passive resistors

A.5.2.1 *Resistor area.* The area that a passive resistor occupies can be controlled by modifying the strip width W and length L to obtain the same number of squares of resistive material. The total resistance of a resistive strip is given by the following equation:

$$R = R_{\square} \cdot \frac{L}{(W - DW)} \tag{27}$$

where $R_{\square}$ is the sheet resistance in ohms per square of resistive material ($\Omega/\square$), and DW is the width reduction (such that the effective width is $W_{eff} = W - DW$).

For a total resistance R to occupy a given area $A_R = W \cdot L$, L must be:

$$L = \frac{-DW + \sqrt{DW^2 + 4 \cdot \frac{A_R \cdot R_{\square}}{R}}}{2 \cdot \frac{R_{\square}}{R}} \tag{28}$$[16]

W can be obtained by using the following equation:

$$W = \frac{R_{\square} \cdot L}{R} + DW \tag{29}$$

A.5.2.2 *Resistor aspect ratio.* Folding the resistor into m unitary strips can be used to obtain a different aspect ratio. Given W, L, the strip separation S,

[16] As long as $L > L_{min}$ where L_{min} is the process-minimum length, Eq. (28) is valid as long as $A_R > \frac{R_{\square}}{R} L_{min}^2 + L_{min} DW$.

and a desired aspect ratio A_{ratio}, m can be obtained as the nearest positive integer number to:

$$\frac{-(S \cdot A_{ratio}) \pm \sqrt{S^2 \cdot A_{ratio}^2 + 4 \cdot L \cdot A_{ratio} \cdot (S+W)}}{-2A_{ratio} \cdot (S+W)} \quad (30)$$

5.3 Passive capacitors

Capacitor area. The capacitance of a rectangular capacitor is defined as $C = C_A \cdot X \cdot Y + 2 \cdot C_P \cdot (X+Y)$, where C_A is the capacitance per unit of area, C_P is the capacitance per unit of perimeter of the capacitive structure, and X, Y are the two sides of the rectangular capacitor. For a specified area, A_C, X can be obtained by using the following equation

$$X = \frac{C - 2C_A \pm \sqrt{C^2 + A_C^2 \cdot C_A^2 - 2A_C \cdot C_A - 8A_C \cdot C_P^2}}{4C_P} \quad (31)$$

Providing that X fulfills the following inequality:

$$X_{min} \leq X \leq \frac{C - 2C_P \cdot X_{min}}{C_A \cdot X_{min} + 2C_P} \text{ [17]} \quad (32)$$

where X_{min} is the process minimum size of the capacitor, Y is then given by:

$$Y = \frac{C - 2C_P \cdot X}{C_A \cdot X + 2C_P} \quad (33)$$

A.5.3.1 *Capacitor aspect ratio.* An intended aspect ratio A_{ratio} can be achieved by assigning X the value:

$$X = \frac{-C_P \cdot (A_{ratio} + 1) + \sqrt{C_P^2 \cdot (A_{ratio} + 1)^2 + C \cdot C_A \cdot A_{ratio}}}{A_{ratio} \cdot C_A} \quad (34)$$

Provided that X fulfills Eq. (32), Y can be obtained through Eq. (33).

[17] The upper bound is to ensure that $Y \leq X_{min}$.

References

[Aaron56] M. Aaron, "The use of least squares in system design," *IRE Transactions on Circuit Theory*, vol. 3, no. 4, pp. 224-231, December 1956.

[Abid99] *Integrated circuits for wireless communications*, A. Abidi, P. R. Gray, and R. G. Meyer, Eds. New York: IEEE Press, 1999.

[Agar04] A. Agarwal, H. Sampath, V. Yelamanchili, and R. Vemuri, "Accurate estimation of parasitic capacitances in analog circuits," in Proc. of Design, Automation and Test in Europe Conference and Exhibition, 2004. Proceedings, pp. 1364-1365, 2004.

[Ahuja82] B. K. Ahuja, "Implementation of active distributed RC anti-aliasing/smoothing filters," *IEEE J. Solid-State Circuits*, vol. 17, no. 6, pp. 1076-1080, December 1982.

[Alva88] A. R. Alvarez, B. L. Abdi, D. L. Young, H. D. Weed, J. Teplik, and E. R. Herald, "Application of statistical design and response surface methods to computer-aided VLSI device design," *IEEE Trans. Computer-Aided Design*, vol. 7, no. 2, pp. 272-288, February 1988.

[Apan98] Z. V. Apanovich and A. G. Marchuk, "Top-down approach to technology migration for full-custom mask layouts," in *Proc. of Int. Conf. on VLSI Design*, 1998, pp. 48-52.

[Arno00] G. Arnout, "SystemC standard," in *Proc. of Asia and South Pacific Design Automation Conf.*, 2000, pp. 573-577.

[Arora96] N. D. Arora, K. V. Raol, R. Schumann, and L. M. Richardson,

"Modeling and extraction of interconnect capacitances for multilayer VLSI circuits," *IEEE Trans. Computer-Aided Design*, vol. 15, no. 1, pp. 58-67, 1996.

[Bäck93] T. Bäck and H.-P. Schwefel, "An overview of evolutionary algorithms for parameter optimization," *Evolutionary Computation*, vol. 1, no. 1, pp. 1-23, MIT Press, Spring 1993.

[Bagg92] B. Baggini, G. Coppero, G. Gazzoli, L. Sforzini, F. Maloberti, and G. Palmisano, "An integrated circuit for GSM mobile communications," *Analog Integrated Circuits and Signal Processing*, vol. 2, no. 3, pp. 197-206, September 1992.

[Barke88] E. Barke, "Line-to-ground capacitance calculation for VLSI: a comparison," *IEEE Trans. Computer-Aided Design*, vol. 7, no. 2, pp. 295-298, February 1988.

[Barn90] T. J. Barnes, "SKILL: a CAD system extension language," in *Proc. of ACM/IEEE Design Automation Conf.*, 1990, pp. 266-271.

[Basa93] B. Basaran, R. A. Rutenbar, and L. R. Carley, "Latchup-aware placement and parasitic-bounded routing of custom analog cells," in *Proc. of ACM/IEEE Int. Conf. on Computer-Aided Design*, 1993, pp. 415-421-

[Basa96] B. Basaran and R. A. Rutenbar, "An O(n) algorithm for transistor stacking with performance constraints," in *Proc. of ACM/IEEE Design Automation Conf.*, 1996, pp. 221-226.

[Bast96a] J. Bastos, M. Steyaert, B. Graindourze, and W. Sansen, "Matching of MOS transistors with different layout styles," in *Proc. of IEEE Int. Conf. on Microelectronic Test Structures*, 1996, pp. 17-18.

[Bast96b] J. Bastos, *Matching characterization for precision analog design*. Ph.D. Thesis, Katholieke Universiteit Leuven, 1996.

[Been93] G. Beenker, J. Conway, G. G. Schrooten, and A. Slenter, "Analog CAD for consumer ICs," in *Analog Circuit Design*, J. Huijsing, R. van der Plassche, and W. Sansen, Eds. Dordrecht: Kluwer Academic Publishers, 1993.

[Bext93] V. M. zu Bexten, C. Moraga, R. Klinke, W. Brockherde, and K.-G. Hess, "ALSYN: flexible rule-based layout synthesis for analog IC's," *IEEE J. Solid-State Circuits*, vol. 28, no. 3, pp. 261-268, March 1993.

[Bhat04] S. Bhattacharya, N. Jangkrajarng, R. Hartono, and C. J. R. Shi, "Correct-by-construction layout-centric retargeting of large analog designs," in *Proc. of ACM/IEEE Design Automation Conf.*, 2004, pp. 139-144.

[Borel99] J. Borel, "Design automation in MEDEA: present and future," *IEEE Micro*, vol. 19, no. 5, pp. 71-79, September-October 1999.

[Bruce96] J. D. Bruce, H. W. Li, M. J. Dallabetta, and R. J. Baker, "Analog layout using ALAS!," *IEEE J. Solid-State Circuits*, vol. 31, no. 2, pp. 271-274, February 1996.

[Cade05] *Cadence® analog design environment user guide. Product version 5.1.41*, Cadence Design Systems Inc., 2005.

[Candy92] J. C. Candy and G. C. Temes, "Oversampling methods for A/D and D/A conversion," in *Oversampling ΣΔ converters*. New York: IEEE Press, 1992, pp. 1-25.

[Cast00] R. Castro-López, M. Delgado-Restituto, F. V. Fernández, and A. Rodríguez-Vázquez, "Reusability methodology for mixed-signal IC designs at layout and schematic levels," in *Proc. of Design of Circuits and Integrated Systems Conf.*, 2000, pp. 492-497.

[Cast01] R. Castro-López, F. V. Fernández, M. Delgado-Restituto, and A. Rodríguez-Vázquez, "Retargeting of mixed-signal blocks for SoCs," in *Proc. of Design, Automation, and Test in Europe Conf.*, 2001, pp. 772-773.

[Cast02a] R. Castro-López, F. V. Fernández, M. Delgado-Restituto, F. Medeiro, and A. Rodríguez-Vázquez, "Generation of technology-portable flexible analog blocks," in *Proc. of IEEE Int. Symp. on Circuits and Systems*, 2002, vol. 2, pp. II-61-II-64.

[Cast02b] R. Castro-López, F. V. Fernández, F. Medeiro, and A. Rodríguez-Vázquez, "Generation of technology-independent retargetable analog blocks," *Analog Integrated Circuits and Signal Processing*, vol. 33, no. 2, pp. 157-70, Kluwer Academic Publishers, November 2002.

[Cast03] R. Castro-López, J. Ruíz-Amaya, R. Romay, J. M. de la Rosa, R. del Río, F. Medeiro, F. V. Fernández, B. Pérez-Verdú, and A. Rodríguez-Vázquez, "Description languages and tools for the behavioural simulation of sigma-delta modulators: a comparative survey," in *Proc. of Forum on Specification & Design Languages*, 2003, pp. 121-132.

[Cast04] R. Castro-López, O. Guerra, F. Medeiro, and A. Rodríguez-

Vázquez, "Synthesis of a wireless communication analog back-end based on a mismatch-aware symbolic analysis," *Analog Integrated Circuits and Signal Processing*, vol. 40, no. 3, pp. 215-233, September 2004.

[Chang97] H. Chang, E. Charbon, U. Choudhury, A. Demir, E. Felt, E. Liu, E. Malavasi, A. Sangiovanni-Vincentelli, and I. Vassiliou, *A top-down constraint-driven design methodology for analog integrated circuits*. Boston: Kluwer Academic Publishers, 1997.

[Char92] E. Charbon, E. Malavasi, U. Choudhury, A. Casotto, and A. Sangiovanni-Vincentelli, "A constraint-driven placement methodology for analog integrated circuits," in *Proc. of IEEE Custom Integrated Circuits Conf.*, 1992, pp. 28.2.1-28.2.4.

[Char94] E. Charbon, E. Malavasi, D. Pandini, and A. Sangiovanni-Vincentelli, "Imposing tight specifications on analog IC's through simultaneous placement and module optimization," in *Proc. of IEEE Custom Integrated Circuits Conf.*, 1994, pp. 525-528.

[Chou90a] U. Choudhury and A. Sangiovanni-Vincentelli, "Constraint-based channel routing for analog and mixed analog/digital circuits," in *Proc. of ACM/IEEE Int. Conf. on Computer-Aided Design*, 1990, pp. 198-201.

[Chou90b] U. Choudhury and A. Sangiovanni-Vincentelli, "Constraint generation for routing analog circuits," in *Proc. of ACM/IEEE Design Automation Conf.*, 1990, pp. 561-566.

[Chou90c] U. Choudhury and A. Sangiovanni-Vincentelli, "Use of performance sensitivities in routing analog circuits," in *Proc. of IEEE Int. Symp. on Circuits and Systems*, 1990, vol.1, pp. 348-351.

[Chou91] U. Choudhury and A. Sangiovanni-Vincentelli, "An analytical-model generator for interconnect capacitances," in *Proc. of IEEE Custom Integrated Circuits Conf.*, 1991, pp. 8.6/1-8.6/4.

[Chou93] U. Choudhury and A. Sangiovanni-Vincentelli, "Automatic generation of parasitic constraints for performance-constrained physical design of analog circuits," *IEEE Trans. Computer-Aided Design*, vol. 12, no. 2, pp. 208-224, February 1993.

[Chou95] U. Choudhury and A. Sangiovanni-Vincentelli, "Automatic generation of analytical models for interconnect capacitances," *IEEE Trans. Computer-Aided Design*, vol. 14, no. 4, pp.

470-480, April 1995.

[Cohn91] J. M. Cohn, R. A. Rutenbar, and L. R. Carley, "KOAN/ ANAGRAM II: new tools for device-level analog placement and routing," *IEEE J. Solid-State Circuits*, vol. 26, no. 3, pp. 330-342, March 1991.

[Comp04] *Virtuoso® schematic composer user guide. Product version 5.1.41*, Cadence Design Systems Inc., 2004.

[Comp05] *Component Description Format User Guide. Product version 5.1.41*, Cadence Design Systems Inc., 2005.

[Cong97] J. Cong, L. He, A. B. Kahng, D. Noice, N. Shirali, and S. H.-C. Yen, "Analysis and justification of a simple, practical 2 1/2-D capacitance extraction methodology," in *Proc. of ACM/IEEE Design Automation Conf.*, 1997, pp. 627-632.

[Conw92] J. D. Conway and G. G. Schrooten, "An automatic layout generator for analog circuits," in *Proc. of European Design Automation Conf.*, 1992, pp. 513-519.

[Corm01] T. H. Cormen, C. E. Leiserson, R. L. Rivest, C. Stein, *Introduction to Algorithms, Second Edition*. The MIT Press, 2001.

[Cost87] G. I. Costache, "Finite element method applied to skin-effect problems in strip transmission lines," *IEEE Microwave Theory Tech.*, vol. 35, no. 11, pp. 1009-1013, November 1987.

[Daems02] W. Daems, *Symbolic analysis and modeling of analog integrated circuits*. Ph.D. Thesis, Katholieke Universiteit Leuven, 2002.

[Deb02] K. Deb, A. Pratap, S. Agarwal, and T. Meyarivan, "A fast and elitist multiobjective genetic algorithm: NSGA-II," *IEEE Trans. Evol. Compu.*, vol. 6, no. 2, pp. 182-197, April 2002.

[Degr87] M. G. R. Degrauwe, O. Nys, E. Dijkstra, J. Rijmenants, S. Bitz, B. L. A. G. Goffart, E. A. Vittoz, S. Cserveny, C. Meixenberger, G. van der Stappen, and H. J. Oguey, "IDAC: an interactive design tool for analog CMOS circuits," *IEEE J. Solid-State Circuits*, vol. 22, no. 6, pp. 1106-1116, December 1987.

[Dengi97] E. A. Dengi and R. A. Rohrer, "Hierarchical 2-D field solution for capacitance extraction for VLSI interconnect modeling," in *Proc. of ACM/IEEE Design Automation Conf.*, 1997, pp. 127-132.

[Dengi98] E. A. Dengi and R. A. Rohrer, "Boundary element method macromodels for 2-D hierarchical capacitance extraction," in

Proc. of ACM/IEEE Design Automation Conf., 1998, pp. 218-223.

[Dess99] M. Dessouky, A. Greiner, and M.-M. Louërat, "CAIRO: A hierarchical layout language for analog circuits," in *Proc. of Int. Conf. Mixed Design of Integrated Circuits and Systems*, 1999, pp. 105-110.

[Dess01a] M. Dessouky, *Design for reuse of analog circuits. Case study: very low-voltage delta-sigma modulator.* Ph.D. Thesis, University of Paris VI, 2001.

[Dess01b] M. Dessouky, A. Kaiser, M.-M. Louërat, and A. Greiner, "Analog design for reuse-case study: very low-voltage ΔΣ modulator," in *Proc. of Design, Automation, and Test in Europe Conf.*, 2001. pp. 353-360.

[Dess00a] M. Dessouky and M.-M. Louërat, "A layout approach for electrical and physical design integration of high-performance analog circuits," in *Proc. of IEEE Int. Symp. on Quality Electronic Design*, 2000, pp. 291-298.

[Dess00b] M. Dessouky, M.-M. Louërat, and J. Porte, "Layout-oriented synthesis of high performance analog circuits," in *Proc. of Design, Automation, and Test in Europe Conf.*, 2000, pp. 53-57.

[Dias92] V. F. Dias, V. Liberali, and F. Maloberti, "Design tools for oversampled data converters: needs and solutions," *Microelectronics Journal*, vol. 23, no. 8, pp. 641-650, Elsevier, December 1992.

[Dier82] W. H. Dierking and J. D. Bastian, "VLSI parasitic capacitance determination by flux tubes," *IEEE Circuits Systems Mag.*, pp. 11-18, March 1982.

[Diva05] *Diva® reference. Product version 5.0*, Cadence Design Systems Inc., 2005.

[Dong01] Y. Dong and A. Opal, "An overview on computer-aided analysis techniques for sigma-delta modulators," in *Proc. of IEEE Int. Symp. on Circuits and Systems*, 2001, vol. 5, pp. 423-426.

[Donn94a] S. Donnay, K. Swings, G. Gielen, W. Sansen, W. Kruiskamp, and D. Leenaerts, "A methodology for analog design automation in mixed-signal ASICs," in *Proc. of the European Design and Test Conf.*, 1994, pp. 530-534.

[Donn94b] S. Donnay, K. Swings, G. Gielen, and W. Sansen, "A methodology for analog high-level synthesis," in *Proc. of the*

IEEE Custom Integrated Circuits Conf., 1994, pp. 373-376.

[Duque93] J. F. Duque-Carrillo, "Continuous-time common-mode feedback networks for fully-differential amplifiers: a comparative study," in *Proc. of IEEE Int. Symp. on Circuits and Systems*, 1993, vol. 2, pp. 1267-1270.

[Durh93] A. M. Durham and W. Redman-White, "Integrated continuous-time balanced filters for 16-b DSP interfaces," *IEEE J. Solid-State Circuits*, vol. 28, no. 7, pp. 835-839, July 1993.

[Edif04] *Guide to EDIF 300 translators. Product version 5.1.41*, Cadence Design Systems Inc., 2004.

[ElTu89] F. El-Turky and E. E. Perry, "BLADES: an artificial intelligence approach to analog circuit design," *IEEE Trans. Computer-Aided Design*, vol. 8, no. 6, pp. 680-692, June 1989.

[Fari04] M. Farina, K. Deb, and P. Amato, "Dynamic Multiobjective Optimization Problems: Test Cases, Approximations, and Applications," *IEEE Trans. Evol. Compu.*, vol. 8, no. 5, pp. 425-442, October 2004.

[Felt93] E. Felt, E. Malavasi, E. Charbon, R. Totaro, and A. Sangiovanni-Vincentelli, "Performance-driven compaction for analog integrated circuits," in *Proc. of IEEE Custom Integrated Circuits Conf.*, 1993, pp. 17.3.1-17.3.5.

[Fern97] *Symbolic analysis techniques: applications to analog design automation*. F. Fernández, A. Rodríguez-Vázquez, J. L. Huertas, and G. Gielen, Eds. New York: IEEE Press, 1997.

[Fern03] F. Fernández, F. Medeiro, R. del Río, R. Castro-López, B. Pérez-Verdú, and A. Rodríguez-Vázquez, "Design methodologies for sigma-delta converters," in *CMOS telecom data converters*, A. Rodríguez-Vázquez, F. Medeiro, E. Janssens, Eds. Boston: Kluwer Academic Publishers, 2003, pp. 15-1, 15-38.

[Fran99a] J. Franca, N. Horta, M. Pereira, J. Vital, R. Castro-López, M. Delgado-Restituto, F. Fernández, A. Rodríguez-Vázquez, J. Ramos, and P. Santos, "RAPID-retargetability for reusability of application-driven quadrature D/A interface block design," in *Proc. of IEEE Int. Conf. on Electronics, Circuits, and Systems*, 1999, vol. 3, pp. 1679-1683.

[Fran99b] K. Francken and G. Gielen, "Methodology for analog technology porting including performance tuning," in *Proc. of IEEE Int. Symp. on Circuits and Systems*, 1999, vol. 1, pp.415-418.

[Frie96] V. Friedman, K. R. Lakshmikumar, D. L. Price, T. N. Le, and J. Kumar, "A baseband processor for IS-54 cellular telephony," *IEEE J. Solid-State Circuits*, vol. 31, no. 5, pp. 646-655, May 1996.

[Garr88] D. J. Garrod, R. A. Rutenbar, and L. R. Carley, "Automatic layout of custom analog cells in ANAGRAM," in *Proc. of ACM/IEEE Int. Conf. on Computer-Aided Design*, 1998, pp. 544-547.

[Gatti89] U. Gatti, F. Maloberti, and V. Liberali, "Full stacked layout of analogue cells," in *Proc. of IEEE Int. Symp. on Circuits and Systems*, 1989, vol. 2, pp. 1123-1126.

[Giel89] G. Gielen, H. Walscharts, and W. Sansen, "ISAAC: a symbolic simulator for analog integrated circuits," *IEEE J. Solid-State Circuits*, vol. 24, no. 6, pp. 1587-1597, December 1989.

[Giel90] G. Gielen, H. Walsharts, and W. Sansen, "Analog circuit design optimization based on symbolic simulation and simulated annealing," *IEEE J. Solid-State Circuits*, vol. 25, no. 3, June 1990.

[Giel00] G. Gielen and R. A. Rutenbar, "Computer-aided design of analog and mixed-signal integrated circuits," *Proc. IEEE*, vol. 88, no. 12, pp. 1825-1854, December 2000.

[Giel01] G. Gielen, M. del Mar Hershenson, K. Kundert, P. Magarshack, A. Matsuzawa, R. A. Rohrer, and P. Yang, "When will the analog design flow catch up with digital methodology?," in *Proc. of ACM/IEEE Design Automation Conf.*, 2001, pp. 419.

[Gilb02] B. Gilbert, "Design for manufacture," in *Trade-offs in analog circuit design. The designer's companion*, C. Toumazou, G. Moschytz, and B. Gilbert, Eds. Boston: Kluwer Academic Publishers, 2002.

[Glos03] Semiconductor Industry Association, *Glossary of terms*. Available: http://www.semichips.org/ind_glossary.cfm, 2004.

[Goran95] P. N'Goran and A. Kaiser, "A building block approach to the design and simulation of complex current-memory circuits," *Analog Integrated Circuits and Signal Processing*, vol. 7, no. 3, pp. 189-199, Kluwer Academic Publishers, May 1995.

[Gray01] P. R. Gray, P. J. Hurst, S. H. Lewis, and R. G. Meyer, *Analysis and design of analog integrated circuits*, 4th ed. New York: John Wiley & Sons, 2001.

[Greg86] R. Gregorian and G. C. Temes, *Analog MOS integrated circuits for signal processing*. New York: John Wiley & Sons, 1986.

[Grei94] A. Greiner and F. Petrot, "Using C to write portable CMOS VLSI module generators," in *Proc. of European Design Automation Conf.*, 1994, pp. 676-681.

[Guer98] O. Guerra, J. D. Rodríguez-García, E. Roca, F. V. Fernández, and A. Rodríguez-Vázquez, "A simplification before and during generation methodology for symbolic large-circuits analysis," *Proc. of IEEE Int. Conf. on Electronic, Circuits, and Systems*, 1998, vol. 3, pp. 81-84.

[Guer02a] O. Guerra, E. Roca, F. V. Fernández, and A. Rodríguez-Vázquez, "Approximate symbolic analysis of hierarchically decomposed analog circuits," *Analog Integrated Circuits and Signal Processing*, vol. 31, no. 2, pp. 131-145, Kluwer Academic Publishers, May 2002.

[Gyur89] R. S. Gyurcsik and J.-C. Jeen, "A generalized approach to routing mixing analog and digital signal nets in a channel," *IEEE J. Solid-State Circuits*, vol. 24, no. 2, pp. 436-442, April 1989.

[Hall87] J. Hall, D. Hocevar, P. Yang, and M. McGraw, "SPIDER - A CAD system for modeling VLSI metallization patterns," *IEEE Trans. Computer-Aided Design*, vol. 6, no. 6, pp. 1023-1031, November 1987.

[Hamo03] M. Hamour, R. Saleh, S. Mirabbasi, and A. Ivanov, "Analog IP design flow for SoC applications," in *Proc. of IEEE Int. Symp. on Circuits and Systems*, 2003, pp. 676-679.

[Harb86] M. Harbour and J. Drake, "Calculation of multiterminal resistances in integrated circuits," *IEEE Trans. Circuits Systems*, vol. 33, no. 4, pp. 462-465, April 1986.

[Harj89] R. Harjani, R. A. Rutenbar, and L. R. Carley, "OASYS: A Framework for Analog Circuit Synthesis," *IEEE Trans. Computer-Aided Design*, vol. 8, no. 12, pp. 1247-1265, December 1989.

[Harv92] J. P. Harvey, M. I. Elmasry, and B. Leung, "STAIC: an interactive framework for synthesizing CMOS and BiCMOS analog circuits," *IEEE Trans. Computer-Aided Design*, vol. 11, no. 11, pp. 1402-1417, November 1992.

[Hasp90] J. J. J. Haspeslagh, D. Sallaerts, P. P. Reusens, A. Vanwelsenaers, R. Granek, and D. Rabaey, "A 270-kb/s 35-mW modulator IC for GSM cellular radio hand-held terminals," *IEEE J. Solid-State Circuits,* vol. 25, no. 6, pp. 1450-1457, December 1990.

[Hast01] A. Hastings, *The art of analog layout*. New Jersey: Prentice Hall, 2001.

[Hend93] R. K. Henderson, L. Astier, A. El Khalifa, and M. G. R. Degrauwe, "A spreadsheet interface for analog design knowledge capture and re-use," in *Proc. of IEEE Custom Integrated Circuits Conf.*, 1993, pp. 13.3.1-13.3.4.

[Hend94] R. K. Henderson, M. Hinners, P. Nussbaum, and L. Astier, "Capture and re-use of analog simulation knowledge," in *Proc. of IEEE Custom Integrated Circuits Conf.*, 1994, pp. 357-360.

[Hers98] M. del Mar Hershenson, S. P. Boyd, and T. H. Lee, "GPCAD: a tool for CMOS op-amp synthesis," in *Proc. of ACM/IEEE Int. Conf. Computer-Aided Design*, 1998, pp. 296-303.

[Hjal03] E. Hjalmarson, *Studies on design automation of analog circuits - The design flow.* Ph.D. Thesis, Linköpings Universitet, 2003.

[Holl92] J. H. Holland, "Genetic algorithms," *Scientific American*, pp. 44-50, July 1992.

[Holl01] T. Hollman, S. Lindfors, M. Lansirinne, J. Jussila, and K. A. I. Halonen, "A 2.7-V CMOS dual-mode baseband filter for PDC and WCDMA," *IEEE J. Solid-State Circuits*, vol. 36, no. 7, pp. 1148-1153, July 2001.

[Horo83] M. Horowitz and R. Dutton, "Resistance extraction for mask layout data," *IEEE Trans. on Computer-Aided Design*, vol. 2, no. 3, pp. 145-150, July 1983.

[Horta91] N. Horta, J. Franca, and C. Leme, "Framework for architecture synthesis of data conversion systems employing binary-weighted capacitor arrays," in *Proc. of IEEE Int. Symp. on Circuits and Systems*, 1991, vol. 3, pp. 1789-1792.

[Hspi04] *HSPICE® simulation and analysis user guide. Release V-2004.03*. Synopsys, Inc., 2004.

[Hu93] C. Hu, "Future CMOS scaling and reliability," *Proc. IEEE*, vol. 81, no. 5, pp. 682-689, May 1993.

[Itrs99] *Int. Technology Roadmap for Semiconductors. Edition 1999.* Available: http://public.itrs.net/files/1999_SIA_Roadmap/Design.pdf, 1999.

[Itrs01] *Int. Technology Roadmap for Semiconductors. Edition 2001.*

Available: http://public.itrs.net/Files/2001ITRS/Home.htm, 2001.

[Itrs03] *Int. Technology Roadmap for Semiconductors. Edition 2003*. Available: http://public.itrs.net/Files/2003ITRS/Home.htm, 2003.

[Jing01] X. Jingnan, J. Serras, M. Oliveira, R. Belo, M. Bugalho, J. Vital, N. Horta, and J. Franca, “IC design automation from circuit level optimization to retargetable layout,” in *Proc. of IEEE Int. Conf. on Electronics, Circuits and Systems*, 2001, vol. 1, pp. 95-98.

[Jusuf90] G. Jusuf, P. R. Gray, and A. L. Sangiovanni-Vincentelli, “CADICS-cyclic analog-to-digital converter synthesis,” in *Proc. of ACM/IEEE Int. Conf. on Computer-Aided Design*, 1990, pp. 286-289.

[Kao01] W. H. Kao, C.-Y. Lo, M. Basel, and R. Singh, “Parasitic extraction: current state of the art and future trends,” *Proc. IEEE*, vol. 89, no. 5, pp. 729-739, May 2001.

[Keat99] M. Keating and P. Bricaud, *Reuse methodology manual*, 2nd ed. Boston: Kluwer Academic Publishers, 1999.

[Koch03] R. J. Koch and F. Dielacher, “Analog IP - stairway to SoC heaven?,” in *Proc. of IEEE Int. Solid-State Circuits Conf.*, 2003, pp. 1-2.

[Koh90] H. Y. Koh, C. H. Sequin, and P. R. Gray, “OPASYN: A compiler for CMOS operational amplifiers”. *IEEE Tran. Computer-Aided Design*, vol. 9, no. 2, pp. 113-125, February 1990.

[Kras99] M. Krasnicki, R. Phelps, R. A. Rutenbar, and L. R. Carley, “MAELSTROM: efficient simulation-based synthesis for custom analog cells,” in *Proc. of ACM/IEEE Design Automation Conf.*, 1999, pp. 945-950.

[Kuhn87] J. Kuhn, “Analog module generators for silicon compilation,” *VLSI Systems Design*, vol. 8, no. 5, pp. 74-80, CMP Publications, May 1987.

[Kund00] K. Kundert, H. Chang, D. Jefferies, G. Lamant, E. Malavasi, and F. Sendig, “Design of mixed-signal systems-on-a-chip,” *IEEE Trans. Computer-Aided Design*, vol. 19, no. 12, pp. 1561-1571, December 2000.

[Laar87] P. J. M. Laarhoven and E. H. L. Aarts, *Simulated annealing: theory and applications*. Dordrecht: D. Reidel, 1987.

[Laks86] K. R. Lakshmikumar, R. A. Hadaway, and M. A. Copeland, "Characterisation and modeling of mismatch in MOS transistors for precision analog design," *IEEE J. Solid-State Circuits*, vol. 21, no. 6, pp. 1057-1066, December 1986.

[Laks91] K. R. Lakshmikumar, D. W. Green, K. Nagaraj, K.-H. Lau, O. E. Agazzi, J. R. Barner, R. S. Shariatdoust, G. A. Wilson, T. Le, M. R. Dwarakanath, J. G. Ruch, J. Kumar, T. Ali-Vehmas, J. J. Junkkari, and L. Siren, "A baseband codec for digital cellular telephony," *IEEE J. Solid-State Circuits*, vol. 26, no. 12, pp. 1951-1958, December 1991.

[Lamp95] K. Lampaert, G. Gielen, and W. Sansen, "A performance-driven placement tool for analog integrated circuits," *IEEE J. Solid-State Circuits*, vol. 30, no. 7, pp. 773-780, July 1995.

[Lamp99] K. Lampaert, G. Gielen, and W. Sansen, *Analog layout generation for performance and manufacturability.* Boston: Kluwer Academic Publishers, 1999.

[Lane89] W. A. Lane and G. T. Wrixon, "The design of thin-film polysilicon resistors for analog IC applications," *IEEE Trans. Electron Devices*, vol. 36, no. 4, pp. 738-744, April 1989.

[Leen01] D. Leenaerts, G. Gielen, and R. A. Rutenbar, "CAD solutions and outstanding challenges for mixed-signal and RF IC design," *Proc. of ACM/IEEE Int. Conf. on Computer Aided Design*, 2001, pp. 270-277.

[Leme91] C. Leme, A. Yufera, L. Paris, N. Horta, A. Rueda, T. Oses, J. Franca, and J. L. Huertas, "Flexible silicon compilation of charge redistribution data conversion systems," in *Proc. of IEEE Midwest Symp. on Circuits and Systems*, 1991, vol. 1, pp. 403-406.

[Leng88] T. Lengauer, "The combinatorial complexity of layout problems," in *Physical design automation of VLSI systems.* Menlo Park: Benjamin-Cummings Publishing Company, 1988, pp. 461-497.

[Levi92] M. Levitt, "Economic and productivity considerations in ASIC test and design-for-test," in *Proc. of IEEE Computer Society Int. Conf.*, 1992, pp. 440-445.

[Libe93] V. Liberali, V. F. Dias, M. Ciapponi, and F. Maloberti, "TOSCA: a simulator for switched-capacitor noise-shaping A/D converters," *IEEE Trans. Computer-Aided Design*, vol. 12, no. 9, pp. 1376-1386, September 1993.

[Liu02] H. Liu, A. Singhee, R. A. Rutenbar, and L. R. Carley, "Remembrance of circuits past: macromodeling by data mining in large analog design spaces," in *Proc. of ACM/IEEE Int. Design Automation Conf.*, 2002, pp. 437-442.

[Liu03] D. Liu, S. Sidiropoulos, and M. Horowitz, "A framework for designing reusable analog circuits," in *Proc. of ACM/IEEE Int. Conf. on* Computer Aided Design, 2003, pp. 375-380.

[Makr95] C. A. Makris and C. Toumazou, "Analog IC design automation: Part II— Automated circuit correction by qualitative reasoning," *IEEE Trans. Computer-Aided Design*, vol. 14, no. 2, pp. 239-254, February 1995.

[Mala90] E. Malavasi, U. Choudhury, and A. Sangiovanni-Vincentelli, "A routing methodology for analog integrated circuits," in *Proc. of ACM/IEEE Int. Conf. on Computer-Aided Design*, 1990, pp. 202-205.

[Mala93] E. Malavasi and A. Sangiovanni-Vincentelli,"Area routing for analog layout," *IEEE Trans. Computer-Aided Design*, vol. 12, no. 8, pp. 1186-1197, August 1993.

[Mala95] E. Malavasi and D. Pandini, "Optimum CMOS stack generation with analog constraints," *IEEE Trans. Computer-Aided Design*, vol. 14, no. 1, pp. 107-122, January 1995.

[Mala96] E. Malavasi, E. Charbon, E. Felt, and A. Sangiovanni-Vincentelli, "Automation of IC layout with analog constraints," *IEEE Trans. Computer-Aided Design*, vol. 15, no. 8, pp. 923-942, August 1996.

[Malo94] F. Maloberti, "Layout of analog and mixed analog-digital circuits," in *Design of Analog-Digital VLSI Circuits for Telecommunications and Signal Processing*, J. Franca and Y. Tsividis, Eds. Englewood Cliffs: Prentice Hall, 1994, pp. 341-367.

[Man80] H. De Man, "DIANA as a Mixed Mode Simulator for MOS LSI Sampled Data Circuits," *Proc. of IEEE Int. Symp. on Circuits and Systems*, 1980, pp. 435-438.

[Marp90] D. Marple, M. Smulders, and H. Hegen, "Tailor: a layout system based on trapezoidal corner stitching," *IEEE Trans. Computer-Aided Design*, vol. 9, no. 1, pp. 66-90, January 1990.

[Math02a] *Using MATLAB version 6.5*, The MathWorks Inc., 2002.

[Math02b] *Using Simulink version 5*, The MathWorks Inc., 2002.

[Math02c] *Writing S-functions version 5*, The MathWorks Inc., 2002.

[Maul93] P. C. Maulik, L. R. Carley, and D. J. Allstot, "Sizing of cell-level analog circuits using constrained optimization techniques," *IEEE J. Solid-State Circuits*, vol. 28, no. 3, pp. 233-241, March 1993.

[Maul95] P. C. Maulik, N. van Bavel, K. S. Albright, and X.-M. Gong, "An analog/digital interface for cellular telephony," *IEEE J. Solid-State Circuits*, vol. 30, no. 3, pp. 201-209, March 1995.

[Mayo90] R. N. Mayo, M. H. Arnold, W. S. Scott, D. Stark, and G. T. Hamachi, "1990 DECWRL/Livermore Magic release," WRL Research Report 90/7, September. 1990.

[McNu94] M. J. McNutt, S. LeMarquis, and J. L. Dunkley, "Systematic capacitance matching errors and corrective layout procedures," *IEEE J. Solid-State Circuits*, vol. 29, no. 5, pp. 611-616, May 1994.

[Mead80] C. A. Mead and L. A. Conway, *Introduction to VLSI systems*. Reading: Adison-Wesley, 1980.

[Mede94] F. Medeiro, R. Rodríguez-Macías, F. V. Fernández, R. Domínguez-Castro, J. L. Huertas, and A. Rodríguez-Vázquez, "Global design of analog cells using statistical optimization techniques," *Analog Integrated Circuits and Signal Processing*, vol. 6, no. 3, pp. 179-195, Kluwer Academic Publishers, November 1994.

[Mede95] F. Medeiro, B. Pérez-Verdú, A. Rodríguez-Vázquez, and J. L. Huertas, "A vertically integrated tool for automated design of ΣΔ modulators," *IEEE J. Solid-State Circuits*, vol. 30, no. 7, pp. 762-772, July 1995.

[Mede99] F. Medeiro, B. Pérez-Verdú, and A. Rodríguez-Vázquez, *Top-down design of high-performance sigma-delta modulators*. Boston: Kluwer Academic Publishers, 1999.

[Ment90] *Mentor Graphics Lx User's Guide*. Mentor Graphics Corporation, 1990.

[Mino95] P. Minogue, "A 3 V GSM codec," *IEEE J. Solid-State Circuits*, vol. 30, no. 12, pp. 1411-1420, December 1995.

[Mitra92] S. Mitra, S. K. Nag, R. A. Rutenbar, and L. R. Carley, "System-level routing of mixed-signal ASICs in WREN," in *Proc. of ACM/IEEE Int. Conf. on Computer-Aided Design*, 1992, pp. 394-399.

[Mitra94] S. Mitra, R. A. Rutenbar, L. R. Carley, and D. J. Allstot, "Substrate-aware mixed-signal macro-cell placement in

WRIGHT," in *Proc. of IEEE Custom Integrated Circuits Conf.*, 1994, pp. 529-532.

[Moga89] M. Mogaki, N. Kato, Y. Chikami, N. Yamada, and Y. Kobayashi, "LADIES: an automatic layout system for analog LSI's," in *Proc. of ACM/IEEE Int. Conf. on Computer-Aided Design*, 1989, pp. 450-453.

[Moore65] G. E. Moore, "Cramming more components onto integrated circuits," *Electronics*, vol. 38, no. 8, April 1965.

[Moore75] G. E. Moore, "Progress in digital integrated circuit," *IEEE Int. Electron Devices Meeting Technology Digest*, p. 11, December 1975.

[Morie93] T. Morie, H. Onodera, and K. Tamaru, "A system for analog circuit design that stores and re-uses design procedures," in *Proc. of IEEE Custom Integrated Circuits Conf.*, 1993, pp. 13.4.1-13.4.4.

[Nabo92] K. Nabors and J. White, "Multipole-accelerated capacitance extraction algorithms for 3-D structures with multiple dielectrics," *IEEE Trans. Circuits Systems I*, vol. 39, no. 11, pp. 946-954, November 1992.

[Naka91] Y. Nakamura, T. Miki, A. Maeda, H. Kondoh, and N. Yazawa, "A 10-b 70-MS/s CMOS D/A converter," *IEEE J. Solid-State Circuits*, vol. 26, no. 4, pp. 637-642, April 1991.

[Naik99] R. Naiknaware and T. S. Fiez, "Automated hierarchical CMOS analog circuit stack generation with intramodule connectivity and matching considerations," *IEEE J. Solid-State Circuits*, vol. 34, no. 3, pp. 304-317, March 1999.

[Neff95] R. Neff, *Automatic synthesis of CMOS digital/analog converters*. Ph.D. Thesis, University of California at Berkeley, 1995.

[Neo04a] NeoCircuit® homepage: http://www.cadence.com/products/custom_ic/neocircuit/index.aspx, 2004.

[Neo04b] NeoCell® homepage: http://www.cadence.com/products/custom_ic/neocell/index.aspx, 2004.

[Ning88] Z.-Q. Ning and P. M. Dewilde, "SPIDER: capacitance modelling for VLSI interconnections," *IEEE Trans. Computer-Aided Design*, vol. 7, no. 12, pp. 1221-1228, December 1988.

[Nye88] W. Nye, D. C. Riley, A. Sangiovanni-Vincentelli, and A. L. Tits, "DELIGHT.SPICE: an optimization-based system for the design of integrated circuits," *IEEE Trans. Computer-Aided*

Design, vol. 7, no. 4, pp. 501-519, April 1988.

[Ocean05] *Ocean Reference. Product Version 5.1.41*, Cadence Design Systems Inc., 2005.

[Ocho96] E. Ochotta, R. Rutenbar, and L.R. Carley, "Synthesis of high-performance analog circuits in ASTRX/OBLX," *IEEE Trans. Computer Aided-Design*, vol. 15, pp. 273-294, March 1996.

[Ocho98] E. S. Ochotta, T. Mukherjee, R. A. Rutenbar, and L. R. Carley, *Practical synthesis of high-performance analog circuits*. Boston: Kluwer Academic Publishers, 1998.

[Ohr02] S. Ohr and L. Marchant, "PANEL: analog intellectual property: now? or never?," in *Proc. of ACM/IEEE Design Automation Conf.*, 2002, pp. 181-182.

[Ohr03] S. Ohr, *Interest in analog IP outpaces execution*. Available: http://www.eedesign.com/articleshowArticle.jhtml?articleId=17408178, 2003.

[OLea91] P. O'Leary, "Practical aspects of mixed analogue and digital design," in *Analogue-Digital ASICS*, R. S. Soin, F. Maloberti, and J. Franca, Eds. London: Peter Peregrinus, 1991, pp. 213-237.

[Onod90] H. Onodera, H. Kanbara, and K. Tamaru, "Operational-amplifier compilation with performance optimization," *IEEE J. Solid-State Circuits*, vol. 25, no. 2, pp. 466-473, April 1990.

[Onod92] H. Onodera and K. Tamaru, "Analog circuit placement – Branch-and-bound placement with shape optimization," in *Proc. of IEEE Custom Integrated Circuits Conf.*, 1992, pp. 11.5.1-11.5.6.

[Onoz95] A. Onozawa, K. Chaudhary, and E. S. Kuh, "Performance driven spacing algorithms using attractive and repulsive constraints for submicron LSI's," *IEEE Trans. Computer-Aided Design*, vol. 14, no. 6, pp. 707-719, June 1995.

[Otten82] R. H. J. M. Otten, "Automatic floorplan design," in *Proc. of ACM/IEEE Design Automation Conf.*, 1982, pp. 261-267.

[Oust84] J. K. Ousterhout, "Corner stitching: a data-structure technique for VLSI layout tools," *IEEE Trans. on Computer-Aided Design*, vol. 3, pp. 87-100, December 1984.

[Owen95] B. R. Owen, R. Duncan, S. Jantzi, C. Ouslis, S. Rezania, and K. Martin, "BALLISTIC: an analog layout language," in *Proc. of IEEE Custom Integrated Circuits Conf.*, 1995, pp. 41-44.

[Peas96] R. A. Pease, J. D. Bruce, H. W. Li, and R. J. Baker, "Comments on 'Analog layout using ALAS!' [and reply]," *IEEE J. Solid-State Circuits*, vol. 31, no. 9, pp. 1364-1365, September 1996.

[Pelg89] M. J. M. Pelgrom, A. C. J. Duinmaijer, and A. P. G. Welbers, "Matching properties of MOS transistors," *IEEE J. Solid-State Circuits*, vol. 24, no. 5, pp. 1433-1439, October 1989.

[Perez00] F. M. Pérez-Montes, F. Medeiro, R. Domínguez-Castro, F. V. Fernández, and A. Rodríguez-Vázquez, "XFridge: a SPICE-based, portable, user-friendly cell-level sizing tool," in *Proc. of Design, Automation, and Test in Europe Conf.*, 2000, p. 739.

[Perl93] R. L. Schwartz and T. Phoenix, *Learning Perl, 4th edition*. Sebastopol, CA.: O'Reilly, 1993.

[Phel00a] R. Phelps, M. J. Krasnicki, R. A. Rutenbar, L. R. Carley, and J. R. Hellums, "Anaconda: simulation-based synthesis of analog circuits via stochastic pattern search," *IEEE Trans. Computer-Aided Design*, vol. 19, no. 6, pp. 703-717, June 2000.

[Phel00b] R. Phelps, M. J. Krasnicki, R. A. Rutenbar, L. R. Carley, and J. R. Hellums, "A case study of synthesis for industrial-scale analog IP: redesign of the equalizer/filter frontend for an ADSL CODEC," in *Proc. of ACM/IEEE Design Automation Conf.*, 2000, pp. 1-6.

[Plas01] G. Van der Plas, G. Debyser, F. Leyn, K. Lampaert, J. Vandenbussche, G. Gielen, W. Sansen, P. Veselinovic, and D. Leenarts, "AMGIE-A synthesis environment for CMOS analog integrated circuits," *IEEE Trans. Computer-Aided Design*, vol. 20, no. 9, pp. 1037-1058, September 2001.

[Plet86] T. Pletersek, J. Trontelj, I. Jones, and G. Shenton, "High-performance designs with CMOS analog standard cells," *IEEE J. Solid-State Circuits*, vol. 21, no. 2, pp. 215-222, April 1986.

[Porte97] J. Porte, "COMDIAC: compilateur de dispositifs actifs, reference manual," (in French) Ecole Nationale Supérieure des Télécommunications, Paris, September 1997.

[Prie97] J. A. Prieto, A. Rueda, J. M. Quintana, and J. L. Huertas, "A performance-driven placement algorithm with simultaneous Place&Route optimization for analog ICs," in *Proc. of the European Design and Test Conf.*, 1997, pp. 389-394.

[Prie01] J. A. Prieto, *GELSA: un colocador flexible para circuitos integrados analógicos*. (in Spanish) Ph.D. Thesis, University of Seville, 2001.

[Ramet88] S. Ramet, "A low-distortion anti-aliasing/smoothing filter for sampled data integrated circuits," *IEEE J. Solid-State Circuits*, vol. 23, no. 5, pp. 1267-1270, October 1988.

[Ranj04] M. Ranjan, W. Verhaegen, A. Agarwal, H. Sampath, R. Vemuri, and G. Gielen, "Fast, layout-inclusive analog circuit synthesis using pre-compiled parasitic-aware symbolic performance models," in *Proc. of Design, Automation, and Test in Europe Conf.*, 2004, pp. 604-609.

[Rant02] C. R. C. De Ranter, G. Van der Plas, M. Steyaert, G. Gielen, and W. Sansen, "CYCLONE: automated design and layout of RF LC-oscillators," *IEEE Trans. Computer-Aided Design*, vol. 21, no. 10, pp. 1161-1170, October 2002.

[Rapa96] T. S. Rappaport, *Wireless communications, principles and practice*. New Jersey: Prentice Hall, 1996.

[Rash01] P. Rashinkar, P. Paterson, and L. Singh, *System-on-a-chip verification*. Kluwer Academic Publishers, 2001.

[Raza01] B. Razavi, *Design of analog CMOS integrated circuits*. New York: McGraw-Hill, 2001.

[Rein02] M. Reinhardt, *Automatic layout modification*. Boston: Kluwer Academic Publishers, 2002.

[Rijm89] J. Rijmenants, J. B. Litsios, T. R. Schwarz, and M. G. R. Degrauwe, "ILAC: an automated layout tool for analog CMOS circuits," *IEEE J. Solid-State Circuits*, vol. 24, no. 2, pp. 417-425, April 1989.

[Rubin87] S. M. Rubin, *Computer aids for VLSI design*. Reading: Addison-Wesley, Inc., 1987.

[Rueh73] A. E. Ruehli and P. A. Brennan, "Efficient capacitance calculations for three-dimensional multiconductor systems," *IEEE Trans. Microwave Theory Tech.*, vol. 21, no. 2, pp. 76-82, February 1973.

[Rute02] R. A. Rutenbar and J. M. Cohn, "Layout tools for analog ICs and mixed-signal SoCs: a survey," in *Computer-Aided Design of Analog Integrated Circuits and Systems*, R. A. Rutenbar, G. Gielen, and B. A. Antao, Eds. Piscataway: John Wiley & Sons Inc., 2002, pp. 365-372.

[Saber02] *Analyzing designs using SABER designer*, Synopsys Inc., 2002.

[Saint02] C. Saint and J. Saint, *IC layout basics. A practical guide*. New York: McGraw-Hill, 2002.

[Saku83] T. Sakurai and K. Tamaru, "Simple formulas for two- and three-dimensional capacitances," *IEEE Trans. Electron Devices*, vol. 30, no. 2, pp. 183-185, February 1983.

[Samp03] H. Sampath and R. Vemuri, "MSL: A high-level language for parameterized analog and mixed signal layout generators," in *Proc. of 12th Int. IFIP VLSI Conf.*, 2003, pp. 416-421.

[Sayed02] D. Sayed and M. Dessouky, "Automatic generation of common-centroid capacitor arrays with arbitrary capacitor ratio," in *Proc. of Design, Automation, and Test in Europe Conf.*, 2002, pp. 576-580.

[Scha90] R. Schaumann, M. S. Ghausi, and K. R. Laker, *Design of analog filters. Passive, active RC, and switched-capacitor*. Englewood Cliffs: Prentice-Hall, 1990.

[Scott85] W. S. Scott and J. K. Ousterhout, "Magic's circuit extractor," in *Proc. of ACM/IEEE Design Automation Conf.*, 1985, pp. 286 - 292.

[Sedra78] A. S. Sedra and P. P. Brackett, *Filter theory and design: active and passive*. Champaign: Matrix Publishers, 1978.

[Seep99] *Reuse techniques for VLSI design,* R. Seepold and A. Kunzmann Eds. Boston: Kluwer Academic Publishers, 1999.

[Seep01] *Virtual components design reuse,* R. Seepold and N. Martínez Eds. Boston: Kluwer Academic Publishers, 2001.

[Seidl87] A. Seidl, M. Svoboda, J. Oberndorfer, and W. Rosner, "CAPCAL - A 3-D capacitance solver for support of CAD systems," *IEEE Trans. Computer-Aided Design*, vol. CAD-7, no. 5, pp. 644-649, March 1987.

[Shi96] W. Shi, "A fast algorithm for area minimization of slicing floorplans," *IEEE Trans. Computer-Aided Design*, vol. 15, no. 12, pp. 1525-1532, December 1996.

[Shing86] M. T. Shing and T. C. Hu, "Computational complexity of layout problems," in *Layout design and verification*, T. Ohtsuki, Ed. Amsterdam: Elsevier Science Publishers, 1986, pp. 267-294.

[Sini01] P. P. Siniscalchi, J. K. Pitz, R. K. Hester, S. M. DeSoto, M. Wang, S. Sridharan, R. L. Halbach, D. Richardson, W. Bright, M. M. Sarraj, J. R. Hellums, C. L. Betty, and G. Westphal, "A CMOS ADSL codec for central office applications," *IEEE J. Solid-State Circuits*, vol. 36, no. 3, pp. 356-365, March 2001.

[Skill04] *SKILL language reference. Product version 06.30*, Cadence Design Systems Inc., 2004.

[Stan94] B. R. Stanisic, N. K. Verghese, R. A. Rutenbar, L. R. Carley, and D. J. Allstot, "Addressing substrate coupling in mixed-mode ICs: simulation and power distribution synthesis," *IEEE J. Solid-State Circuits*, vol. 29, no. 3, pp. 226-238, March 1994.

[Stan96] B. R. Stanisic, R. A. Rutenbar, and L. R. Carley, *Synthesis of power distribution to manage signal integrity in mixed-signal ICs*. Norwell: Kluwer Academic Publishers, 1996.

[Stoc83] L. Stockmeyer, "Optimal orientation of cells in slicing floorplan designs," *Information and Control*, vol. 57, no. 2-3, pp. 91-101, Academic Press, May-June 1983.

[Su93] D. K. Su, M. J. Loinaz, S. Masui, and B. A. Wooley, "Experimental results and modeling techniques for substrate noise in mixed-signal integrated circuits," *IEEE J. Solid-State Circuits*, vol. 28, no. 4, pp. 420-430, June 1993.

[Sun97] W. Sun, W. Wei-Ming Dai, and W. Hong, "Fast parameter extraction of general interconnects using geometry independent measured equation of invariance," *IEEE Trans. Microwave Theory Tech.*, vol. 45, no. 5, pp. 827-836, May 1997.

[Suya90] K. Suyama, S.-C. Fang, and Y. Tsividis, "Simulation of mixed switched-capacitor/digital networks with signal-driven switches," *IEEE J. Solid-State Circuits*, vol. 25, no. 6, pp. 1403-1413, December 1990.

[Tang02] H. Tang and A. Doboli, "Employing layout-templates for synthesis of analog systems," in *Proc. of IEEE Midwest Symp. on Circuits and Systems*, 2002, pp. 505-508.

[Tang03] H. Tang, H. Zhang, and A. Doboli, "Layout-aware analog system synthesis based on symbolic layout description and combined block parameter exploration, placement and global routing," in *Proc. of IEEE Computer Society Annual Symp. on VLSI*, 2003, pp. 266-271.

[Tayl85] C. D. Taylor, G. N. Elkhouri, and T. E. Wade, "On the parasitic capacitances of multilevel parallel metallization lines," *IEEE Trans. Electron Devices*, vol. ED-32, no. 11, pp. 2408-2414, November 1985.

[Temes67] G. C. Temes and D. A. Calahan, "Computer-aided network optimization - The state-of-the-art," *Proc. IEEE*, vol. 55, pp. 1984, December 1967.

[Tokh96] V. M. Tokhomirov, "The Evolution of Methods of Convex Optimization," *The American Mathematical Monthly*, vol. 103, no.1, pp. 65-71, Mathematical Association of America, January 1996.

[Toum95] C. Toumazou and C. A. Makris, "Analog IC design automationn: Part I— Automated circuit generation: new concepts and methods," *IEEE Trans. Computer-Aided Design,* vol. 14, pp. 218-238, February 1995.

[Tsiv96] Y. Tsividis, *Mixed analog-digital VLSI devices and technology.* New York: McGraw-Hill, 1996.

[Uebb86] R. H. Uebbing and M. Fukuma, "Process-based three-dimensional capacitance simulation – TRICEPS," *IEEE Trans. Computer-Aided Design,* vol. CAD-5, no. 1, pp. 215-220, January 1986.

[Vach03] A. Vachoux, C. Grimm, and K. Einwich, "Analog and mixed signal modelling with SystemC-AMS," in *Proc. of IEEE Int. Symp. on Circuits and Systems,* 2003, vol. 3, pp. 914-917.

[Vanc01] P. Vancorenland, G. Van der Plas, M. Steyaert, G. Gielen, and W. Sansen, "A layout-aware synthesis methodology for RF circuits," in *Proc. of ACM/IEEE Int. Conf. on Computer-Aided Design,* 2001, pp. 358-362.

[Vand96] R. J. Vanderbei, *Linear programming: foundations and extensions.* Boston: Kluwer Academic Publishers, 1996.

[Vand03] J. Vandenbussche, G. Gielen, and M. Steyaert, *Systematic design of analog IP blocks.* Boston: Kluwer Academic Publisher, 2003.

[VanP90] P. M. VanPeteghem and J. F. Duque-Carrillo, "A general description of common-mode feedback in fully-differential amplifiers," in *Proc. of IEEE Int. Symp. on Circuits and Systems,* 1990, vol. 4, pp. 312-320.

[Venk86] J. Venkataraman, S. M. Rao, A. R. Djordjevic, T. K. Sarkar, and Y. Naiheng, "Analysis of arbitrarily oriented microstrip transmission lines in arbitrarily shaped dielectric media over a finite ground plane," *IEEE Trans. Microwave Theory Tech.,* vol. 33, no. 10, pp. 952-960, October 1985.

[Veri00] *Verilog-AMS language reference manual. Version 2.0,* Open Verilog Int., 2000.

[Veri01] *IEEE standard Verilog hardware description language,* IEEE Std 1346-2001, 2001.

[Vhdl99] *IEEE standard VHDL analog and mixed-signal extensions,* IEEE Std 1076.1-1999, 1999.

[Vhdl00] *IEEE standard VHDL language reference manual*, IEEE Std 1076-2000, 2000.

[Virt00a] *Virtuoso® layout editor user guide. Product version 4.4.6*, Cadence Design Systems Inc., 2000.

[Virt00b] *Virtuoso® parameterized cell reference. Product version 4.4.6*, Cadence Design Systems Inc., 2000.

[Virt01] *Virtuoso® relative object design user guide. Product version 4.4.6*, Cadence Design Systems Inc., 2001.

[Vital93] J. Vital, N. Horta, N. S. Silva, and J. Franca, "CATALYST: A highly flexible CAD tool for architecture-level design and analysis of data converters," in *Proc. of European Conf. on Design Automation*, 1993, pp. 472-477.

[Vitt85] E. A. Vittoz, "The design of high-performance analog circuits on digital CMOS chips," *IEEE J. Solid-State Circuits*, vol. 20, no. 3, pp. 657-665, June 1985.

[Vsia99] *Analog/Mixed-signal VSI extension specification 1 version 2.0 (AMS 1 2.0)*, Analog/Mixed-Signal Development Working Group (VSI Alliance™), 1999.

[Vsia01] *Model* taxonomy version *2.1 (SLD 2 2.1)*, VSI Alliance™, 2001.

[Wei84] C. Wei, R. F. Barrington, J. R. Mautz, and T. K. Sarkar, "Multiconductor transmission lines in multilayered dielectric media," *IEEE Trans. Microwave Theory Tech.*, vol. 32, no. 4, pp. 439-450, April 1984.

[Welch03] B. B. Welch, K. Jones, and J. Hobbs, *Practical programming in Tcl and Tk*. New Jersey: Prentice Hall, 2003.

[Wimer88] S. Wimer, I. Koren, and I. Cederbaum, "Floorplans, planar graphs, and layouts," *IEEE Trans. Circuits Systems*, vol. 35, no. 3, pp. 267-278, March 1988.

[Wolf99] M. Wolf and U. Kleine, "Reliability driven module generation for analog layouts," in *Proc. of IEEE Int. Symp. on Circuits and Systems*, 1999, vol. 6, pp. 412-415.

[Wong86] D. F. Wong and C. L. Liu, "A new algorithm for floorplan design," in *Proc. of ACM/IEEE Conf. on Design Automation*, 1986, pp. 101-107.

[Yagh88] H. Yaghutiel, S. Shen, A. Sangiovanni-Vincentelli, and P. R. Gray, "Automatic layout of switched-capacitor filters for custom applications," in *Proc. of IEEE Int. Solid-State Circuits*

Conf., 1998, pp. 170-171 & 353-354.

[Yúfe91] A. Yúfera, A. Rueda, and J. L. Huertas, "Flexible capacitor and switch generators for automatic synthesis of data convertors," in *Proc. of IEEE Int. Symp. on Circuits and Systems*, 1991, vol. 5, pp. 3162-3165.